The work of the ILO in the employment field aims to encourage and assist member states to adopt and implement policies and programmes designed to promote full, productive and freely chosen employment and to reduce poverty.

In response to the pressing challenges of the 1990s, including the rapidly growing interdependence of national economies, increasing reliance on market mechanisms, technological innovation and environmental concerns, the ILO has refocused its work in the field of employment. Specific attention is now devoted to assessing the impact of the above challenges on employment, migratory pressures, unemployment and poverty, examining ways of coping with them, and understanding the linkages between macro- and micro-economic policies. Greater emphasis is put on policy-oriented research and technical advisory services. This is to ensure a swift response to requests of assistance, further enhanced through the worldwide network of 14 ILO multidisciplinary advisory teams.

Through these activities the ILO is helping national authorities and social partners take advantage of the opportunities created by current global changes, so as to achieve more and better employment. They also provide a basis for an ongoing dialogue with other international organizations on employment and the social dimensions of growth and development in the world economy.

This publication is an outcome of such work, and is intended to disseminate information of relevance to a variety of countries and a wide audience.

The Macmillan Series of ILO Studies

Iftikhar Ahmed (*editor*)
BIOTECHNOLOGY: A Hope or a Treat?

Iftikhar Ahmed and Jacobus A. Doeleman
BEYOND RIO: The Environmental Crisis and Sustainable Livelihoods in the Third World

Richard Anker and Catherine Hein (*editors*)
SEX INEQUALITIES IN URBAN EMPLOYMENT IN THE THIRD WORLD

W.R. Böhning
STUDIES IN INTERNATIONAL LABOUR MIGRATION

Enyinna Chuta and Carl Liedholm
EMPLOYMENT AND GROWTH IN SMALL-SCALE INDUSTRY

Alain de Janvry, Samir Radwan, Elisabeth Sadoulet and Erik Thorbecke (*editors*)
STATE, MARKET AND CIVIL ORGANIZATIONS: New Theories, New Practices and their Implications for Rural Development

Ghazi M. Farooq and George B. Simmons (*editors*)
FERTILITY IN DEVELOPING COUNTRIES

David J.C. Forsyth
TECHNOLOGY POLICY FOR SMALL DEVELOPING COUNTRIES

Dharam Ghai, Azizur Rahman Khan, Eddy Lee and Samir Radwan (*editors*)
AGRARIAN SYSTEMS AND RURAL DEVELOPMENT

Jeffrey James and Susumu Watanabe (*editors*)
TECHNOLOGY INSTITUTIONS AND GOVERNMENT POLICIES

Vali Jamal and John Weeks
AFRICA MISUNDERSTOOD OR WHATEVER HAPPENED TO THE RURAL–URBAN GAP?

Nicolas Jéquier and Yao-Su Hu
BANKING AND THE PROMOTION OF TECHNOLOGICAL DEVELOPMENT

Azizur Rahman Khan and Dharam Ghai
COLLECTIVE AGRICULTURE AND RURAL DEVELOPMENT IN SOVIET CENTRAL ASIA

Ng Sek-Hong and Victor Fung-Shuen Sit
LABOUR RELATIONS AND LABOUR CONDITIONS IN HONG KONG

A.S. Oberai
POPULATION GROWTH, EMPLOYMENT AND POVERTY IN THIRD-WORLD MEGA-CITIES

Guy Standing
UNEMPLOYMENT AND FEMALE LABOUR

Wouter van Ginneken and Christopher Brown (*editors*)
APPROPRIATE PRODUCTS, EMPLOYMENT AND TECHNOLOGY

Susumu Watanabe (*editor*)
MICROELECTRONICS AND THIRD-WORLD INDUSTRIES

A.B. Zahlan
ACQUIRING TECHNOLOGICAL CAPACITY

Beyond Rio

The Environmental Crisis and Sustainable Livelihoods in the Third World

Edited by

Iftikhar Ahmed
Employment Strategies and Policies Branch
International Labour Office
Geneva

and

Jacobus A. Doeleman
Senior Lecturer in Economics
University of Newcastle
Australia

A study prepared for the International Labour Office

© International Labour Organization 1995

This book is published in *The Macmillan Series of ILO Studies*
All rights reserved. No reproduction, copy or transmission of
this publication may be made without written permission.

No paragraph of this publication may be reproduced, copied or
transmitted save with written permission or in accordance with
the provisions of the Copyright, Designs and Patents Act 1988,
or under the terms of any licence permitting limited copying
issued by the Copyright Licensing Agency, 90 Tottenham Court
Road, London W1P 9HE.

Any person who does any unauthorised act in relation to this
publication may be liable to criminal prosecution and civil
claims for damages.

First published 1995 by
MACMILLAN PRESS LTD
Houndmills, Basingstoke, Hampshire RG21 6XS
and London
Companies and representatives
throughout the world

ISBN 0-333-63176-5

A catalogue record for this book is available
from the British Library.

10 9 8 7 6 5 4 3 2 1
04 03 02 01 00 99 98 97 96 95

Printed and bound in Great Britain by
Antony Rowe Ltd, Chippenham, Wiltshire

Published in the United States of America 1996 by
ST. MARTIN'S PRESS, INC.,
Scholarly and Reference Division
175 Fifth Avenue, New York, N.Y. 10010

ISBN 0-312-12712-X

The designations employed in ILO publications, which are in conformity with
United Nations practice, and the presentation of material therein do not imply the
expression of any opinion whatsoever on the part of the International Labour
Office concerning the legal status of any country, area or territory or of its
authorities, or concerning the delimitation of its frontiers.
The responsibility for opinions expressed in studies and other contributions rests
solely with their authors, and publication does not constitute an endorsement by
the International Labour Office of the opinions expressed in them.
Reference to names of firms and commercial products and processes does not
imply their endorsement by the International Labour Office, and any failure to
mention a particular firm, commercial product or process is not a sign of
disapproval.

To our wives Selina and Jacquie

Contents

Notes on the Contributors	xi
Editors' Preface	xvii
Acknowledgements	xix

1 Introduction: In Quest of Sustainable Livelihood   1
Iftikhar Ahmed

Interpreting Sustainable Development	1
Multidisciplinarity in Sustainable Development Interpretation	2
'Green' versus 'Brown' Issues in Sustainable Livelihoods	2
Destruction versus Creation of Sustainable Livelihoods	3
Jobs and the Environment	4
Failures: Markets, Policies, Institutions and Central Planning	6
Methodology: Innovative Approaches for Sustainable Livelihoods	7
Design of the Volume	8

PART ONE CONCEPTUAL APPROACHES

2 Environmental Degradation at Different Stages of Economic Development   13
Theodore Panayotou

Inequality: Environmental Degradation and Development	13
Theoretical Underpinnings	14
Empirical Tests	19
Policy Implications	28
Implications for Employment, Technology and External Assistance	32
Conclusion	36

3 Sustainable Livelihoods and Environmentally Sound Technology: Theoretical and Conceptual Approaches   37
Charles Perrings

Sustainable Development	38
The Notion of Environmental Soundness	44

Subcategories: Sustainable Livelihood (SL) and Environmentally Sound Technology (EST)	48
Making the Concepts Operational	54
The Decision Process	58
Concluding Remarks	62

PART TWO EMPIRICAL STUDIES

4 Technology, Environment and Employment in Third World Agriculture — 69
Anil Markandya

The Conceptual Framework	69
The Context of the Developing Countries	71
The Linkages between Technology Transfer, Employment Creation and Environmental Degradation	76
Government Responses to Environmental Damage, and the Role of Multilateral Institutions	87
General Policy Implications	92

5 Incentives for Sustainable Development in Sub-Saharan Africa — 95
Charles Perrings

The Ecological and Socioeconomic Setting	95
Climate Change and Ecosystem Resilience	100
Population Growth and Environmental Degradation	103
Property Rights and the Overutilization of Resources	106
Risk Management and Innovation in the Rural Economy	111
Price Incentives	116
Income and Asset Distribution	122
Environmental Degradation, Technology and Employment	125
Concluding Remarks	130

6 Quantification of the Trade-Offs in Zambia between Environment, Employment, Income and Food Security — 133
Dodo J. Thampapillai, Phiri T. Maleka and John T. Milimo

Aims and Scope	133
The Study Area	135
Method of Analysis: Formulation of Models	138
Application of the Models and Evaluation of the Options	141

Contents ix

	Evaluation of Environmental Quality	150
	Conclusions	156
7	**Soil Conservation and Sustainable Development in the Sahel: A Study of Two Senegalese Villages** *Elise H. Golan*	**159**
	The Analytical Framework	159
	The Peanut Basin	161
	Data Collection	164
	Policy Analysis	170
	Conclusions	191
8	**The Green Revolution, Biotechnology and Environmental Concerns: A Case Study of the Philippines** *R. Sathiendrakumar and Keith Norris*	**195**
	The Green Revolution in the Philippines	195
	Chemical Fertilizer Usage and the Associated Environmental Problems	204
	Alternatives to Chemical Fertilizers	211
	Summary and Conclusions	217
9	**Technology–Environment–Employment Linkages and the Rural Poor of Bangladesh: Insights from Farm-Level Data** *Mohammad Alauddin, Mustafa K. Mujeri and Clement A. Tisdell*	**221**
	State of the Rural Environment and Trends	222
	The Survey Area	224
	Agriculture in an Unforested Area on the Floodplains	226
	Agriculture in a Forested Area: Case Studies from Chittagong Hill Districts	246
	Concluding Comments	253
10	**Sustainable Development and Employment: Forestry in Malaysia** *James E. Jonish*	**257**
	Deforestation	257
	Forest Management: Concept and Issues	258
	Malaysia's Forestry Sector	263

	Malaysian Forest Policy Alternatives	272
	Conclusions	284
11	**Agrarian Structure and Sustainable Livelihoods of Tribal People in Indian Forestry** *Vijay Shankar Vyas*	**289**
	Poverty and Environmental Decay	289
	Poor People in a Poor Region	291
	Process of Impoverishment and Environmental Degradation	296
	Determinants of Change	298
	Conclusion: The Lessons Drawn	311

PART THREE   CONCLUSION AND POLICY SYNTHESIS

12	**Sustainable Livelihood and Employment: Pragmatic Approaches** *Iftikhar Ahmed*	**317**
	Is Development Possible without Environmental Degradation?	317
	Sustainable Livelihood	318
	Are Environment, Development and Employment Interrelated?	334
	Policy Instruments	354
	Institutional Dimensions	356
	Concluding Remarks	358

Bibliography	363
Index	383

Notes on the Contributors

Iftikhar Ahmed is a Development Economist with the ILO's Employment Department in Geneva and has worked on agricultural and rural development. Previously, he was a Post-Doctoral Associate at the Iowa State University of Science and Technology, United States, a Visiting Fellow at the Institute of Development Studies, University of Sussex, United Kingdom, and Associate Professor of Economics, Dhaka University, Bangladesh. He is the author of *Technological Change and Agrarian Structure: A Study of Bangladesh* (Geneva, ILO, 1981), co-editor of *Farm Equipment Innovations in Eastern and Central Southern Africa* (Aldershot, United Kingdom, Gower, 1984), editor of *Technology and Rural Women: Conceptual and Empirical Issues* (London, George Allen & Unwin, 1985), co-editor (with Vernon W. Ruttan) of *Generation and Diffusion of Agricultural Innovations: The Role of Institutional Factors* (Aldershot, United Kingdom, Gower, 1988) and editor of *Biotechnology: A Hope or a Threat?* (London, Macmillan, 1992).

Mohammad Alauddin is Senior Lecturer in Economics, University of Queensland, Australia. He previously taught at Rajshahi University, Bangladesh, and at the University of Melbourne, Australia. He was Visiting Faculty Fellow, Southwest Fisheries Science Center, National Marine Fisheries Science, La Jolla, United States. He is the author (jointly with C.A. Tisdell) of *The Green Revolution and Economic Development: The Process and its Impact in Bangladesh* (London, Macmillan, 1991) and many articles published in international journals, including *Journal of Development Economics*, *Journal of Development Studies*, *Applied Economics*, *Oxford Bulletin of Economics and Statistics* and *Economic Development and Cultural Change*.

Jacobus Doeleman is Senior Lecturer in Economics at the University of Newcastle, New South Wales, Australia. Specializing in environmental economics, he has acted as Environmental Consultant for the Australian Government and, recently, spent two years as Environmental Economist with the International Labour Office in Geneva. His writings comprise an introductory monograph on environmental economics and a range of shorter publications in the same field.

Elise H. Golan is a Research Associate at the Natural Resources and Environmental Research Center at the University of Haifa, Israel. Previously (1990–92) she held a post-doctoral fellowship awarded by the Israeli Higher Education Council. She received her PhD from the Department of Agricultural and Resource Economics at the University of California at Berkeley, United States. Her work to date has focused on conservation policy and resource-tenure systems in the Sahel region of western Africa.

James E. Jonish is Professor of Economics and Deputy Director of the International Center for Arid and Semiarid Land Studies at Texas Tech University in Lubbock, Texas, United States. He is the author of more than 80 papers, chapters and reports in the area of natural resource and environmental economics and human resource economics. His main research topics of interest include environmental-energy and environmental-water resource concerns.

Phiri T. Maleka (died May 1993) was a Natural Resources Economist. He was a Research Fellow at the Rural Development Studies Bureau of the University of Zambia, where he carried out a number of studies and published a number of papers in international journals, especially on food security. Previously he was with the Planning Division of the Ministry of Agriculture and Water Development, where he became Senior Economist and Head of the Sectoral and Policy Analysis Section.

Anil Markandya is an Environmental and Resource Economist at the Harvard Institute for International Development, Harvard University, United States. He has been working in this field for over 20 years and has made several contributions in the areas of environmental valuation discounting, the environmental economics of the ozone layer, and biodiversity and economic regulations. In 1991 he was awarded the Mazzoti Prize for contributions to ecology and was elected a Fellow of the Royal Society of Arts, United Kingdom. He has acted as a consultant for several national governments as well as most international organizations and has published more than 50 books and articles in the areas of environmental economics and welfare economics.

John T. Milimo is a Social Anthropologist at the University of Zambia, where he has been Director of the Rural Development Studies Bureau since 1982. He has taught as Visiting Fulbright Professor at the University of Alabama, United States. He is co-editor of the following books: *Adopting Improved Farm Technology* (with R. Celis and S. Wanmali), *The Dynamics of Agricultural Policy and Reform in Zambia* (with A.P. Wood, S.A. Stuart and D.M. Warren), and *Agricultural Policies in Frontline States* (with B. Chisanga and C. Mwila). He is currently taking a keen interest in the use of participatory research methods and is co-author of the World Bank-sponsored *Participatory Poverty Assessment in Zambia*.

Mustafa K. Mujeri is an economist currently working on monitoring adjustment and poverty in Bangladesh with the Centre on Integrated Rural Development for Asia and the Pacific (CIRDAP), Dhaka. Previously, he worked as an expert and macroeconomic planner with the Bangladesh Planning Commission and UNDP/UNDTCD, and was a Visiting Lecturer at the Department of Economics, University of Queensland, Australia and an Associate Professor of Economics, Rajshahi University, Bangladesh. He has contributed widely to professional journals and books on development and environmental economics.

Keith Norris is Professor of Economics at Murdoch University, Perth, Australia. He is author of several books, including the *Economics of Australian Labour Markets* (third edition, 1993).

Theodore Panayotou is a Fellow at the Harvard Institute for International Development and Lecturer in the Department of Economics at Harvard University, United States. A specialist in environmental and resource economics, environmental policy analysis and development economics, Dr Panayotou has advised governments and institutes in Asia, Africa and Eastern Europe, as well as numerous other national and international institutions, on the interactions between the natural resource base and economic development. He served for a decade as Visiting Professor and Resident Adviser in southeast Asia, and recently co-wrote a multivolume environmental policy study at the Thailand Development Research Institute in Bangkok. The author of several books, including *Not by Timber Alone: Economics and Ecology for Sustaining Tropical Forests* (with Peter S. Ashton: Island Press, 1992) and *Green Markets: The Economics of Sustainable Development* (ICS Press, 1993), and numerous articles and monographs, he is currently at work on another book, *Natural Resources, Environment and Development: Economics, Policy and Management*. Dr Panayotou received the 1991 Distinguished Achievement Award of the Society for Conservation Biology for his wide-ranging efforts in using economic analysis as a tool for conservation.

Charles Perrings is Professor of Environmental Economics and Environmental Management at the University of York, United Kingdom, and Director of the Biodiversity Programme at the Beijer International Institute for Ecological Economics, Stockholm, Sweden. He was formerly Professor of Economics at the University of California, Riverside, United States, and the University of Botswana, and has taught at the University of Auckland, New Zealand. His research interests intersect the problems of environmental management and development, especially in low-income countries, and he has published extensively in both areas.

R. Sathiendrakumar is a Lecturer in Economics at Murdoch University, Western Australia. He currently holds a Visiting Fellowship in the Department of Economics and Statistics, National University of Singapore. He holds a PhD degree in Economics from the University of Newcastle, New South Wales, Australia and a Master's Degree in Economics from Manchester University, United Kingdom. He has published widely in international journals in the areas of environment, fisheries and tourism. His current research interests are in environmental economics, natural resource economics, the economics of tourism, fisheries economics and microeconomic policy.

Dodo J. Thampapillai is a Senior Lecturer in Environmental Economics at Macquarie University, Sydney, Australia. During 1993, he was a Visiting Professor in Environmental Economics at the Swedish University of Agricultural Sciences at Uppsala. He has been a Visiting Associate Professor in Economics at Simon Fraser University, Canada, and has held visiting positions at the Australian National University, and the Christian Albrechts Universität of Kiel, Germany. He has written several publications, including textbooks, in the areas of environmental economics and development economics.

Clement A. Tisdell is Professor of Economics at the University of Queensland, Australia. He is also an Honorary Professor of the Institute of Economic Research, People's University of China, Beijing, and has held visiting appointments at Princeton University, Stanford University and the East–West Center, United States, at the University of York, United Kingdom, and elsewhere. His books include *Science and Technology Policy: Priorities of Governments, Natural Resources, Growth and Development, Economics of Environmental Conservation and Economic Development in the Context of China*. He is co-editor of *Economic Development and Environment: A Case Study of India*.

Vijay Shankar Vyas is the Director of the Institute of Development Studies, Jaipur, India. He was previously the Director of the Agro-Economic Research Centre for Western India; Member of the Agriculture Prices Commission, Government of India; Chair, Professor, and later Director of the Indian Institute of Management, Ahmedabad; and Senior Advisor in the Agriculture and Rural Development Department of the World Bank. He has served as a consultant with United Nations organizations, the World Bank, the Asian Development Bank and various institutions in India and abroad. He has written extensively on poverty and sustainable development. Currently he is editing a volume on *Public Interventions for Poverty Alleviation: Experiences of Indian States*. He was invited to deliver the Elmhurst Memorial Lecture at the Tokyo Conference of the International Association of Agricultural Economists in 1991.

Editors' Preface

Despite the prolific international debate on sustainable livelihood there does not seem to be any clear understanding of the operational significance of this concept. This volume attempts to raise more questions rather than to offer answers on this vital issue, which is one of the cornerstones of Agenda 21 adopted by the 1992 UNCED Rio Earth Summit.

Since the issue of sustainable livelihoods basically relates to the survival of the rural poor, the bulk of the case-studies in this volume has to deal with the issues of natural resource management. This volume has shed more light on the causes and consequences of the destruction of sustainable livelihoods, particularly of indigenous people, than it has on their creation. However, it is clear that the concern for equity is central to any strategy for the promotion of sustainable livelihoods.

The volume, in our view, is essentially designed to stimulate discussion by posing provocative questions on whether environmental sustainability depends on socioeconomic sustainability. In this respect the volume shows that there is no simple straightforward answer to the question of how environment, development and employment are interrelated.

Largely drawing on experience from industrialized countries the volume confirms the optimism on the vast job potential of the emerging environmental industry expressed in the book *Environment, Employment and Development* (edited by A.S. Bhalla, Geneva, ILO, 1992) which was an ILO contribution to the Rio Summit. However, the analysis of the viability of livelihoods goes beyond speculation, where the volume deals concretely with the possible employment implications of clean production and waste management in developing countries. The volume concludes with a discussion on the appropriate policy instruments and institutional reforms essential for promoting sustainable development. A pragmatic approach is necessary if the twin objectives of sustainable development and job creation are to be pursued simultaneously.

This volume was completed within the framework of the ILO research programme on environment and employment as an ILO contribution to the implementation of Agenda 21 adopted by the Rio Earth Summit. This global review is being followed by a more focused regional ILO study entitled *Sustainable Development and Poverty Alleviation in Sub-Saharan Africa* by Charles Perrings.

Iftikhar Ahmed   Jacobus A. Doeleman

Acknowledgements

Ajit Bhalla, Massoud Karshenas, Philippe Garnier and Amarjit Oberai provided valuable comments on the text. We are grateful to Ron Kirkman for his editorial scrutiny of the manuscript and careful verification of the bibliography.

We are thankful to Susan Saidi and Sue Fewings who helped with the tedious work of typing the manuscript and transferring it to the word processor.

1 Introduction: In Quest of Sustainable Livelihood

Iftikhar Ahmed

INTERPRETING SUSTAINABLE DEVELOPMENT

The surge in the literature on sustainable development since the 1987 Brundtland Report (WCED, 1987), boosted by the adoption of Agenda 21 at the 1992 Rio Earth Summit, has led to a more comprehensive interpretation of this concept. Yet the breadth, diversity and the growing complexity of its interpretation has baffled policy-makers struggling to design innovative policies which could even partially satisfy the growing number of sustainable development criteria.[1]

The purpose of this volume is not to enter into a debate on the interpretation of this concept nor even to attempt a survey of the literature on the subject. No matter how sustainable development is interpreted, the concept is still considered ambiguous as there is no agreement yet on what exactly is to be sustained (Redclift, 1992). Is the goal of sustainable development to sustain levels of production, consumption, the resource base itself or the livelihood derived from it? In Agenda 21 the concept of sustainable livelihood is applied to the least well-off members of society. Sustainable livelihood requires the conservation of the productive potential of the produced and natural capital on which the entire community depends currently and in the future, and a distribution of income that meets the needs of the dependent groups (elaborated in Chapter 3).

Since the empirical case studies in this volume deal essentially with sustainable development in natural resource-based economies, sustainability could be interpreted in somewhat simpler and more operational terms of sustainable resource use (Pezzey, 1992). This also raises questions of who gains access to and controls the use of the natural resources upon which their livelihood depends (Friedman and Rangan, 1993). In this sense sustainability is linked to the notions of social justice and equity, both between generations and within generations as well as within nations and between nations (Colchester, 1992).

For instance, sustainability requires that total capital stocks are held at least constant, so that future generations have the same capability to develop as current generations. Inequality within a nation may again cause unsustainability if the poor, who rely on natural resources more heavily than the rich, deplete the natural resources faster because they have no access to any alternative resources for their livelihood. Similarly, the pattern of demand for traded natural resources or natural resource-intensive products in rich countries may cause unsustainable management of those resources in poor countries (Pearce and Warford, 1993).

MULTIDISCIPLINARITY IN SUSTAINABLE DEVELOPMENT INTERPRETATION

Another question arising in the interpretation of sustainable development is that of multidisciplinarity. Fortunately, there appears to be a consensus among economists, ecologists and sociologists about the role of their respective disciplines in any comprehensive analysis of sustainable development. However, the economists' perception of sustainable development is believed to be narrower than the overall understanding of environmentally sound sustainable development. According to the economists, the economic objectives are growth and efficiency; the social objectives concern employment creation, equity and poverty alleviation; and the ecological objectives relate to the questions of natural resource management (Serageldin, 1993).[2] In order to avoid overgeneralization, this volume deliberately follows the economists' narrower but more focused interpretation of sustainable development which, as we shall see, permits a more operational and in-depth treatment of the subject of sustainable livelihoods.

'GREEN' VERSUS 'BROWN' ISSUES IN SUSTAINABLE LIVELIHOODS

One must also recognize the divergences in the environmental issues facing rural and urban populations (Dixon, 1993). The rural environmental problems could be categorized as 'green' issues, as they are concerned with the use, control, accessibility and management of natural resources. In urban centres and industrial locations the environmental problems could be classified as 'brown' issues, as they relate to air, water and noise pollution and waste disposal (sewage and solid waste) resulting from urbanization (congestion) and industrialization (e.g. manufacturing, power generation, oil refining, transport, etc.). The brown issues of industrial production can again be distinguished in terms of the so-called 'end-of-pipe'

Introduction: In Quest of Sustainable Livelihood

environmental clean-up and pollution prevention approaches (Almeida, 1993). The former is often automatically assumed to create extra jobs, while the latter, it is believed uncritically, could require sacrificing employment.

While Chapters 2 and 12 partially deal with the 'brown' issues, the main emphasis of a volume concerned with sustainable livelihood is obviously on the 'green' issues, as the majority (over two-thirds) of the world's poor people, including the very poorest, live in rural areas (World Bank, 1990a). They are almost totally dependent for their livelihood on environmental resources and on their own labour and rarely on land or capital assets, to which they have little access.

Furthermore, from the perspective of sustainable livelihood we need to know not only how much poverty there is, but also where it is. To establish priorities for action on sustainable livelihoods, it is important to analyse the distributional and environmental implications of the high incidence of poverty in ecologically fragile areas of the globe.

One-half of the world's poor derive their living from the least resilient and most threatened environmental areas of the world, which are typified by *tropical forests* where soils are acidic and subject to serious erosion once deforestation occurs, *upland areas* where soil erosion is a serious risk, and *arid* and *semi-arid zones* where soils are light and easily eroded by wind (Pearce and Warford, 1993). In contrast, only about 15 per cent of the world's poor live in the peripheries of cities.

Therefore, it is clear that by emphasizing primarily, but not exclusively, the 'green' issues of the ecologically fragile areas inhabited by the poor of the world, this volume addresses the problems of sustainable livelihood more directly. For example, it is often uncritically assumed that the poor have no alternative to consuming environmental resources simply because the poor are highly dependent on natural resources for their survival. As we shall see, the plundering of the planet's natural resources on a massive scale has more to do with deliberate policies to benefit the privileged and with the biases in institutional structure favouring the rich.

DESTRUCTION VERSUS CREATION OF SUSTAINABLE LIVELIHOODS

The above analysis would suggest that the concern for equity in sustainable livelihoods is greater than that for sustainable development. This volume makes a distinction between the creation and the destruction of sustainable livelihoods. The destruction of sustainable livelihoods is almost entirely concerned with the annihilation of indigenous populations living in total harmony with nature. The combined effects of distorted incentive structures, institutional inequalities, vested interests, infrastructural developments, trade deficits, debt burdens and consumption patterns in affluent societies are examined for their relative and cumulative impact on the relentless destruction of sustainable livelihoods, particularly of the indigenous populations inhabiting the tropical forests (Chapters 4, 9, 10, 11 and 12).

The prospects for the creation of and strategies for the preservation of sustainable livelihoods are taken up in Chapter 12, on the basis of case study experiences and on some recently available evidence. The conditions under which environmental sustainability and socioeconomic sustainability could be mutually supportive are also taken up in Chapter 12.

JOBS AND THE ENVIRONMENT

While much has been said about the sacrifice in employment that inevitably accompanies environmental conservation in Third World countries, the immature literature on this question (particularly projecting job creation potential) is still very speculative (Bhalla, 1992; Bhalla, 1994). For instance, the assumed conflict between jobs and the environment is dramatized by the somewhat empty slogan of 'save a logger, kill an owl', implying that measures to protect endangered species from extinction in the United States will cost the jobs of tens of thousands of loggers (Renner, 1991). On the other hand, the preservation of endangered species and biodiversity in other parts of the globe, e.g. in Zambia (Chapter 6), Malaysia (Chapter 10) and the Amazon forests, Kenya and Costa Rica (Chapter 12), could enhance the sustainability of future incomes and employment while at the same time reducing environmental destruction.

It has been argued, often with inadequate supporting evidence, that the use of pollution abatement policies and technologies in industry and transport and increased investments in environmental protection activities such as waste recycling, reduction and management could actually generate jobs (Renner, 1991; Bhalla, 1994). As we shall see from evidence summarized in Chapter 12, this is true not only for industrial and urban pollution control but also for sound natural resource management, e.g. in environmental protection in the Philippines uplands and in semi-arid Kenya. Is employment-intensive environmental protection equally important where communities have abandoned environmental conservation as a result of labour shortages, as in the Latin American highlands? Chapter 12 shows

Introduction: In Quest of Sustainable Livelihood

that the concern for job creation with environmental conservation does not necessarily have a universal appeal even within the entire developing world.

Therefore, the concluding chapter of this volume attempts to synthesize, from the available literature on developing countries, the rather limited, partial, fragmented and scattered empirical evidence on the connection between employment and the environment covering both the 'green' and the 'brown' issues. Concrete quantitative data generated by some of the empirical case studies, e.g. on Zambia (Chapter 6) and Malaysia (Chapter 10), have also been taken into account in demonstrating with rigorous econometric and regression analysis the value of a range of policy options actually available for simultaneously maximizing sustainable employment and minimizing environmental degradation. As we shall see, sustainable development strategies may require minor sacrifices in current levels of employment and incomes in order to steady the stream of future employment and income flows from existing natural capital stock.

The most polluting industries (e.g. paper and pulp, chemical products, petroleum refinery, iron and steel, nonferrous metals, leather tanning and textiles) appear to be the fastest-growing industries in developing countries (Almeida, 1993). Has this pattern of growth accompanied by environmental pollution stimulated employment expansion in developing countries? Does this trend reflect the flight of dirty industries from the environmentally conscious North to pollution havens in the South? Will the introduction of stricter environmental standards (for flattening the environmental Kuznets curve) in these polluting industries stunt growth and destroy jobs in the Third World? Have environmental compliance costs really been 'job-killers' in the industrialized countries? What are the consequences for sustainable growth and jobs of 'end-of-pipe' environmental clean-up and pollution prevention approaches?

A quantitative assessment of the overall and net employment implications of specific sectoral environmental measures is made difficult when one has to take into account the indirect employment effects created through cross-sectoral linkages (e.g. when assessed in an inter-industry input–output framework). This question is concretely brought up in Chapter 12 in a country context. Substantial quantitative work at the macro and sectoral levels has been undertaken for advanced countries on the employment dimensions of sustainable development, using the most sophisticated statistical and econometric tools of analysis and macroeconomic models (ECOTEC, 1994a; Majocchi, 1994; Sprenger, 1994), the findings of which are analysed in Chapter 12 in order to examine their relevance for developing countries. Of course, one has to keep in mind the myriad assumptions made on the structure of the models used, the policy instruments applied, the changes in consumer behaviour expected and the production responses by firms assumed to be rational.

As we shall see, environmental action affects employment levels not only quantitatively but also qualitatively. It is important to identify (as we do in Chapter 12) the kind of skills that will be in greater demand on account of environmental job creation as well as the spatial distribution of jobs, of interest for regions with the highest unemployment rates.

FAILURES: MARKETS, POLICIES, INSTITUTIONS AND CENTRAL PLANNING

The literature appears to point to a series of policy failures, market failures, institutional (property rights) failures and planning failures at the core of sustainable development problems (Karshenas, 1992, Chapter 5; Pearce and Warford, 1993). For instance, resource degradation may be caused by price failure when the prices of resources do not even reflect the private marginal cost of production, let alone marginal social costs which include external costs. Many resources do not even have markets. Prices have to be established for them by assigning property rights to the free resource. Under central planning, problems arose from the overlap between the state as the main polluter and the state as the regulator, which has been compared to entrusting a poacher with the job of a gamekeeper (Pearce and Warford, 1993). This was also true of environmentally sensitive industries such as steel, chemicals, energy, water and petrochemicals run by government bureaucracies prior to their recent privatization in Argentina (*Financial Times* (London), 27 October 1993).

Instead of plunging into a detailed examination of the ramifications for sustainable development of such failures (which is already available in the literature), this volume empirically focuses on distortions in incentive structures and the most glaring policy and property rights failures which have contributed simultaneously to environmental degradation and the destruction of sustainable livelihoods (primarily in Chapters 3, 5 and 12). The analysis is not only limited to identifying policy distortions (e.g. input price subsidies, and fiscal and monetary incentives) which need to be eliminated to stem environmental degradation, but also deals with incentives which ought to be introduced more aggressively in order to protect the environment, generate employment and offer the promise of sustainable livelihoods at the same time. In this respect the appropriateness and the relative effectiveness of the 'command and control' approaches and market-based incentives are evaluated and compared in relation to both the 'green' and the 'brown' issues (Chapters 4 and 12).

Are the types and forms (direct and indirect) of policy instruments designed to tackle 'brown' problems of environment and employment equally applicable to solving similar problems with respect to the 'green' issues? Can the package of environmental policy instruments be adapted to bring about greater convergence between the environmental and employment goals?

METHODOLOGY: INNOVATIVE APPROACHES FOR SUSTAINABLE LIVELIHOODS

Highly sophisticated econometric techniques, macroeconomic models and quantitative tools of analysis have been developed for analysing the direct and indirect employment implications of environmental policies in advanced countries (ECOTEC, 1994a; ECOTEC, 1994b; Majocchi, 1994; Sprenger, 1994). The theoretical framework and methodology adopted for analysing the environment–employment linkages for developing countries will be completely different from those applied for advanced countries, for several reasons. Firstly, the emphasis on sustainable livelihoods for developing countries requires a focus on the 'green' issues concerning the sustainable use of natural resources, while the thrust of environmental issues in advanced countries is primarily towards the 'brown' issues covering energy use and pollution problems associated with transport, industry and infrastructure, where it is far easier to identify specific environmental and market-based economic policy instruments and generate the required statistics to evaluate the quantitative and qualitative impact on employment. Secondly, advanced countries have already generated a wealth of statistical data such as is required for rigorous econometric analysis. Indeed, much progress has been made with quantitative macroeconomic models and inter-sectoral input–output techniques for the analysis of the employment implications of environmental policies and programmes, particularly for the countries of the European Union. For developing countries, which lack hard data, a pragmatic mix of macro, micro, sectoral and cross-sectoral approaches will have to be applied, often piecing together partial fragmented data hidden in country case studies. Thirdly, the analysis of sustainable livelihood, which is of greater concern for the indigenous people of Third World countries, has to focus on inequalities in institutional structures (e.g. property rights, landownership distribution and land legislation).

As a result of the above differences, evaluating the employment implications of environmental policies is much more complex, difficult and challenging for developing countries than for advanced countries. However, the methodologies used and the results obtained from the analysis of the employment–environment relationship in advanced countries will still be useful in providing clues for deducing the likely scenarios for developing countries. Such an extension might facilitate the indirect assessment of the possible employment effects, both quantitative and qualitative, if similar environmental policies were adopted in developing countries. This would also require a great deal of methodological innovativeness in tailoring the limited scattered evidence to testing hypotheses on some critical aspects of sustainable livelihoods in developing countries.

DESIGN OF THE VOLUME

In the light of the above background information, the conceptual and methodological issues and hypotheses raised in this volume deal with policy and analysis based on empirical case studies from three developing continents of Africa, Asia and Latin America. The analysis is further facilitated by a substantive body of quantitative data on the trends in employment creation from environmental action in advanced countries to secure insights and clues on the likely impact on employment from the adoption of similar sustainable development strategies in Third World countries.

Although the book is divided into three distinct parts, many of the issues and themes raised in respect of employment, sustainable livelihood, environment, policies and institutions in each part are very much interrelated. The chapters in Part One attempt to provide the analytical and policy framework for the empirical country case studies as well as regional overview chapters. They build on the conceptual and methodological issues raised in this chapter. Chapter 2 systematically examines the relationship between environment and development by statistically verifying the existence and measuring the intensity of the environmental Kuznets curve. Chapter 3 constitutes the core of this volume by providing the hypotheses and theoretical framework for evaluating sustainable livelihood. It spells out the criteria and conditions that need to be fulfilled in relation to the concepts of sustainable livelihood and environmentally sound technologies.

Part Two begins with an in-depth survey of the use and effectiveness of a diverse range of policy options for sustainable development by focusing on the agricultural sector of Third World countries. It is followed by a regional overview for sub-Saharan Africa (Chapter 5), which critically examines the contribution of incentive structures to the overuse of natural resources and challenges the view that the poor are exclusively responsible for following a suicidal path by rapidly descending a 'downward spiral' of poverty and environmental degradation.

The rest of the empirical case studies in Part Two can be classified as follows: (a) quantitative estimates of the exact trade-offs between employment, environmental quality, food security and incomes using two separate scenarios (Chapters 6 and 7); (b) trade-offs between current and future streams of incomes and employment for sustainable use of natural resources (Chapter 10); (c) the significance of enforcing property rights for the sustainable livelihood of indigenous people (Chapters 9 and 10); and (d) linkages between poverty, landownership distribution, free access to common property and environmental degradation (Chapters 8, 9 and 10).

Chapter 12 is not merely a summary of the findings of the volume, although it highlights the major issues raised. More importantly, it reinforces the conclusions and findings with a substantial amount of entirely new empirical data and the most recent qualitative data gathered not only from developing countries but also from North America and the European Union countries. It concludes the volume with a short synthesis of policy issues.

Notes

1. See, for example, Pezzey (1992) for a collection of nearly 50 different interpretations of this concept.
2. The broader interpretation of environmentally sound sustainable development encompasses a much larger and more complex set of elements in the objective function. The economic objectives consist of growth, equity and efficiency; the social objectives consist of empowerment, participation, social mobility, social cohesion, cultural identity and institutional development; and the ecological objectives consist of ecosystem integrity, carrying capacity, biodiversity and global issues (Serageldin, 1993).

Part One
Conceptual Approaches

2 Environmental Degradation at Different Stages of Economic Development

T. Panayotou

INEQUALITY: ENVIRONMENTAL DEGRADATION AND DEVELOPMENT

Three decades ago, Simon Kuznets (1965, 1966) advanced the then intriguing hypothesis that in the course of economic development income disparities at first rise and then begin to fall. This inverted U-shape relationship between income inequality and income per capita, which is supported by considerable statistical evidence, came to be known as the Kuznets curve.

The idea that things may have to get worse before they get better appears to have a more general applicability. Casual observation suggests that environmental degradation also rises at first and then falls in the course of economic development. For example, the cities of the newly industrializing countries, be they Seoul, Bangkok or Mexico City, are far more polluted than they were 20 or 30 years ago, with their pollution levels rising at rates that match or exceed those of economic growth; while cities in industrialized countries such as Japan and the United States, and in Western Europe, are cleaner today than they were 20 or 30 years ago.

In this study, we hypothesize a Kuznets-type inverted U-shape relationship between the rate of environmental degradation and the level of economic development. At low levels of development both the quantity and the intensity of environmental degradation are limited to the impacts of subsistence economic activity on the resource base and to limited quantities of biodegradable wastes. As economic development accelerates with the intensification of agriculture and other resource extraction and the take-off of industrialization, the rates of resource depletion begin to exceed the rates of resource regeneration, and waste generation increases in quantity and toxicity. At higher levels of development, structural change towards information-intensive industries and services, coupled with increased environmental awareness, the enforcement of environmental regulations, better technology and higher environmental expenditures, result in a levelling-off and gradual decline of environmental degradation.

The purpose of this study is to test empirically the hypothesis that an environmental Kuznets curve (referred to henceforth as an EK-curve) exists and to derive its policy implications for employment, technology transfer and development assistance. We test the hypothesis using cross-section data on deforestation and air pollution from a sample of developing and developed countries.

By way of a preview, the acceptance of an EK-curve would seem to suggest that the observed environmental deterioration in developing countries is a temporary phenomenon associated with their stage of development. Therefore, efforts to pressure or assist developing countries to improve their environmental performance may be based on the wrong premises and hence unnecessary, ineffective, or both. This, however, may also turn out to be the wrong conclusion if there are ecological thresholds beyond which environmental degradation is irreversible. Such thresholds are more likely to exist in today's developing countries, most of which are located in the tropics. Tropical resources such as forests, fisheries and soils are known to be more fragile and less resilient than temperate resources. If severely depleted or degraded during the take-off stage of economic development, tropical resources may take an inordinate length of time and a prohibitively high cost to restore at a later stage. Since these resources have a positive option and existence value (see Krutilla and Fisher, 1985) for high-income countries today and for developing countries in the future, a case could be made for assistance to developing countries to help them to flatten their EK-curve so as to avoid, or at least to limit, irreversible environmental damage. Yet, direct environmental assistance and the transfer of command-and-control regulations may not be the most cost-effective means of attaining a sustainable improvement in the environmental performance of developing countries. More rapid (rather than slower) economic development of the 'right' kind with appropriate market-based incentives for sound environmental behaviour may help developing countries to leapfrog ecological thresholds, thus attaining higher incomes, reducing inequalities and improving environmental quality.

THEORETICAL UNDERPINNINGS

The state of natural resources and the environment in a country depends on five main factors: (a) the level of economic activity or size of the economy; (b) the sectoral structure of the economy; (c) the vintage of the technology; (d) the demand for environmental amenities; and (e) the conservation and environmental expenditures and their effectiveness. The larger the size of the economy as measured by GNP, *other things being equal*, the more rapid the depletion of natural resources and the higher the level of pollution generated. The type and level of resource depletion and pollution also depend on the sectoral structure of the economy: economies that depend heavily on agriculture and other primary industries tend to suffer from rapid rates of resource depletion, such as deforestation and soil erosion, and low rates of industrial pollution. As countries become industrialized, the rural resource depletion problems are gradually being transformed into urban pollution and congestion problems. This trend tends to be modified, however, by two factors: (a) the unbalanced structural change of employment relative to output, which keeps undue large numbers of rural people dependent on declining resource sectors with consequent forest encroachment and unsustainable resource use; and (b) the environmental shadow cast on the rural resource base by the industrial urban sector both in terms of demand of raw materials and in terms of pollutants (e.g. acid rain damaging forests and crops).

There is a fairly close relationship between the level of development (as measured by GDP per capita), the share of the industrial sector in GDP and the structure of industry. In low-income countries, the share of industry in GDP is small (less than that of agriculture) and the sector is dominated by agroprocessing and light assembly. In middle-income countries, the industry's share approaches or exceeds one-third of GDP and the sector is dominated by the heavy steel, pulp and paper, cement and chemical industries. The relationship, however, is not monotonic. In higher-income countries, the share of the industry stabilizes or declines somewhat, and the sector is dominated by sophisticated technology industries (such as electrical machinery and electronics) and services. Industrial emissions vary positively with the size of the industrial sector and the share of the chemical and heavy industries. In later stages of development, the share of the industrial sector and, within industry, the share of chemicals and heavy industries level off and begin to decline gradually, while the share of information technologies and services continues to rise. These structural changes alone may explain the inverted relationship between emissions and level of economic development.

Countries with the same industrial structure, however, may generate different levels of industrial emissions and wastes if their capital stock and production technology are of different quality or vintage. Outdated, old or ill-maintained industrial plants and machinery tend to be less efficient in energy and material use and to produce, as a result, higher levels of wastes and emissions than new and better-maintained industrial plants.

The choice of technology and inputs is influenced by relative prices and by the policy and regulatory framework. A country that allows the duty-free import of 'clean' technologies is expected to have lower pollution levels than a country that imposes heavy duties on imported machinery. Countries that subsidize energy, electricity, water and raw materials should expect higher pollution levels than countries which implement full cost pricing.

There are three forces at work here: (a) substitution in favour of the underpriced input; (b) inefficient and wasteful use of the underpriced input; and (c) reduced incentive for input-minimizing technology, in respect of both implementation and development. Industrial protection and investment promotion tend to favour an industrial structure that favours capital, energy and pollution-intensive industries.

Pollution intensity or emissions per unit of output are further influenced by the strictness and intensity of enforcement of environmental regulations or by the effective price at which the environment is made available to the industry for the disposal of waste. Other factors being equal, countries with lax pollution control regulations would experience more industrial emissions than countries which effectively control or charge for the use of the environment as a waste sink.

The level of industrial emissions varies not only across countries but also over time in any given country, because of growth, structural transformation and policy change. Industrial growth has counteracting impacts on the level of emissions. On the one hand, industrial growth implies more output and presumably more inputs of materials and energy and hence more emissions and wastes. On the other hand, industrial growth also implies the introduction of later vintage capital, more efficient technologies and higher value added, all of which help to reduce emissions per unit of economic output. To the extent that old vintage capital and inefficient technologies and industries are withdrawn and replaced by new ones, the absolute level of pollution may also be reduced. In Japan, for example, economic growth in the past 15 years has been accompanied by a significant reduction in energy use and related emissions. Germany's tradable pollution permit system (see Organisation for Economic Co-operation and Development (OECD), 1989) also aims at sustaining industrial growth and reducing emissions by replacing old plants with new ones of the same kind but with substantially lower emissions. The impact of industrial growth on the level of emissions, we find, depends on the stage of industrialization, being positive in the earlier stages and negative in the later stages.

Economic structure is a dynamic feature that changes with the level of development. The changes in economic structure of a single economy over time are also reflected in the cross-section variations of structure among countries at different stages of development. Its changes parallel cross-section variations among countries. The faster economic growth is, the faster the structural change that propels industry from a minor to a dominant sector of the economy, and from light, through heavy, to technologically sophisticated industry.

Macroeconomic and industrial policies also change over time. Economic and trade liberalization may eliminate ailing or inefficient industries and restructure public enterprises that are often major sources of industrial pollution; they may also open up the floodgates for the import of polluting industries from countries with stricter environmental regulations. Depending on the stage of development, liberalization policies, such as a reduction in the level of protection or the devaluation of an overvalued currency, may or may not lead to a reduction in industrial emissions in the short-to-medium run. In the long run, however, policies that enhance competitiveness and economic efficiency help both to promote growth and to reduce industrial pollution by reducing waste, by recycling by-products and by upgrading the capital stock. Thus, while the general trend is for industrial emissions first to grow with industrialization and then to begin to decline, government policy can retard or speed up the process of structural change and technological development, thereby modifying the relationship between industrial emissions and economic development.

Over time, as incomes grow, pollutants accumulate and people can afford to become more environmentally conscious, and environmental regulations are tightened and more strictly enforced. Thus, another set of forces that underlie the EK-curve is: (a) the share of environmental expenditures in the government and industry budgets at different levels of development; and (b) the demand for environmental amenities by the population at different levels of income. In early stages of development when poverty is still pervasive, tax collection ineffective and environmental awareness low, few or no funds are allocated to environmental protection. This is partly a response to the very low demand for environmental amenities by those with low levels of income, a factor which accounts for the low level of private defensive expenditures against environmental pollution. Environmental amenity is an income-elastic 'commodity' that does not constitute a significant part of the consumer's budget until fairly high levels of income have been attained. Despite the low levels of public expenditures in environmental protection and the almost non-existent defensive expenditures on the part of the private sector (with the exception of the élite), environmental degradation remains limited at low levels of development because of the low levels of largely biodegradable waste and the relatively unimpaired natural assimilative capacities of the environment.

As the development process takes off, resource depletion accelerates and environmental pollution begins to accumulate at an increasing rate as the natural assimilative capacity of the environment becomes overloaded with pollutants. In contrast, environmental protection expenditures grow only slowly, because of lags in environmental awareness, in the change of preferences with rising incomes, and in the rise of the environmental movement.

As a country approaches the status of a newly industrializing economy (NIE), environmental quality reaches its lowest point because of the cumulative effects of rapid growth and the delayed effect of past accumulations of pollutants, while environmental clean-up expenditures are still limited and their impact not yet felt. As the higher level of income and wealth is consolidated economically (whilst being threatened ecologically by the impact of resource depletion and pollution on productivity, on health, on property values and on the quality of life), the demand for environmental amenities (being income elastic) rises, creating a perceptible disequilibrium *vis-à-vis* the dwindling supply of environmental amenities. As a result, economic, social and political pressures are built up to institute and enforce environmental regulations and to increase budgetary allocations to environmental protection and clean-up. This results not only in a reduction of emissions but also in structural change adversely affecting heavy polluters such as the steel, cement and chemical industries and in favour of more sophisticated, less polluting industries, thus reinforcing other continuing structural changes. These pressures are exerted through a number of channels, including joining the environmental movement, voting for 'green' or pro-environment parties, boycotting polluting industries, and expressing a preference for 'green' products. Thus, in the later stages of development, environmental quality improves.

Since the size of the economy, the change in economic and industrial structure, the vintage of technology, the demand for environmental amenities and the level of environmental expenditures are all a function of the level of development, it is reasonable to hypothesize a relationship between environmental degradation (deforestation, erosion, pollution) and GNP per capita. Furthermore, given the dynamics of structural change, technological development and consumption expenditure explained above, it is reasonable to hypothesize that this relationship is non-linear and has an inverted U-shape.

The experience of Japan, the United States and Western Europe, and more recently of the NIE of Hong Kong, the Republic of Korea,

Singapore and Taiwan, China, appears to conform to such a U-shape relationship between environmental quality and economic growth. The countries making up the next wave of newly industrializing economies (Brazil, Indonesia, Malaysia, Mexico and Thailand) may have reached their lowest point and are gearing up for an upturn (witness the logging ban and the introduction of lead-free petrol in Thailand and the serious efforts being made, at last, to deal with urban pollution in Mexico City).

EMPIRICAL TESTS

In the absence of long time series on environmental degradation for individual countries, we test the hypothesis of an EK-curve using cross-section data from a sample of developed and developing countries for which data are available. Not having available a single environmental degradation index, we select 'deforestation' as a representative variable for natural resource depletion and sulphur dioxide (SO_2), nitrogen oxides (NO_x) and solid particulate matter (SPM) for industrial and energy-related pollutants. There is an asymmetry between resource depletion and environmental pollution as treated here: deforestation rates are compensated by reforestation and therefore they reflect net effects. Pollution rates, on the other hand, concern emission rather than concentration levels. The latter depend on emissions and geographical location, atmospheric conditions and dispersion. If we were trying to explain variations in environmental quality among countries, the use of emissions rather than concentrations would have been inappropriate. However, since our explanatory variable is the production of pollutants, ignoring atmospheric conditions is justifiable since there is no reason to believe that developed and developing countries differ in any systematic way in the dispersion of pollutants. For this reason, pollutant emission data may confirm the postulated inverted U-shaped curve between environmental quality and income, albeit with a less steep right-hand side because of increasing levels of pollution abatement as incomes rise.

Deforestation and level of development

In countries with low population densities and low levels of economic activity, forests represent the dominant form of land use (with the exception of deserts, savannas and permafrost regions). As the population grows and economic activity expands, forests are being cut to obtain construction materials, fuelwood and land for cultivation. Yet the level of forest exploitation is still sufficiently low to permit natural regeneration, and no undesirable permanent forest loss takes place. For example, slash-and-burn or shifting cultivation with long fallow periods does not result in permanent deforestation. However, as the rural population increases and the economy opens up to international trade, deforestation may result from overharvesting for fuelwood and timber and from land clearing for agriculture, unless more attractive alternatives (such as an improvement in agricultural productivity, alternative fuels and non-farm employment) are made available. Forests are particularly vulnerable during the take-off process of industrialization, when the rural sector is heavily taxed to generate a surplus for industrial growth and while a protected industry generates very few jobs for the induced 'surplus' rural labour. This results in forest encroachment as exemplified by the cases of the Philippines and Thailand. Government policies may further exacerbate this trend, as the colonization of the Amazon region and the transmigration to the outer Indonesian islands demonstrate.

Once industrialization takes off, agricultural productivity rises, non-agricultural employment spreads, incomes grow, inequality levels off, the dependence on the land and on forests gradually declines and rates of deforestation begin to fall. Thailand reached this stage in the late 1980s (at income levels of US$1,000 per capita). Marginal land that was formerly cultivated is now left fallow or abandoned and is being progressively recolonized by forests (assuming for the moment that there has been no irreversible damage to the soil). At the same time, a combination of government-sponsored reforestation projects and the establishment of private plantations and national parks eventually reverses the trend into a net forest gain, as has happened in much of Western Europe and parts of North America. (The assumption of reversibility will be relaxed later on.)

To test the hypothesis that deforestation follows an EK-curve, we postulate deforestation as a function of income per capita and population density:[1]

$$\ln \text{DEF} = \alpha_1 \ln \text{INC} + \alpha_2 \ln \text{POP} + \tfrac{1}{2} \alpha_{11} (\ln \text{INC})^2 + \alpha_{12} (\ln \text{INC})(\ln \text{POP}) + \tfrac{1}{2} \alpha_{22}(\ln \text{POP})^2.$$

where: DEF = rate of deforestation (+1);
INC = income per capita;
POP = population density.

Environmental Degradation and Economic Development

The translog functional formulation permits us to test the hypothesis of a non-linear, inverted U-shape relationship between deforestation and income per capita. We would have failed to reject the hypothesis if:

$$\epsilon = \frac{\partial \ln \text{DEF}}{\partial \ln \text{INC}} > 0 \quad \text{for } 0 < \text{INC} < X$$
$$\epsilon < 0 \quad \text{for INC} > X$$

where: ϵ is the elasticity of deforestation with respect to income per capita and X is the level of income corresponding to the maximum 'predicted' level of deforestation;

and: $\dfrac{\partial^2 \text{DEF}}{\partial (\text{INC})^2} < 0 \quad$ throughout.

Using mid-to-late-1980s figures from 41 tropical, mostly developing countries and ordinary least squares regression techniques, we obtained the following parameter estimates for the equation above:

ln DEF = 0.035 ln INC − 0.0385 ln POP − 0.00467 (ln INC)² +
          (3.14)   (3.08)   (3.57)

        0.00108 (ln POP)² + 0.0459 (ln INC) / (ln POP)
        (1.50)   (3.69)

Adj. R^2 = 0.75; degrees of freedom = 35

The variations in the explanatory variables, income per capita and population explain 75 per cent of the variation of deforestation among tropical countries. Both variables, their square terms and their interaction are statistically significant at the 5 per cent significance level. The hypothesis of an inverted U-shaped relationship between deforestation and income is not rejected. Our results lend support to the hypothesis that resource depletion, at least in the case of tropical deforestation, exhibits a U-shaped relationship with the level of development as represented by income per capita, since:

$\epsilon > 0$ for INC < X

$\epsilon < 0$ for INC > X

where: $\epsilon = \partial \ln \text{DEF}/\partial \ln \text{INC} = 0.0627 - 0.00934 \ln \text{INC}$
evaluated at the geometric mean of POP;

and where the mode (the turning-point income) of the inverted U-curve is:

$$X = e^{\ln \text{INC}} = e^{6.713} = 823$$

and: $\partial^2 \ln \text{DEF}/\partial (\ln \text{DEF})^2 = -0.00934 < 0$ throughout.

The elasticity of deforestation with respect to income per capita (ϵ) varies according to the level of income per capita. At the geometric mean of income, $\epsilon = 0.19$. At the minimum income value in the sample (US$126 for Ethiopia), $\epsilon = 1.12$ and at the maximum income in the sample (US$3,226 for Venezuela), $\epsilon = -0.81$.

The estimated EK-curve (Figure 2.1) indicates a rate of deforestation of 1.3 per cent per annum beginning at US$100 per capita and rising to reach a peak level of 3.5 per cent per annum at an income of US$823 per capita. Thereafter, the rate of deforestation drops slowly below 3 per cent as income rises above US$2,000 per capita. Net deforestation continues to decline slowly, falling under 2 per cent at income levels above US$4,000 and reaches zero only when income levels exceed US$12,000 per capita. Thereafter, deforestation becomes negative, i.e. *net* reforestation takes place. At a per capita income of US$20,000 per capita (e.g. Switzerland), net reforestation reaches 1.4 per cent per annum.

However, since the above parameters were obtained with a sample of developing (tropical) countries alone, extrapolations of the results to the developed countries might be risky. For this reason, we ran a second model with an extended sample that includes 27 developed countries in addition to the 41 developing countries. Using ordinary least square

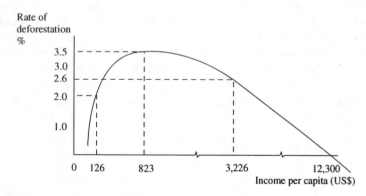

Figure 2.1 Environmental Kuznets curve (EK-curve): tropical deforestation

regression techniques, we obtained the following estimates for the translog deforestation function:

ln DEF = 0.0133 ln INC − 0.0181 ln POP − 0.0013 (ln INC)^2 +
(1.71)   (1.75)   (1.95)

0.0011 (ln POP)^2 + 0.00096 (ln INC) (ln POP) +
(1.51)   (1.74)

0.0189 (TROPIC)
(4.13)

Adj. R^2 = 0.63; degrees of freedom = 60

In this model, the variations in population and income, their interactions and a dummy variable for tropical countries explain 63 per cent of the variation in deforestation. All variables are statistically significant at the 90 per cent confidence level, except for the square of population. The dummy variable representing tropical countries is significant at the 99 per cent confidence level and positive, suggesting a significantly higher deforestation rate in tropical developing countries than in temperate developed countries. On the average, deforestation rates in tropical countries are 1 per cent higher than in temperate countries at the same level of income and population.

With the inclusion of developed countries, the level of income per capita at which the rate of deforestation begins to decline rises from US$823 to US$1,200. With the new estimates, the rate of net reforestation for a country with approximately Switzerland's income per capita (US$20,000) is predicted to be about 1.8 per cent, while in a poor country such as Ethiopia (US$125), the net deforestation rate is predicted to be 1.0 per cent, which is only slightly lower than the rate predicted with 'the model based on a sample of developing countries only'.

Environmental pollution and level of development

The level of environmental pollutants such as SO_2, NO_x and SPM varies over time and across countries with the quantity and type/source of energy consumed, which itself varies with: (a) the level and type of industrial activity; (b) the number and quality (vintage) of vehicles; (c) domestic electricity consumption; and (d) pollution control and abatement (induced by environmental regulations, economic incentives and public sector environmental expenditures). Clearly, the levels of industrial activity, the number of vehicles and the amount of consumption of electricity are directly related to the stage of economic development and level of income. Hence, the consumption of energy and the generation of emissions will rise with the level of development (represented here by income per capita). On the other hand, the *type* of industrial activity, the vintage of the capital stock, the age and quality of vehicles and the quality of fuels and other energy sources tend to improve with the level of economic development. While there are more sources of pollution at higher levels of development, each source generates fewer emissions per unit of output. This is partly the result of introducing new technology as obsolete capital stock is replaced, and partly the result of improved economic efficiency. As the country ascends the ladder of industrialization, it gradually exhausts the opportunities for import substitution behind protective tariff walls and increasingly moves towards export-oriented industrialization. This, in turn, requires a reduction of protection and a realignment of domestic prices with world prices, factors which encourage improved efficiency in input use and hence reductions of waste per unit of output. On the other hand, the expansionary effect of export-oriented industrialization means a greater output and possibly greater aggregate level of emissions and wastes than under the earlier inward-looking, import-substituting industrialization.

At a later stage of development a combination of two factors ensures that not only emissions per unit of output but also aggregate emissions begin to decline. These are: (a) increased regulation of emissions (through ambient and effluent standards and/or economic incentives) in response to arising demand for income-elastic environmental amenities, including clean air; and (b) advanced structural change away from industry to services and within industry from energy-intensive, highly polluting industry to technology- and knowledge-intensive industries such as electronics, which generate lower levels of pollution per unit of GDP.

In order to test the hypothesis that emissions first rise with the level of economic development and later decline when a certain level of income is attained, we focus on three air pollutants, SO_2, NO_x and SPM. These pollutants are expressed in per capita terms at a point in time in a cross-section of countries and are related to income per capita at that particular point in time. The sample used in the analysis includes both developed and developing countries in the late 1980s. For income per capita we used figures from World Bank (1988). National figures for emissions were not available for most countries. Available figures refer to ambient levels in urban areas of selected countries in different years.

National-level data on emissions of SO_2, NO_x and SPM for developed countries are available from the OECD *State of the Environment* (SO_2 and

NO$_x$ for the late 1980s and SPM for 1987). For developing countries such data are not available. As a rough approximation, we constructed data for emissions of SO$_2$, NO$_x$ and SPM from data on the consumption of petroleum, coal and natural gas, which account for over 90 per cent of man-made emissions.

Our use of uniform conversion factors for developing countries obscures differences in fuel-specific technology among these countries, and to some degree it distorts the relationship between emissions and levels of development. It is not, however, tantamount to representing variations in emissions by variations in energy use, since the mix of fuels also varies between countries, as does the level of emissions, with coal being the most polluting and natural gas the least polluting energy source. Thus, of the three factors that determine the level of emissions, energy consumption, fuel mix and technology, our emission surrogate captures the first two and obscures the latter.

As countries embark on rapid industrialization, energy use per capita increases and tends to shift towards dirtier fuels, such as coal or lignite, that are often available from domestic sources (e.g. China, India and Thailand). At a much higher level of development, countries stabilize or even reduce energy consumption per capita (and per unit of GDP) and shift to cleaner fuels, such as natural gas. All this is captured by our model.

What is not captured is the use of cleaner production technology and the installation of pollution control equipment, such as scrappers to reduce SO$_2$ or filters to reduce SPM. This is admittedly a weakness of our model that cannot be resolved until national level data on ambient air levels become available.

The relationship between emission of SO$_2$ per capita and income per capita was specified to be a log-quadratic function in the following form:

$$\ln (SO_2/N) = \alpha_0 + \alpha_1 \ln INC + \alpha_2 (\ln INC)^2$$

where: SO_2/N = SO$_2$ per capita;
INC = income per capita.

Using late-1980 figures for 55 countries (both developed and developing) and ordinary least square techniques, we obtain the following parameter estimates for the second equation:

$$\ln (SO_2/N) = -35.26 + 8.13 \ln INC - 0.51 (\ln INC)^2$$
$$(-5.8) \quad (5.2) \quad (-5.3)$$

Adj. R^2 = 0.33; degrees of freedom = 51

The hypothesis of an inverted U-shaped relationship between SO_2 emissions per capita and income per capita was not rejected. However, only 33 per cent of the total variations in emissions is explained by income. At an income level of US$300 per capita (the lowest in the sample represented by India), the elasticity of emissions with respect to income was found to be 2.31; that is, a 1 per cent increase in income per capita results in a 2.3 per cent increase in emissions. However, as income rises, the emissions elasticity declines, reaching a unitary value at about US$1,000 (e.g. Thailand): emissions grow proportionally with incomes. The emissions elasticity, with respect to income, continues to decline, reaching zero at about US$3,000 per capita (Portugal). Thereafter, increases in income contribute to a reduction rather than a rise in emissions. At a per capita income of US$5,000 (e.g. Greece), the emissions elasticity is –0.55, i.e. a 1 per cent increase in income results in an 0.55 per cent reduction in emissions. A unitary (but negative) emissions elasticity is reached at income levels of US$8,000 per capita (e.g. Singapore), indicating a proportional reduction in emissions with income growth. Emission reductions become income-elastic at about US$10,000 (e.g. Italy) with the elasticity reaching –2.00 at an income level of US$20,000 per capita (e.g. Switzerland).

Correspondingly, our estimates predict 0.005 tons of emissions per capita for India, 0.032 tons for Thailand, 0.049 for Greece, 0.034 for Singapore, 0.026 for Italy and 0.009 for Switzerland. The implied inverted U-shaped relationship between SO_2 emissions and incomes is depicted in Figure 2.2.

In an alternative formulation that includes a dummy variable for high-income, oil-exporting countries which tend grossly to underprice domestic energy, we were able to increase the explanatory power of the model to 50 per cent. In this new formulation, the critical level of income (at which the emissions elasticity becomes zero) rises by US$900 (from US$2,900 to US$3,800) per capita. Indeed, high-income oil exporters emit 0.25 tons more SO_2 per capita than non-oil exporters at the turning-point of their inverted U-shaped curve.

Similar results were obtained for NO_x emissions modelled as a quadratic function of income per capita:

$$\ln (NO_x/N) = 25.36 + 4.82 \,[\ln INC] + 0.28 \,(\ln INC)^2$$
$$(-5.6) \quad (4.19) \quad\quad\quad (-3.88)$$

Adj. $R^2 = 0.35$; degrees of freedom = 51

All parameter estimates are statistically significant at the 99 per cent confidence level. Again, the hypothesis of an inverted U-shaped relation-

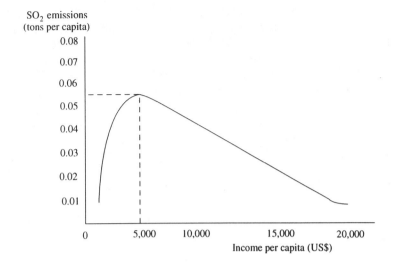

Figure 2.2 An environmental Kuznets curve between SO_2 emissions and income per capita

ship between NO_x emissions per capita and income per capita is not rejected. However, the turning-point of the NO_x curve occurs at a significantly higher income level than that of the SO_2 curve: increases in per capita income begin to result in reductions of emissions per capita at about US$5,500. At the lowest income level (US$300) in our sample, represented by India, the income elasticity of emissions is 1.63, indicating that a 10 per cent increase in income per capita results in a 16 per cent increase in NO_x emissions per capita. In contrast, at the highest income level (US$21,332), represented by Switzerland, a 10 per cent increase in income results in a 7.6 per cent reduction in NO_x emissions.

Finally, we tested the hypothesis that emissions of solid particulate matter (SPM) behave over the course of economic development in a fashion similar to SO_2 and NO_x, i.e. they exhibit an inverted-U relationship with income per capita. While the hypothesis has not been rejected, income per capita and its square term explained only 12 per cent of the variations in emissions, a clearly unsatisfactory result. SPM emissions rise with income to reach a maximum of 0.01 tons per capita at an income level of US$4,500 (e.g. Spain) and begin to decline thereafter reaching a level of 0.005 tons per capita at the level of income currently enjoyed by Switzerland (US$21,332).

The above results, albeit preliminary and subject to confirmation with better data and further analysis, tend to lend credence to the hypothesis of an environmental Kuznets curve in the sphere of air pollution. Our results are corroborated by the findings of other recent studies. Grossman and Krueger (1991) analysed the relationship between SO_2 and SPM concentrations and income levels in major urban centres. They found that SO_2 and SPM concentrations are positively correlated with per capita income levels lower than US$5,000 (1985) and negatively correlated with income levels greater than US$5,000.

These figures are surprisingly close to our findings, which put the turning-point in the range between US$3,800 and US$5,500. Similarly, a World Bank study by Shafik and Bandyopadhyay (1992) found a quadratic relationship for SO_2 with a turning-point around US$3,670 per capita, and for suspended particle matter with a turning-point of US$3,280 per capita. Again, these findings corroborate the results of the present study, despite the difference in data and formulation (the use of data on ambient levels in selected urban areas between 1972 and 1988).

POLICY IMPLICATIONS

Our analysis, both theoretical and empirical, suggests an inverted U-shape relationship between at least two forms of environmental degradation and economic development. While more appropriate data and further analysis are necessary, studies carried out simultaneously and independently have found similar results, which lends support to the hypothesis of the EK-curve. Our findings indicate that the turning-point for deforestation occurs much earlier (at US$800–1,200 per capita) than the turning-point for emissions, US$3,800–5,500. This is as one would expect. Deforestation for either agricultural expansion or surplus extraction (via timber exports) takes place at an earlier stage of development than heavy industrialization. Environmental degradation overall (combined resource depletion and pollution) is worse at levels of income per capita under US$1,000. Between US$1,000 and US$3,000, both the economy and environmental degradation undergo dramatic structural change from rural to urban, from agricultural to industrial. A second structural transformation begins to take place as countries surpass a per capita income of US$10,000 and begin to shift from energy-intensive heavy industry into services and information/technology-intensive industry.

The non-rejection or provisional acceptance of the EK-curve hypothesis has important policy implications. First, it implies a certain inevitability of environmental degradation along a country's development path, especially during the take-off process of industrialization. Second, it suggests that as

Environmental Degradation and Economic Development

the development process picks up, when a certain level of income per capita is reached, economic growth turns from an enemy of the environment into a friend, although for quite a while growth helps only to undo the damage done in earlier years. Nevertheless, economic growth appears to be a powerful way for improving environmental quality in developing countries – even more powerful than our results seem to suggest, because we have not captured the effect of the income-elastic expenditure on pollution abatement and the increased supply of environmental amenities. If economic growth is good for the environment, policies that stimulate growth (such as trade liberalization, economic restructuring and price reform) ought also to be good for the environment. This in turn would tend to suggest that the environment needs no particular attention, in terms of either domestic environmental policy or international pressure or assistance; resources can best be focused on achieving rapid economic growth to move quickly through the environmentally unfavourable stage of development to the environmentally favourable range of the Kuznets curve.

However, there are several reasons why this policy may not be optimal. First, the positively sloping part of the curve, where growth worsens rather than improves, may take several decades to cross, in which case the present value of higher future growth and cleaner future environment may be more than offset by high current rates of environmental damage. Therefore, efforts to mitigate emissions and resource depletion in the earlier stages of development may be justified on purely economic grounds. Second, it may be less costly to prevent or abate certain forms of environmental degradation today rather than in the future; for example, it might be less costly to treat and safely dispose of hazardous waste as it is being produced than to leave it scattered and difficult to collect and treat later on once economic growth has both increased the demand for a cleaner environment and the available resources to accomplish it.

Third, and perhaps more importantly, certain types of environmental degradation allowed at an earlier stage of development may be physically irreversible at a later stage. Tropical deforestation, the loss of biological diversity, the extinction of species, the destruction of unique natural sites are either physically irreversible or prohibitively costly to reverse. In terms of pollution, lead emissions from the burning of leaded fuels and radioactive contamination from nuclear accidents or the operation of unsafe reactors may also have irreversible effects. For example, a study prepared jointly by the United States Agency for International Development and the Environmental Protection Agency (USAID/EPA, 1990) found that lead emissions from the burning of leaded petrol are responsible for high levels of lead in the blood of school-age children and a consequent slow-down of their mental development and school performance that is equivalent to the loss of five IQ points; the resulting loss of earning capacity of the affected children and the loss of development potential for the country are not easily reversed by switching to unleaded petrol at later stages of development.

This leads us to the fourth reason why an exclusive focus on policies and investments that directly promote economic growth may not be sufficient. Certain forms of environmental degradation, such as soil erosion, watershed destruction, loss of resilience to natural disasters, sedimentation of irrigation and hydroelectric reservoirs, damage to human health and productivity, loss of working time owing to congested traffic, and respiratory diseases, all constrain economic growth. Therefore, environmental degradation may need to be attacked directly through environmental policies and investments in order to remove obstacles to economic growth itself, from which a second generation of environmental benefits are expected.

To sum up, while an EK-curve is an empirical reality and an inevitable result of structural change accompanying economic growth, it is not necessarily optimal. In the presence of ecological thresholds that might be crossed irreversibly (see Figure 2.3), and of complementarities between environmental protection and economic growth (see Figure 2.4), a deep EK-curve (implying high rates of resource depletion and pollution per unit of incremental GDP per capita) is neither economically nor environmentally optimal, because more of both could be obtainable with the same resources, if better managed. Perhaps the source and pattern of economic growth are as important as the rate of economic growth, if not more so.

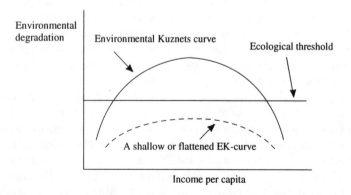

Figure 2.3 Environmental Kuznets curve (EK-curve) and ecological thresholds

Environmental Degradation and Economic Development

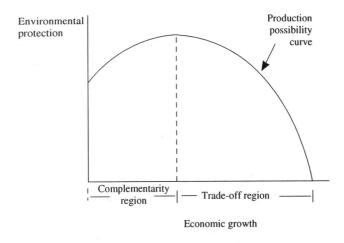

Figure 2.4 Complementarities and trade-offs between environmental protection and economic growth

Economic growth that results from the subsidized use of energy and raw materials or from the liquidation of natural capital imposes a higher cost on the environment than does growth that derives from efficient resource use, labour-intensive industries and services and information technology. Economic growth that takes place in an environment where property rights over resources are not well defined and where secure and environmental costs are not accounted for and internalized imposes an excessively high environmental cost on society, which ultimately undermines the sustainability of growth itself.

The reverse danger is also possible, i.e. when developing countries prematurely adopt the strict environmental standards of developed countries, and attempt to attain them overnight through strict end-of-pipe emission standards and requirements for the mandatory instalment of waste treatment facilities and pollution abatement equipment. This over-ambitiousness is a prescription for retarding economic growth without necessarily improving the environment. It simply stretches, as it were, the EK-curve without making it shallower. A superior alternative that flattens the EK-curve without retarding growth is the adoption of more appropriate environmental standards for the country's level of development, or a gradual phasing-in of international standards, pursued through flexible market-based instruments or economic incentives/disincentives rather than rigid command-and-control regulations. A consistent structure of economic instruments such as environmental taxes, pollution charges, user fees,

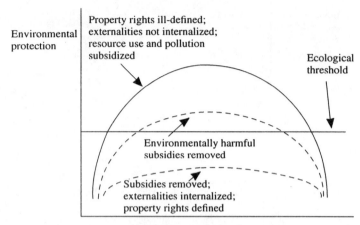

Figure 2.5 Flattening out the Environmental Kuznets curve (EK-curve) through removing environmental harmful subsidies, internalizing externalities and ensuring the clear definition and enforcement of property rights over natural resources

deposit refund systems, tradable permits and transferable development rights would direct new investments and eventually the structure of the economy away from energy-intensive, highly polluting industries towards more environmentally benign agricultural, industrial and service sectors. If gradually phased in, a combination of flexible and revenue-neutral economic instruments can ensure that environmental management does not constrain economic growth but only helps to direct it towards a more sustainable trajectory. With full-cost pricing of resources that reflects growing scarcities and full internalization of externalities, the EK-curve would not disappear but would be a shallow one (Figure 2.5). The persistence of the shallow EK-curve underlies a changing structure of demand and falling effective discount rates with growing incomes.

IMPLICATIONS FOR EMPLOYMENT, TECHNOLOGY AND EXTERNAL ASSISTANCE

Certain forms of environmental degradation, such as forest encroachment, the cultivation of marginal lands and erodible soils, urban squatting and deteriorating urban slums, are related to the prevalence of underemployment, landlessness and illiteracy. Ideally, one would expect an abundance of low-cost labour to be absorbed by the growing industrial and service sectors. This does not happen generally because of a combination of subsidized capital intensity, minimum wage laws and lack of essential skills. Particularly damaging is the use of technology that does not correspond to a country's resource endowment or relative factor prices, e.g. investment promotion with general tax and credit incentives results in capital- and energy-intensive industries while a country with abundant low-cost labour has a competitive advantage in labour-intensive activities that help both to alleviate poverty and to limit the damage to the environment. As structural changes in the demand for labour lag behind the structural change of output, underemployed farmers and labourers colonize forests, fishing grounds, pastures and other public or open access lands in search of a livelihood, causing considerable damage to the resource base and to the environment, simply because they lack the capital, technology and security of tenure to exploit resources sustainably. Others crowd the cities, where they squat on public lands, burden an already inadequate infrastructure, scatter solid waste in empty lots and waterways and cause congestion and unsanitary conditions in urban slums. This situation results in a steeper and lengthier left leg of the EK-curve and a greater likelihood that ecological thresholds will be crossed. A cost-effective solution calls for the removal of direct and indirect subsidies, the reduction of protection for capital-intensive industries that employ little labour and the promotion of more labour-intensive activities, coupled with increased investment in education and training. Titling untitled agricultural land and allocating water rights to farmers and other rural dwellers would provide them with access to capital markets, thereby increasing on-farm investments that improve agricultural productivity and help to arrest the erosion of the resource base.

Much has been said about the need for technology transfer to enable developing countries to achieve sustainable development. The idea derives from the observation that production technologies in developed countries are cleaner than those in developing countries and that a wide range of pollution abatement technologies are also available in developed countries. Since most developing countries lack the financial resources to import these technologies at commercial cost, the case has been made that developed countries should transfer these technologies to LDCs on concessionary terms. Even at low cost, the transfer of such technologies to developing countries could prove to be a double-edged sword. On the one hand, less polluting technologies may reduce emissions per unit of output. On the other hand, such technologies, designed to be profitable under conditions of high-cost labour and low-cost capital, might be poorly suited to countries with reversed relative prices. If this mismatch is ignored, the transfer of inappropriate technologies, even at concessionary cost, may further reduce employment and intensify resource depletion and environmental degradation.

The net impact of technology transfer might vary depending on the industry. In the power and transport sectors, more efficient technologies are certain to have net environmental benefits. This may not be so in other industries such as construction, textiles, agriculture and services. Rather, what developing countries appear to need more than technology transfer is the transfer of information, know-how and skills to enable them to design their own technologies or modify existing ones.

Developing countries are not only short of technology; they are also short of financial resources. A number of industrialized countries are experiencing a net outflow of financial resources to service their external debt. While the case for debt rescheduling, and, in some cases, debt relief, has been made and implemented in a number of countries, the EK-curve suggests two additional types of external funding: (a) environmentally sound development financing; and (b) incremental cost financing of local projects with global environmental benefits.

The scope for development financing arises from the need to speed up economic growth, which offers a way out of the vicious circle of mutually reinforcing poverty and environmental degradation. The traditional sources of external funds (grants and loans) have been bilateral and multilateral donor agencies. Though a qualified success, development financing has not been either adequate or fully cognizant of its environmental implications, both positive and negative. For example, a study of the environmental impacts of the World Bank's structural adjustment programme for Thailand (Panayotou and Sussangkarn, 1992) found that the programme helps to reduce pollution per unit of output by improving economic efficiency, but that its expansionary impact resulted in higher aggregate pollution because it did not provide for the internalization of environmental costs. Moreover, resource sectors with secure property rights benefited from increased conservation while sectors with ill-defined or insecure property rights suffered from increased resource depletion. In this respect, the Land Titling Loan that accompanies the Bank's structural adjustment programme may be regarded as an effort to flatten out the EK-curve even though its primary purpose is productivity growth. Future structural adjustment programmes and other policy-related lending can help to flatten out the EK-curve significantly by explicitly providing for secure property rights over resources and for economic instruments to internalize environmental externalities.

Similarly, project lending must fully internalize the environmental impacts of projects such as irrigation systems, and price natural resources such as water at their full scarcity value and environmental costs, where the latter includes the cost of mitigating environmental impacts and resettling affected people. For example, the Nam Pong reservoir in Thailand, which has not provided for the protection of its watershed, for the resettlement of displaced people and for economic water pricing, lies on a 'steep EK-curve' while the Dumoga Bone Irrigation System *cum* National Park in Sulawesi, Indonesia, lies on a much 'shallower EK-curve' for having done all of the above.

Incremental cost financing for environmental projects that cannot be justified on conventional economic grounds or that do not generate sufficient domestic benefits to cover costs has two related justifications: (a) to avert irreversible environmental losses that may occur in earlier stages of development because of a high discount rate and a low demand for environmental amenities at low income levels; and (b) to make possible the realization of global environmental benefits that would not be realized otherwise, such as the conservation of biodiversity, the protection of the ozone layer and the reduction of greenhouse gases. Since global environmental assets are mainly of interest to developed countries, which have high incomes and low discount rates, they are also, in a sense, the very assets for which developing countries will have a high demand, once they reach comparable income levels; and they are assets which they will wish they had saved. However, in so far as the loss of environmental assets is not irreversible or recovery prohibitively costly, a minimum-regret policy can be moderated by restoration at a latter phase in the development process.

Incremental financing of local projects in developing countries by the Global Environment Facility (GEF) could also be considered an effort to flatten out the EK-curve of developing countries. The main problems with the GEF are its limited funding, a project-by-project approach and a limited opportunity to utilize market forces. A much larger effort, at the level of US$125 billion per annum, was estimated by the 1992 United Nations Conference on Environment and Development (UNCED) as the level of financing needed for Agenda 21, a programme of policies and projects designed to help developing countries to shift to a sustainable development path. Such a path can be understood in the context of the present study as a path along a shallow Kuznets curve, much shallower than is the case at present. (For financing mechanisms to fund Agenda 21, see Panayotou, 1992.)

CONCLUSION

Like inequality, environmental degradation tends to become worse before it becomes better along a country's development path. While some deterioration is inevitable as part of the immutable structural changes that accompany economic growth, the EK-curve need not be as steep as appears to be the case in many developing countries. Part of the steepness of the inverted U-shaped relationship between environmental degradation and growth is due to policy distortions such as subsidies on energy and agrochemicals, the protection of industry and the underpricing of natural resources which are both economically and environmentally destructive. Developing country governments can help to flatten out their EK-curve by: (a) eliminating policy distortions; (b) internalizing environmental costs to the activities that generate them; and (c) defining and enforcing property rights over natural resources. Development assistance agencies can help further in flattening out a developing country's EK-curve by making environmental protection an integral part of their project and policy financing. Finally, developed countries can help developing countries, through creative financing mechanisms such as the Global Environment Facility, to conserve natural resources, including biodiversity, which generate global benefits which might be lost irreversibly during the earlier stages of economic development.

Note

1. The translog formulation adopted has the advantages of allowing for interaction efforts between variables and of permitting direct reading of elasticities.

3 Sustainable Livelihoods and Environmentally Sound Technology: Theoretical and Conceptual Approaches[1]

Charles Perrings

The programme of action drawn up following the 1992 Rio Earth Summit, Agenda 21 (UNCED, 1993), introduces a number of currently very imprecise concepts. The policy and other implications of these concepts are accordingly not obvious. Much work has already gone into the operationalization of the various proposals contained in Agenda 21, but considerable ambiguity remains about certain of the key concepts and the programmes to which they relate. This chapter focuses on two: 'sustainable livelihoods' and 'environmentally sound technology' (hereafter SL and EST). These are subcategories of the general concepts 'sustainable development' and 'environmental soundness', respectively. Hence the study also considers both the content and the operational relevance of these more general concepts.

There are three steps in the operationalization of concepts such as SL and EST:

(a) identification of the content of the concept;
(b) identification of its role in economic decision problem(s); and
(c) application of the concept to specific decision problems.

This chapter is concerned with the first two steps only. Specific applications are beyond its scope. Nor does it seek to analyse current policies in terms of these concepts. Indeed, it is not immediately obvious if and how SL and EST currently enter either the objective function or the constraint set of economic decision-makers. The study does, however, seek to derive a set of principles that might be used in the application of these two concepts in different cases.

SUSTAINABLE DEVELOPMENT

Sustainable development as understood in the Brundtland Report (WCED, 1987) is the leitmotif of Agenda 21. Principles 3 and 5 of the Rio Declaration on Environment and Development affirm that development should equitably meet the needs of future generations whilst eliminating poverty within this generation. Principle 4 affirms that environmental protection should be an integral part of the development process. All three principles accord closely with the Brundtland concept of sustainable development. Where Agenda 21 differs from the Brundtland Report and earlier discussions of sustainable development is in the assumption that sustainability is a criterion that may be used to assess all economic behaviour and the use of all resources – that sustainability of the whole implies sustainability of each of the parts. Agenda 21 seeks to promote sustainability not just of the development process but also of each aspect of the development process. It is asserted, for example, that 'sustainable development must be achieved at every level of society' (UNCED, 1993). This requires not only the sustainable 'development' of all communities within a society, but also the sustainable 'livelihood' of individuals within those communities, and the environmental soundness of each process.

We shall consider the divisibility of sustainable development in the context of SL. To begin, however, it is useful to characterize the more general concept of sustainability. This has been extensively discussed in the literature and it is neither feasible nor desirable to offer a comprehensive review of what has become a very diverse set of contributions, as was stressed in the introductory chapter (Chapter 1). However, for the purpose of this chapter it is necessary to distil a small number of common elements from this literature, which are shared by Agenda 21. Any infinite-horizon economic process may be said to be sustainable (from the perspective of the present) if the welfare of society is non-declining in terms of the present structure of preferences (Pezzey, 1989). Every contribution to the literature explicitly or implicitly accepts this proposition. It is obvious from this that sustainability as an economic concept implies some judgement about intergenerational equity. An economic process is not sustainable if future generations are left worse off than the present generation by the standards of the present generation, i.e. if the intergenerational distribution of income is less than egalitarian. This is, of course, explicit in Agenda 21.

It is well understood that a necessary condition for the real consumption expenditure of future generations to be no less than that of the present generation is that the value of the capital stock should be non-declining

(Solow, 1974). True (Hicks/Lindahl) income is, in fact, defined by the maximum amount which may be spent on consumption in one period without reducing real consumption expenditure in future periods: that is, income is the level of real consumption expenditure that leaves society as 'well off' at the end of a period as at the beginning, and society can only be as well off at the end of the period if it is not consuming its productive assets. Since productive assets include the resources of the natural environment, it follows that a necessary condition for the protection of the consumption possibilities open to future generations is that the value of the produced or man-made capital stock plus the value of the resources of the natural environment should be non-declining (see Hartwick, 1977, 1991; Solow, 1986; Daly and Cobb, 1989; Mäler, 1990; Pearce and Turner, 1990; Pearce and Mäler, 1991; Perrings, Folke and Mäler, 1992).

Not all treatments of sustainability explicitly discuss intergenerational equity and the value of the asset base. This is the case with Agenda 21. There is no formal statement concerning intergenerational equity beyond the Brundtland requirement that development should be such that the needs of future generations may be met. Nor is there any formal statement concerning the value of environmental assets. Instead, there is a general emphasis on the need to protect the environmental resources to be passed on to future generations. However, it is clear that sustainability does require conservation of the value of the aggregate capital stock, and that this does include environmental assets. Indeed, a large part of the problem in operationalizing this requirement is that for many environmental assets there are no markets, and so no prices. Any valuation of the asset base requires both a proper specification of the resource base to include all relevant environmental resources, and the proper valuation of those resources. Neither exercise is trivial.

On the identification of the asset base, there is a general consensus that the maintenance of the value of the asset base allows for some substitution between natural and produced capital, but there is also a consensus that the degree of substitutability is limited. This is because of the properties of biotic and abiotic cycles, the multifunctionality of biotic resources, and the irreversibility of many environmental effects. Differences concerning the degree of substitutability between produced and natural capital have led to a distinction in the literature between weak sustainability (involving the maintenance of the value of the aggregate capital stock) and strong sustainability (involving the maintenance of the value of natural capital alone) (Pearce and Turner, 1990). The significance of the distinction will become clear later. For now, what is important is that a flow of income to an individual household or community will be judged to be sustainable only if it involves no net depreciation in the value of the set of all assets/natural assets affected by the income-generating activity.

On the valuation of the asset base, if there exist well-functioning, complete and competitive markets, if there are no policy distortions and if there is full information, the price of individual assets will be a good approximation of their value to society. There are very few environmental assets for which these conditions are satisfied. But even if they are not satisfied, or are only partly satisfied, it is in principle still possible to uncover the private value of assets to individual users. The private value of assets includes not only their value in some current use, but also the value that they may have independent of current use (Krutilla, 1967; Pearce and Turner, 1990), or the value of the option to make use of them in the future (Weisbrod, 1964; Arrow and Fisher, 1974). Most methods of estimating such private value are susceptible to bias in several directions (Kolstad and Braden, 1991) but, subject to this, the aggregation of such values may be taken to approximate the social value of capital.

Where individual resource users base their decisions on a set of private costs that reflects poorly functioning, incomplete and non-competitive markets, a range of policy distortions and less than full information, the resulting allocation will be both inefficient and, in general, unsustainable. A sufficient – though not a necessary – condition for the social value of the stock of all assets (approximated by such estimates) to be non-declining over time is that each and every activity involves no net depreciation of the asset base affected by that activity: that each and every resource user 'makes good' any loss in the value of natural assets in their possession by equivalent investment in substitute assets. This requires that resource users take the social value of assets into account in making their consumption and investment decisions. The problem here lies in the fact that the users of environmental resources are seldom confronted by the social cost of their use of the resource owing to the absence of markets and the existence of distortionary policies, while their own valuation of the resource is biased (often in the same direction) by ignorance, uncertainty, insecurity of tenure and myopia. Indeed, it is argued that this is a major cause of environmental degradation (Perrings, Folke and Mäler, 1992). The issues raised by this problem are discussed below.

Poverty, affluence and the private valuation of environmental resources

The Brundtland Report asserted that the amelioration of poverty and the improvement of the distribution of income and assets were both necessary conditions for the sustainability of development. This has become part of the received wisdom in those organizations responsible for promoting sustainable development (see, for example, World Bank, 1992b), and is reproduced as a guiding principle of Agenda 21. The basis for this widespread acceptance of the idea lies in the empirical observation that there exist a number of positive feedbacks between poverty and environmental degradation. It turns out that poverty is closely linked with the gap between the private and the social valuation of resources, and so with the incentive to overuse those resources.

The difference between private and social valuation of environmental resources is due, in part, to two things: the paucity of information available to private resource users, and the rate at which they discount the future. Both are sensitive to the level of market income. Since information is not costless, knowledge of the wider environmental effects of private activities will tend to be a declining function of the income of resource users (Dasgupta, 1992). In addition, those in poverty tend to discount the future costs of resource use at higher-than-average rates (Perrings, 1989a). Put another way, the private valuation of environmental effects of the poor will be lower than that of the rich not only because of differences in ability to pay, but also because the poor screen out more environmental effects from the decision process.

Poverty also affects the responsiveness of individual resource users to the change in the private costs of resource use. Income effects tend to be more pronounced at lower levels of income, and less certain in their direction. The perverse effects observed in the Giffin Good phenomenon, for example, are associated with extreme poverty. Empirical work on supply responses in the low-income countries has revealed that long-run aggregate supply elasticities in many countries are low, and in some cases even negative (Perrings, 1989b; Rao, 1989; Markandya, 1991). The significance of these findings will be explored later.

While it is argued that poverty depresses the private valuation of natural resources, and so encourages their overuse, per capita demand for environmental resources by the poor will be less than the corresponding demand by the rich. However, since poverty is positively correlated with high rates of population growth, or at least with high rates of fertility, it is also positively correlated with increasing aggregate demand for environmental resources. Once again, the causal connections in the individual decision process are complicated and involve both the information and the discount rate effects already discussed. But the net result is that the poor do not take account of the wider and longer-term effects of fertility decisions (Dasgupta, 1992). Poverty distorts the private valuation of the environmental effects of fertility decisions.

At the other end of the spectrum, it is argued that affluence has exactly the same net impact on aggregate demand for environmental resources. The Rio Declaration includes the injunction to eliminate unsustainable patterns of production and consumption and promote appropriate demographic policies (UNCED, 1993), in which consumption refers largely to consumption by the North. Although the rich do place a higher valuation on environmental resources, and although they may discount the future at lower rates than the poor, they consume many more environmental resources in per capita terms. The high private valuation of environmental resources by the rich, their willingness to pay (WTP) for those resources, is a measure first and foremost of their greater *ability* to pay.

Market failure, policy failure and the private cost of environmental resources

There is a very clear sense in the general literature on sustainability that the often implicit intergenerational equity goals of sustainable development are best served by promoting efficiency in the current allocation of resources. Indeed, this is a common theme in many recent contributions (Repetto, 1989, pp. 69–86; Warford, 1989; World Bank, 1992b). It is also a dominant theme in Agenda 21. Sustainable development, it is argued, will be served by trade liberalization, market development, economic policy reform and a range of other efficiency-promoting innovations. The basis for this argument lies in the fact that the private valuation of resources is sensitive to the market price for those resources or goods and services embodying those resources, and that market prices are distorted by the absence of markets for important environmental effects, a range of institutional constraints and the impact of distortionary economic policies. All these factors are argued to drive a wedge between the private and social costs of environmental resource use.

If these costs differ, decisions on private resource use will tend to be intertemporally inefficient, which implies some loss of welfare. However, an allocation which is intertemporally efficient does not necessarily guarantee the welfare of future generations (Norgaard, 1991). Indeed, depending upon the choice of the social rate of discount, it may be socially efficient to run down the value of the stock of assets. If the social rate of discount is higher than the marginal efficiency of capital measured across the whole of the asset base, this will be the case. The presumption made in Agenda 21 is that a liberalization of markets for products using environmental resources in order to realize efficiency gains will be sustainable, but there is no a priori reason why this should be so. There are good arguments for seeking efficiency gains wherever possible, but there is no reason to believe that trade liberalization, say, will be more environmentally sustainable than trade restriction.

*

The central point of the general literature on sustainable development is that the satisfaction of the intergenerational equity goal implicit in the concept requires the protection of the potential productivity of the asset base which, in economic terms, translates as the value of the asset base. A sufficient condition for this is that every individual behaves in a way that ensures no net depreciation of the asset base. In other words, a sufficient condition is that all agents at least replace the value of any assets, including any environmental assets, they may use. This is not, however, a necessary condition. The necessary condition is that the potential productivity of the overall asset base should be maintained. If some individuals are consuming capital, the overall allocation of resources will still be sustainable, providing that other individuals are investing enough resources to compensate. In general, it will always be the case that individuals are net consumers for at least some part of their lives. The requirement in Agenda 21 that all activities be sustainable needs to be interpreted with this in mind. It is unrealistic to expect that refugees, victims of famine, the disabled, children or senior citizens should behave sustainably. Similarly, it is unrealistic to expect that every resource will be used in a sustainable way: either that it will be used at a rate less than or equal to its natural rate of regeneration, or that the proceeds of its consumption will be diverted to investment in a resource of at least equal value. However, if one thinks in terms of the behaviour of communities, rather than of the individuals within communities, and if one thinks in terms of ecological systems rather than the component organisms of those systems, one comes closer to an operational concept.

If sustainable development requires zero net depreciation of the asset base, it requires that any consumption of assets, including environmental assets, should be compensated through investment. Whether consumption of one asset can be compensated by investment in another asset depends on the degree of substitutability between them. The argument about strong and weak sustainability hinges on this. Given the multifunctionality of many organisms within an ecological system, and given the complementarity between them, there are many ecological goods and services which cannot be 'manufactured'. These need to be safeguarded by suitably constraining patterns of both consumption and investment, and this is where the concept of environmental soundness comes in.

THE NOTION OF ENVIRONMENTAL SOUNDNESS

The notion of environmental soundness is less well developed in the economics literature than the notion of sustainability. There is certainly no body of literature debating the nuances of environmental soundness in the same way as there is for sustainability. Nor is there a literature on the way in which environmental soundness is related to the concept of sustainability (Pereira, 1991). Most of the debate on environmental soundness has taken place outside the discipline of economics. Indeed, the sense in which the term is used in the main text of Agenda 21 derives much more from engineering science than it does from either the biological and economic social sciences. The specific interpretation of the term 'environmentally sound technology' collapses two separate ideas, one from the engineering sciences and one from the biological sciences. The more operational of these ideas is that from the biological sciences, and involves some notion of environmental safety. Indeed, it is this notion that the reports editors choose to emphasize. It is, for example, stated in the preamble to Agenda 21 that, wherever the term 'environmentally sound' is used in the report, it should be interpreted to mean 'environmentally safe and sound', and that this is particularly the case in respect of energy and technology.

The engineering literature on the environmental impact of technology contains references to a number of very closely related concepts: 'clean technology', 'cleaner technology', 'best practicable technology', 'low waste technology' and 'resource-conserving technology'. While the main text of Agenda 21 appears to define EST to mean something very similar, it is important to note that these are not at all the same as environmentally 'safe and sound' technology. Whereas environmentally 'safe and sound' technology uses the safety of the technology with respect to the external environment as the criterion of assessment, the other ideas all use the relative volume of environmental inputs or outputs in some process as the criterion of assessment – without reference to the external environment. In order to clarify what is at issue in the general concept of environmental soundness, the following paragraphs focus on environmental safety.

The valuation of environmental resources depends, to a high degree, on the use they have in either production or consumption. This may be direct (if the resources are used directly) or indirect (usually in the form of the services provided by the ecological systems whose functions they support). The total social value of environmental resources is the combination of this direct and indirect use value, together with any non-use value they may have. The notions of environmental safety and soundness refer primarily to the indirect value of environmental resources.

Sustainable Livelihoods and Environmental Soundness

Human society depends on access to a range of environmental services which are supported by the interaction between the organisms, populations and communities – the ecological systems – of the natural environment. What characterizes ecological systems is that their ability to provide such services is a non-linear function of the mix of biotic and abiotic resources which they comprise. More particularly, there exist threshold values for most resources, below which ecosystems cannot function. If certain resources fall below or exceed their threshold values, the ecosystem will tend to lose resilience or productive potential. Holling (1973, 1986, 1987) has described terrestrial and some marine ecosystem behaviour in terms of the sequential interaction between four system functions: exploitation (represented by those ecosystem processes that are responsible for rapid colonization of disturbed ecosystems); conservation (as resource accumulation that builds and stores energy and material); creative destruction (where an abrupt change caused by external disturbance releases energy and material that have accumulated during the conservation phase); and reorganization (where released materials are mobilized to become available for the next exploitive phase). Ecosystem resilience is measured by the effectiveness of the last two system functions. It describes the ability of the system to satisfy exogenous demands for ecological services, and to respond creatively to exogenous shocks. It can be thought of as a measure of the productive potential of the system.

Ecological systems are able to provide ecological services at given levels of stress, providing that they are resilient with respect to that level of stress. Environmental safety and environmental soundness may accordingly be interpreted in terms of the resilience of ecological systems. More particularly, technologies may be said to be environmentally safe and sound if they do not threaten the supply of essential ecological services by exceeding the thresholds of ecosystem resilience. The essentiality of ecological services depends, in large part, on the degree to which they can be substituted by investment in produced capital. It follows that environmental soundness is related to the weak and strong sustainability debate.

One of the difficulties to be tackled in the operationalization of this concept is that there is still considerable uncertainty about which ecological services are essential, and therefore which thresholds most need to be protected. Some environmental goods and services are obviously indispensable to humanity. These include maintenance of the gaseous quality of the atmosphere, amelioration of climate, operation of the hydrological cycle (including flood controls and drinking-water supply), waste assimilation, recycling of nutrients, generation of soils, pollination of crops, provision of food from the sea, maintenance of the genetic library, and so on

(Ehrlich, 1989). But there may be other services that are less obviously critical, but still essential. Alteration of primary productivity, nutrient availability and hydrological cycles all affect the quality and quantity of ecosystem services exploited by human societies. Landscape transformations at the regional level typically change a range of biogeochemical cycles at the ecosystem level. Emissions of toxic pollutants have similar effects. Such changes affect recycling, feedback loops and internal control mechanisms in the ecosystem. They accordingly affect both the production and the maintenance of ecological services. If the system's internal cycling of nutrients and materials is reduced, it can become both more dependent on external inputs of energy and less resilient.

One of the main threats to ecosystem resilience, and hence to environmental safety and soundness, derives from activities which reduce the functional diversity of ecosystems. Functional diversity, in this context, refers to the range of responses to environmental change, including the space and time scales over which organisms react to each other and the environment (Steele, 1991). Since loss of functional diversity generally implies loss of system resilience, it also implies loss of productive potential. The almost universal tendency of economic development to seek productivity gains through ecological specialization – crudely, the tendency towards monoculture – has the effect of reducing functional diversity, and so resilience. This is typically masked in the short term by the use of exogenous inputs such as imported water, industrial energy, fertilizers, pesticides, and so on. Indeed, this is the substance of much modern environmental management. But it also pushes ecosystems much closer to the thresholds of resilience, and this is where the main problems are reckoned to lie in the longer term (Holling and Bocking, 1990; see also Conway, 1987, and Conway and Barbier, 1990, for an application to agriculture).

*

The key elements in the concepts of sustainable development and environmental safety and soundness are, respectively, the maintenance of the value of the asset base and the protection of thresholds of ecosystem resilience. No development process may be said to be sustainable unless the value of man-made and natural capital together is non-declining. No practice may be said to be environmentally safe and sound if it causes the loss of resilience of those ecosystems on which human life and livelihood depends. What is the link between these concepts? First, if the set of prices on which resources are valued are optimal prices, the condition on the value of the asset base ensures that the value of the flow of services derived will be non-declining; that is, the value productivity of the asset base will be non-declining. Second, if the resilience of the ecosystem is protected, this will conserve what has been described as the potential biophysical productivity of the system.

Both concepts accordingly address the problem of productivity, which ultimately refers to the physical potential of the system in some state of nature. The difference between them lies, in part, in the difference in the perception of the system dynamics in each case. The cyclical dynamics of almost all terrestrial and many marine ecosystems, and the tendency for periodic destruction and renewal within the system, are what lie behind the biologists' focus on the potential of those systems. The economic models behind the key concepts of economic sustainability are not characterized by dynamics of this sort. In reality, economic systems behave much like terrestrial ecosystems. They are characterized by strongly cyclic dynamics which also involve the 'creative' destruction and renewal of assets, and this is recognized in the more recent literature on non-linear economic system dynamics (see, for a review, Rosser, 1991). Hence the notion of potential productivity, secured by conserving certain resilience properties of the system, turns out to be relevant to both concepts. Environmental soundness and sustainable development alike can be conceptualized as the conservation of the resilience of the system concerned: its ability to respond creatively and positively to both stress and shock.

Agenda 21 does not deal with either concept in precisely these terms. But the discussion in Agenda 21 is certainly consistent with this interpretation, and since the protection of system resilience is readily operationalizable, it provides a useful reference point for dealing with SL and EST. The broad focus in Agenda 21 is on the protection of a set of critical ecological functions via the incentives faced by people using the ecosystems which provide those functions – where incentives may mean prices or regulations. This is equivalent to the protection of the resilience of those systems. The main concern in Agenda 21 lies in the fact that current incentives do not safeguard those ecological functions. All resource management activities discussed include activities designed to get the structure of incentives right, and the emphasis on liberalization is at least motivated by the same consideration. The main problems identified in Agenda 21 relate to conditions that inhibit the development of incentives to protect key ecological functions, including both market and government failures. Indeed, the discussion of EST is dominated by the technology transfer issue, which is provoked by the failure of markets for best practice technologies.

SUBCATEGORIES: SUSTAINABLE LIVELIHOOD (SL) AND ENVIRONMENTALLY SOUND TECHNOLOGY (EST)

We now turn to the content of the subcategories: SL and EST. These may be thought of as applications of the general concepts of sustainable development and environmental soundness. As has already been remarked, Agenda 21 proceeds from the assumption that sustainability/soundness of the whole implies sustainability/soundness of the parts. This can be interpreted in one of two ways: either that each household, firm or community should itself be, in some sense, sustainable; or that each household, firm or community should behave in a way that is consistent with the sustainability of the wider system.

Consider these two interpretations separately. Under the first, the protection of the biosphere is taken to imply the preservation of each of its component parts, and sustainability of the economy is taken to imply that the income and assets of each household or firm should be non-declining. This is too stringent a requirement to be useful. Conservation of the productive potential of both environmental and economic systems is not generally compatible with conservation of the productive potential of each component part of the system. Particular communities which thrive at some points in the ecological cycle may crash at other points without prejudice to the resilience of the ecological system. Similarly, individual households and firms, even whole industries and the regional economies they support, may expand the capital they command at some points in the economic cycle and may contract it at others, without prejudice to the resilience of the economic system. Indeed, both restructuring and market liberalization encourage such changes to take place. Preservation of both the economic and the ecological status quo is not an option.

Under the second interpretation, the use of environmental resources by communities, households and firms should not prejudice the resilience of either the economy or the biosphere. This is a more consistent and more easily operationalizable interpretation. It has already been remarked that it is a sufficient condition for assuring both sustainability and environmental soundness. It is still more restrictive than is needed, but is at least intuitive and it has rather natural policy implications. We consider this interpretation below.

Sustainable livelihoods (SL) and Agenda 21

The concept of SL is introduced in Agenda 21 in the context of a programme to combat poverty. While the programme is motivated by a reference to the link between the state of environmental resources and poverty, there is a sense in which the environment is incidental. The objectives of the programme area are to provide everyone with 'the opportunity to earn a sustainable livelihood' by tackling the causes of poverty, hunger, the inequitable distribution of income and low human resource development. The target groups are numerous, and include all of those who are currently disadvantaged: women, children and (presumably) senior citizens, the urban unemployed and the urban poor, the rural poor including smallholders, pastoralists, artisans, fishing communities, landless people, indigenous communities, migrants and refugees. The activities envisaged for the programme are similarly very wide, including the empowerment of communities and a variety of development and aid initiatives.

It is useful to separate these target groups into those who are, by definition, net consumers, and those who are net producers. Many of those targeted in Agenda 21 do not and cannot invest a sufficient proportion of their income to protect the productive potential of the assets on which they (directly or indirectly) depend. Indeed, many survive solely on transfers. If the livelihood of such people is to be sustainable, it must be secured by economic activities that are sustainable – that do not threaten the integrity of the environmental assets on which both donor and recipient depend. SL may be interpreted as the requirement that each community of resource users to which these target groups belong should behave in a way that is consistent with the sustainability of the wider asset base on which that community depends, and that the distribution of income within that community should be such that the needs of all its members are met. This is not a requirement about the value of the assets commanded by each individual. It is a requirement that the productive potential of the assets on which the community depends – including both public and private assets – should be protected.

This has three important implications: (a) that individual decisions about the use of environmental goods should take full account of the future costs to society – the user costs – of the allocation of those goods; (b) that the value of the asset base should be conserved by an appropriate investment strategy; and (c) that the distribution of income within the community should meet the needs of productive and non-productive members alike. Operationalization of the concept of SL requires that private decision-makers be confronted by the social cost of their use of environmental resources, that the value or productive potential of the total asset base be protected, and that the needs of those who are net consumers by reason of age or disability be secured through transfers.

Indeed, it is most useful to read Agenda 21 as a statement of intent about the broad changes that would be needed to make the communities to which the target groups belong both economically and environmentally sustainable. It is then possible to focus on the conditions for the sustainability of the livelihoods of the community: that there exists an appropriate microeconomic decision environment, an appropriate macroeconomic balance between consumption and investment, and an appropriate distribution of income. The important point is that the livelihood of dependent members of the community should not be considered in isolation. It is possible to discuss the net effect of the consumption and investment decisions of one group, given the consumption and investment decisions of the rest of the community, but this may not be very helpful. For example, the consumption of the poor might very well be unsustainable given the profligacy of the rich, but it is not obvious that the 'right' policy is then to cut the consumption of the poor.

Environmentally sound technology (EST) and Agenda 21

ESTs are defined in the main text of Agenda 21 as those which 'protect the environment' by being less polluting, recycling more waste, and disposing of waste in a more sustainable manner than the technologies they replace. ESTs are 'total systems which include know-how, procedures, goods and services, and equipment as well as organizational and managerial procedures'. They are, in addition, 'compatible with nationally determined socioeconomic, cultural and environmental priorities'. By this definition ESTs are the set of all acceptable technologies which are in some sense cleaner than existing technologies. They are, in other words, the set of 'clean technologies', 'cleaner technologies', 'best practicable technologies', 'low waste technologies' and 'resource-conserving technologies' referred to earlier in this chapter. The EST programme includes the promotion of access to and transfer of technology, the improvement of the capacity to develop and manage technology, and the establishment of collaborative arrangements and partnerships (UNCED, 1993, pp. 252ff).

It has already been remarked that this combines both environmental safety and the relative cleanliness of technology in a way that is unhelpful. If EST means cleaner technology, it is very easily satisfied. Any technology which reduces the environmental inputs, waste or emissions per unit of 'economic' output will do. But this is also a source of difficulty in operationalizing the concept. It gives no indication as to the importance of the environmental impact of the technology. There are both qualitative and quantitative problems. For example, nuclear fission is cleaner than coal-fired thermal power generation in respect of sulphur emissions, but not in respect of ionizing radiation. How should one rank them using the criterion of relative cleanliness? A 500 MW coal-fired plant may be dirtier than a 1,000 MW oil-fired plant in terms of emissions per unit of output, but will have a smaller total effect on the environment. Does it imply a more environmentally sound technology? Given the necessity for ESTs to be compatible with socioeconomic, cultural and environmental priorities, the choice of technology involves very much more than its cleanliness relative to the existing technology. There are trade-offs between the depletion of environmental resources and the emission of environmental pollutants, on the one hand, and a range of development objectives, on the other. These trade-offs make it impossible to provide a simple ranking or classification of technologies based on relative cleanliness.

The element of environmental safety introduced in the preamble to Agenda 21 is, as has already been suggested, a much better place to start. Pereira (1991) distinguishes between technologies which are environmentally 'sound', technologies which may not be environmentally 'sound' but which are acceptable, and technologies which are intolerable or unacceptable. The last set of technologies is the only set which is environmentally unsound in the sense of being environmentally unsafe, and it is on these technologies that we need to focus. If EST is to be operationalized, it is important to home in on those characteristics which may be readily tested against some criterion. In terms of the criterion of environmental safety discussed above, technology (or any other factor conditioning human production or consumption activities) may be said to be environmentally safe and sound if its use does not cause the ecological system of interest to lose resilience. In what follows, this is the criterion that is used.

The impact that a technology has is critically dependent both on the scale at which it is applied, and on its context. The parallel with preferences is instructive. Consumer preferences may bias demand for environmental resources in some direction; but whether a change in preferences has a significant effect on ecosystem resilience will depend on the level of consumption of environmental resources, and this will depend on income and the set of relative prices. There is scope for changing demand for environmental resources by changing preferences. The substitution of consumer goods that are perceived to be 'environmentally friendly' is evidence of this. But the consumption of 'environmentally friendly' consumer goods may be just as threatening to the resilience of the system as the consumption of 'environmentally harmful' goods if it occurs at high enough levels.

The safety of production technology, like the safety of consumer preferences, is scale-dependent. The thresholds of ecosystem resilience consist of critical values for either the biotic or abiotic components of the system. This would include, for example, critical populations for particular species, critical levels of acidity or alkalinity, or critical densities for atmospheric pollutants. If a technology requires the extraction of some environmental resource or the emission of some pollutant for which there exist threshold values in the system concerned, whether it is environmentally sound will depend on the level of extraction or emissions relative to those thresholds. It may be that what is environmentally unsafe in one ecosystem will be environmentally safe in another. This is most easily illustrated in the context of an example. Alauddin, Mujeri and Tisdell in Chapter 9 of this volume indicate that the adoption of new agricultural technologies in Bangladesh has increased productivity of the selected crops, but at the cost of a management regime that includes much higher levels of pesticide, herbicide and fertilizer use, and much lower levels of crop diversity. The net effect is that the resilience of the agricultural system is substantially lowered through greater susceptibility to pest infestation and soil degradation. The loss of resilience shows up as an increasing risk that crops will fail if the system is perturbed, and this is critically dependent on the scale of agricultural activity relative to the carrying capacity of the environment. In Chapter 4 Markandya notes that in the agricultural sector this is primarily a function of population growth, in that the application of technologies involving the mining of environmental resources (including the assimilative capacity of the environment) that is safe at one level of population may be wholly unsafe at another.

To operationalize this requires the protection of ecosystem resilience, and this in turn requires that the users of environmental resources face an appropriate set of incentives. To ensure that the technology used is environmentally safe, given all the other factors affecting the economic decision involved, it is once again necessary that the decision-maker be confronted by the true cost of the choice of each of the available technologies. Since the true cost of environmental resource use will depend on the demand for environmental resources relative to the system thresholds or carrying capacity, it follows that the appropriate incentives will be sensitive to this. Pereira (1991) argues that wherever a technology is 'intolerable', the appropriate protection is provided by regulatory instruments which he distinguishes from economic instruments. This is understood to include those instruments conventionally termed 'command-and-control' measures. These include a range of restrictions on the level of resource use or waste emissions. Hunting and harvesting quotas, open and closed seasons, emission caps and safe minimum standards are all examples. What these physical restrictions mean in actuality is that, wherever technologies are intolerable, the cost to the individual of using such technologies should be 'very high': that is, the penalties for exceeding critical ecological thresholds should be such that individual resource users will avoid doing so.[2] The problem with the present system of incentives in most low-income countries is not only that important ecological thresholds are unprotected: it is also that the private costs of technologies, whether they threaten thresholds or not, do not reflect the environmental costs of those technologies. The private costs of older and dirtier technologies are usually less than the private costs of newer and cleaner technologies.

In the literature on EST this last issue has been discussed in the context of the transfer of technology. The problem has been the subject of numerous research programmes in which the main concerns have been the effect of the system of property rights on the ability of users in low-income countries to acquire technologies involving lower social costs. From the perspective of the low-income countries, the difficulty has been argued to lie in the system of patents, which it is claimed have restricted the availability of clean technology in two senses. First, it has allegedly enabled patent holders to extract monopoly profits and so has inflated the cost of clean technologies. Second, it has discouraged patent holders from distributing clean technologies in those countries which have weak protection for the holders of intellectual property rights. There is undoubtedly some justification for both claims, but it is more instructive to look at the demand side. Demand for patented technologies in the high-income countries is largely driven by environmental regulations, and since the existence of environmental regulations involving some safe minimum standard reduces the elasticity of demand for technologies that enable such regulations to be satisfied at higher levels of output, the equilibrium price in countries with environmental regulations will be higher than in countries without such regulations. Hence, from the perspective of the suppliers of environmentally clean technologies, the 'solution' to the problem of low demand in the low-income countries lies in the imposition of environmental regulations in those countries – along with the concessionary financing of purchases (OECD, 1992).

The following section will argue that environmental regulation and the use of safe minimum standards is indeed the most effective way of assuring the resilience of ecological systems. In this sense this chapter supports the conclusions reached by both ILO and OECD studies (see Pereira, 1991; OECD, 1992). But it is worth repeating that whether a technology is sustainable or not depends on the demand for environmental resources under that technology relative to the carrying capacity of the environment: that is, it will depend on the use of environmental resources relative to carrying capacity. 'Clean' technologies may be no more sustainable than 'dirty' technologies at low levels of use.

MAKING THE CONCEPTS OPERATIONAL

Sustainable livelihoods (SL)

A considerable amount of work has already gone into the operationalization of sustainability concepts by agencies such as the World Bank, and the first part of this section draws on this material. Three questions have been raised in the literature. First, since the driving forces behind environmental degradation are the factors that condition private resource use decisions, what is the best way of dealing with these? Second, since sustainability does imply some judgement about intergenerational equity, what is the best means of assuring equity? Third, since sustainability implies the need to avoid crossing thresholds involving irreversible welfare loss, how should these thresholds be protected?

The initial approach to the operationalization of sustainability by the international organizations considered the problem in the context of project spending. The questions raised by UNCED are much wider than this, but the approach is instructive nevertheless. The starting point for work in this area was the observation that conventional cost/benefit analysis of projects involving significant environmental effects failed to test for sustainability for two reasons: the pervasiveness of externalities (including the existence of crowding effects) for which it was impossible to provide reasonable shadow prices, and the difficulty of dealing satisfactorily with the intergenerational equity issues.

The problem of externalities in projects has been tackled through the valuation of non-marketed effects. This has relied on a variety of techniques, but principally on: (a) direct valuation of productivity changes, loss of earnings associated with environmental impacts, and defensive expenditures against environmental degradation; (b) the use of surrogate markets to give proxies for environmental amenities; (c) the inclusion of non-marketed environmental resources in functions explaining the production of marketed goods; and (d) the use of direct estimates of willingness to pay/accept. All these methods have severe limitations in terms of their ability to provide realistic estimates of present and future values of environmental effects of economic activity (Lutz and Munasinghe, 1993). But the approach does serve to highlight the options open to policy-makers to influence private decisions. The cost/benefit framework provides a means of comparing alternative actions given the set of relative input and output prices, the rate at which future costs and benefits are discounted, and the set of constraints within which the action is undertaken. The valuation approach selects relative prices as the vehicle through which to influence the decision. But while intervention to align the private and social cost of resource use is undoubtedly an important component of any strategy for sustainability, it is not the only one. Nor is it the most effective means of protecting ecological thresholds.

There is general consensus in the literature that it is inappropriate to use the discount rate as the primary means of influencing future relative to present costs. Initially, it was thought that securing intergenerational equity required adjustment of the social rate of discount, since this is the measure of societal preference for consumption at present, relative to consumption in the future. However, the observation that the adjustment of discount rates would have uncertain and potentially perverse effects discouraged this approach (Markandya and Pearce, 1988). The alternative subsequently proposed has been a 'sustainability constraint', of which there are two variants. The first is a restriction on the use of environmental resources to protect the stock of natural capital transferred to future generations to assure intertemporal equity in the distribution of income (Norgaard, 1991). This was motivated by an interest in protecting essential natural assets for which there exist no close substitutes (the strong sustainability requirement) (Daly and Cobb, 1989) or about which there is considerable uncertainty (von Amsberg, 1993). The appropriate environmental 'price' in a cost/benefit framework is the shadow price of the environmental constraint. The second variant is what amounts to a sustainability levy, or a compensation component invested over the lifetime of a project sufficient to yield an asset of equal value to the environmental resource (von Amsberg, 1993). This satisfies the requirement for economic sustainability, i.e. that the value of the asset base is preserved, but it is blind to threshold effects and the non-substitutability of produced and natural capital.

This first variant of the sustainability constraint comes very close to the problem of assuring system resilience where there are important scale effects. It is argued that expansion in the demand for environmental resources, driven partly by population growth and partly by economic growth, has pushed many human activities up to and possibly beyond the carrying capacity of the natural systems being exploited (Daly, 1991). Carrying capacity may appear to be a rather static concept, and in this case, where growth is argued to push society beyond carrying capacity, it appears to be a standing target. But it is recognized that carrying capacity may change as both technology and ecosystem functions change. In some cases, it is clear that technological change has increased the human carrying capacity of ecosystems, but in other cases it is equally clear that degradation of the system has had the opposite effect. Such cases make clear the link between carrying capacity and thresholds of resilience, since exceeding the carrying capacity reduces the ability of the system to accommodate the stress of predation. By capping the level of demand for environmental resources, a sustainability constraint can be seen as a means of protecting against the costs of crossing an important threshold.

The operationalization of SL requires the development of a method for changing the cost/benefit analysis conducted not just by the major development agencies, but also by every single user of environmental resources. Rational individuals faced with prices that are less than the social opportunity cost of environmental resources will overuse those resources. This remains the case for prices in many agricultural markets. The divergence between private and social costs in this case is the product of both incomplete markets (many environmental externalities of agriculture are ignored), of the particular structure of property rights (there is often an implicit subsidy on agricultural land offered by traditional land tenure systems), of economic policies (agricultural subsidies are still extremely widespread), and of the international trading system (including agreements such as those reached under GATT or its successor the World Trade Organization (WTO) or the Lomé Convention). There is a case for reconsidering international agreements from an environmental perspective. The problem of implicit environmental subsidies should be on the agenda in the next 'round' of the WTO. But in the short and medium term, operationalizing the sustainability of livelihoods will require domestic policy reform to remove many of the distortions to the private cost of environmental resources.

Price reform may also be one of the more effective means of dealing with the distortions to the private valuation of environmental resources caused by poverty. It has long been observed that poverty amongst the users of natural resources has been a consequence of the control of agricultural prices and the manipulation of agricultural markets (Sen, 1981). However, price reform will not help those with access only to 'uneconomic' holdings of land. The distribution of both assets and income has tended to widen over time in many low-income countries, partly as a result of the erosion of traditional rights of access to the resource base. This is a problem faced especially by female-headed households, who show the highest levels of relative deprivation (UNDP, 1990).

The problem of externalities can be dealt with in part by the assignment of property rights. There currently exists a very wide range of rights conferred by law or custom on the users of environmental resources. The incentive effects of these various rights are not very well understood at present, and a prerequisite to the operationalization of the sustainability of livelihoods would be to establish what these are. This is particularly important where environmental effects have a long gestation period (owing to the nature of the biogeochemical cycles involved). While it is clear that open-access common property does not give the right incentives, it is not at all clear that private rights in perpetuity are the most appropriate alternative.

Environmentally sound technology (EST)

Two key problems need to be tackled in operationalizing EST. On the one hand are the scale dependence and threshold effects. On the other are the information, institutional, property rights and related issues associated with technology transfer. Agenda 21 focuses on the latter. It takes it for granted that sustainable development requires improved access to technology information by low-income countries, and the strengthening of their capacity to evaluate alternative technologies. This last includes the capacity to undertake environmental impact and risk assessments. On the issue of property rights, Agenda 21 is consistent with the current trend towards the strengthening of intellectual property rights established through patents. Although it acknowledges the concern of such countries over the negative effects of the patent system, it explicitly includes recommendations for 'measures to prevent the abuse of intellectual property rights, including rules with respect to their acquisition through compulsory licensing'. To deal with the income effects of the monopoly prices enjoyed by the proprietors of clean technologies, it recommends the subsidized purchase of such technologies, together with concessionary finance for their acquisition.

To some extent these considerations are peripheral to the question of EST. The recent emphasis on intellectual property rights in technology, and the issue of technology transfer, are not driven by environmental concerns. It is more helpful to think of EST as environmentally appropriate technology. As the literature on appropriate technology has made clear, the optimal choice of technology for the production of any set of goods and services is a function of a wide range of factors, including the set of relative prices, the state of knowledge, output levels, and so on. EST represents any feasible and efficient technology which does not threaten the resilience of the ecosystems in which it is employed: that is, it is any feasible and efficient technology which does not approach the thresholds of ecosystem resilience, given the set of relative prices, state of knowledge, output levels, and so on.

For the most part, the cost of technology is reflected in the cost of the capital in which it is embodied. Consequently, the pricing of assets at marginal social cost – where this includes the cost of future environmental external effects – should ensure the adoption of environmentally appropriate technology. For any given technology, environmental sustainability can then be guaranteed by the use of regulatory instruments involving restrictions on the extraction of environmental resources or the emission of environmental pollutants. These will be discussed shortly. For now, what is important about such restrictions is that they imply discontinuities in the private cost functions associated with the environmental resource or pollutant in question. For example, where firms are fined for catching fish above quota, the cost of the catch will be discontinuous at the catch limit.

The form taken by restrictions to protect ecological thresholds under any given technology will depend on the nature of the technology. Most ecological thresholds are conditional, in the sense that whether the emission of some pollutant at a particular level and under a particular technology does cause loss of resilience will depend both on natural conditions and on the other emissions under the same technology.

The set of restrictions to protect ecological thresholds is currently very wide. It includes harvesting and extraction quotas, together with open and closed 'seasons' (in fisheries, hunting, gathering, some forestry), restrictions on the species to be harvested including the size or age of harvestable species (also in extractive industries and sometimes taking the form of gear, e.g. net size restrictions), emission caps or 'bubbles' (in industrial activities generating liquid or gaseous wastes), and safe minimum standards (usually governing the composition of emissions and including the so-called critical loads). These correspond to a parallel set of economic instruments, the 'prices' that are discontinuous around the restriction. These include – on one side of the restriction – royalties and other user charges in extractive industries, effluent charges, environmental taxes/subsidies, deposit refund systems, and environmental performance bonds; and, on the other side of the restriction, fines, surrendered deposits and bonds, and other penalties.

THE DECISION PROCESS

To show the significance of SL and EST for the decision process in a less indicative way, this section draws on a recent analysis of decision-making in the face of environmental discontinuities by Perrings and Pearce (1993).

Sustainable Livelihoods and Environmental Soundness

The total economic cost (TEC) of an environmental resource committed to some use which has effects that are external to the market is:

TEC = C + E

where $C = C(\mathbf{w}, q)$ denotes private cost and $E = E(\mathbf{r}(q))$ denotes external cost, $\mathbf{w}$ being a strictly positive vector of market input costs, $\mathbf{r}$ a non-negative vector of (unpriced) environmental resources and q output. If $C(\cdot)$ is continuous, differentiable and increasing in both q, and $\mathbf{w}$, and if $E(\cdot)$ is increasing in the level of output, the privately optimal level of output is that at which:

$\partial R/\partial q = \partial C/\partial q$

$R = R(p, q)$ denoting revenue, p being the price of output. The socially optimal level of output is that at which:

$\partial \Pi_p/\partial q = (\partial E/\partial \mathbf{r})(\partial \mathbf{r}/\partial q)$

Π_p being marginal net private benefit (private profit), or private revenue minus private cost. This is the point at which marginal net private benefit is equal to marginal external cost. The two conditions are satisfied at the levels of output q^*_p and q^*_s in Figure 3.1. The 'solution' to the problem is to confront private users with the difference between private and social cost at q^*_s: the 'user' or 'polluter pays' principle.

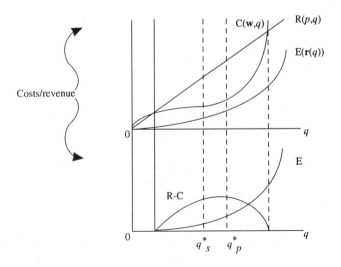

Figure 3.1 Private and social use of environmental resources

However, if there exist threshold values for the resilience of the ecosystems, the external cost function will typically be discontinuous at those thresholds, and the usual conditions for efficiency in the use of the resource may not hold. The social problem in this case is to maximize social net benefit deriving from the exploitation of ecological resources over some time horizon, subject to the effect of the level of economic activity on the environmental resources on which economic activity depends. The approach described here invokes both forms of the sustainability constraint discussed above: that is, it requires both that the value of assets left at the end of the activity is of some defined value, and that some set of ecological thresholds (defined by the essentiality of the ecological services at risk if those thresholds are exceeded) are protected. The set of benefits produced by the activity is described by a stream of income, $Y[\cdot]$. The value of the assets remaining at the end of the activity, time T, is described by the function $W[\cdot]$. The effects caused by the depletion of environmental resources are described in the equations describing the dynamics of the ecological system.

The problem is to maximize the sum of the two components of value, Π_s, through choice of the structure of incentives, $\mathbf{k}$, where

$$\Pi_s = W[\mathbf{r}(T),T]e^{-\delta T} + \int SUP(T,0) Y[\mathbf{r}(t),\mathbf{q}(t),t]e^{-\delta t}dt$$

subject to:

$$\begin{aligned}
O(\mathbf{r},\cdot)(t) &= \mathbf{f}\ [\mathbf{r}(t),\mathbf{q}(t),t] & 0 \leq t \leq T \\
\mathbf{0} &\geq \mathbf{h}\ [\mathbf{r}(t),\mathbf{q}(t),t] & 0 \leq t \leq T \\
\mathbf{q}(t) &= \mathbf{q}\ [\mathbf{k},\mathbf{w}(t),t] & 0 \leq t \leq T \\
\mathbf{w}(t) &= \mathbf{w}\ [\mathbf{q}(t),t] & 0 \leq t \leq T \\
\mathbf{r}(0) &= \mathbf{r}_0
\end{aligned}$$

The vectors $\mathbf{r}$ and $\mathbf{w}$ have already been defined. $\mathbf{r}$ describes the state variables of the system and the equations $O(\mathbf{r},\cdot)(t) = \mathbf{f}[\mathbf{r}(t),\mathbf{q}(t),t]$ describe the evolution of the state variables as a function of both their natural rate of regeneration and the level of economic activity. $\mathbf{q}$ is a vector of economic outputs, and is influenced through the set of economic incentives $\mathbf{k}$. δ is the social rate of discount, and t denotes time. The ecological thresholds are protected by the second form of sustainability constraints referred to in the section 'Making the concepts operational' above, and is reflected in the inequalities: $\mathbf{0} \geq \mathbf{h}[\mathbf{r}(t), \mathbf{q}(t),t]$. These restrict the admissible values of the set of resources, $\mathbf{r}(t)$. If $h_i[\cdot] < 0$ the constraint is non-binding, implying that the resource is in the neighbourhood of its threshold value.

Sustainable Livelihoods and Environmental Soundness

The necessary conditions for an optimum include the following:

$$O(\lambda, \cdot)'(t) = \begin{cases} -Y_r(t)e^{-\delta t} - \lambda'(t)F_r(t) & \mathbf{h}[\cdot] < \mathbf{0} \\ -Y_r(t)e^{-\delta t} - \lambda'(t)F_r(t) - \eta'(t)H_r(t) & \mathbf{h}[\cdot] = \mathbf{0} \end{cases}$$

$$\lambda'(T) = W_r(t)e^{-\delta T}$$

$$0 = \begin{cases} Y_k(t)e^{-\delta t} + \lambda'(t)F_k(t) & \mathbf{h}[\cdot] < \mathbf{0} \\ Y_k(t)e^{-\delta t} + \lambda'(t)F_k(t) + \eta'(t)H_k(t) & \mathbf{h}[\cdot] = \mathbf{0} \end{cases}$$

where $H[\cdot] = Y[\cdot]e^{-\delta t} + \lambda'(t)\mathbf{f}[\cdot] + \eta'(t)\,\mathbf{c}[\cdot]$ is the Hamiltonian for the problem, and λ and η are non-negative vectors of Lagrange multipliers, with $\eta'(t) = 0$ if $\mathbf{h}[\cdot] < \mathbf{0}$ and $\eta'(t) \geq 0$ if $\mathbf{h}[\cdot] = \mathbf{0}$. If the constraint is non-binding, the corresponding multiplier has a zero value. If it is binding, the corresponding multiplier is positive and is in principle equal to the marginal benefit of the relaxation of the constraint.

Let us consider the two types of sustainability constraint separately. The constraints designed to protect ecological thresholds imply the existence of an underlying set of restrictions: quotas, caps, safe minimum standards and so on. It is these which determine the limits imposed on the extraction or emissions. Their 'enforcement' requires the imposition of penalties on resource users who exceed the standards. The effect of such penalties is that expected private marginal costs increase at the standard by the penalty multiplied by the probability that it will be 'enforced'. If the ecological thresholds are not known with certainty, the standard that triggers the penalty should be such as to reduce the 'risk' of overshooting ecological thresholds and the consequential losses to society to acceptable levels (Ciriacy-Wantrup, 1952), the location of the standards relative to the ecological thresholds indicating society's aversion to risk. If society is risk averse, the penalty function will lie inside the threshold, as illustrated in Figure 3.2.

Any technology which is economically feasible for levels of output less than the standard (whether or not it is economically efficient at such levels of output) may be said to be environmentally sound, in the sense that its application will not threaten the resilience of the underlying ecological system. Note that no matter how 'environmentally friendly' a technology may be by the criteria of Agenda 21, it will be unsustainable at some level of output, just as it will be economically infeasible at some set of relative prices. This is the scale issue, and it is worth repeating. Assuring the

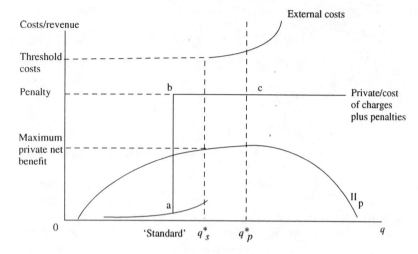

Figure 3.2 Penalty function for breaching standards

environmental soundness of a technology means restricting its application to levels that are environmentally safe.

We now turn to the second form of sustainability constraint: the constraint on the value of capital. Sustainability, in this formulation of the problem, is satisfied through the terminal benefit function, $W[\cdot]$. The social value of the stock of assets at the terminal time, T, is determined on a different basis from its value during the course of the policy. For $t < T$ the value of biological resources is assessed on the basis of the flow of goods and services it yields. $W[\cdot]$ reflects the productive potential of the stock of assets left at the end of the activity. There are several ways of arriving at a value for $W[\cdot]$, and indeed a value of T. The most rigorous attack on this problem to date is probably to be found in von Amsberg (1993).

CONCLUDING REMARKS

This chapter has concentrated on the first two stages in the operationalization of the concepts of SL and EST. No attempt has been made to apply the approach to particular cases. However, this final section both summarizes the main results and offers some general remarks on how the concepts might be applied to particular problems. The first point to be made is that recent work on the concepts of sustainability and environmental soundness highlights two general requirements. The first is a requirement that essential ecological services be protected by restricting the use of environmental resources to 'safe' levels. This is motivated in terms of the need to protect essential natural assets for which there exist no close substitutes (the strong sustainability requirement). In this chapter it is argued that this requirement implies mechanisms to assure the resilience of the ecological systems concerned with respect to the levels of stress and/or shocks to which they are subjected by the economic process. The second is a requirement that those who exploit environmental assets should be confronted by the full future or user cost of their actions, and that the resources generated should be invested so as to conserve the value of the total stock of natural and produced assets.

In considering how these two requirements apply to SL and EST, the first problem to be dealt with lies in the assumption made in Agenda 21 that the sustainability of development is divisible, in the sense that sustainability of the whole implies sustainability of the parts. If sustainability requires only that the value of the aggregate capital stock be maintained over time, is it meaningful to talk of the sustainability of the actions of individual agents? It has been argued here that it is not reasonable to expect that the individual behaviour of refugees, victims of famine, the disabled, children or senior citizens should be sustainable. Nor is it reasonable to expect that all environmental resources will be used in a sustainable way. So in what sense is sustainability divisible? It is certainly possible to characterize the 'sustainability' of any one of the factors conditioning economic decisions, given the remaining factors. It is, for example, possible to characterize the sustainability of preferences or property rights – given relative prices, incomes and so on. It has already been remarked that it is also possible to characterize the sustainability of the consumption and investment decisions of one group of agents, given the consumption and investment decisions of all other groups of agents. But this may not be very helpful. Getting the economic environment right means identifying the most cost-effective policy adjustments, which requires a general rather than a partial view of the problem.

The development of a policy for the promotion of SL requires three preconditions: an appropriate decision environment; an appropriate balance between investment and consumption expenditure; and an appropriate distribution of income. To ensure that individual resource use decisions are consistent with the sustainability of the assets available to the community to which the individual belongs, it is necessary to intervene in the private decision process. It follows that all factors which enter the decision process (and which may be influenced by policy) are natural targets. Apart from those economic instruments which can be used to alter the private costs of resource use (microeconomic and macroeconomic policy), this includes the state of knowledge (information policy), cultural and social values and the preferences to which they are related (education policy), security of tenure, market and institutional conditions (policy on property rights and competition), and so on. The effectiveness of a policy designed to address one parameter in the decision-making process will be conditioned by the remaining parameters, and so by the policies addressing those parameters. Disseminating information on the user cost of groundwater depletion, for example, will be ineffective if the resource is subject to open-access common property rights. Similarly, imposing water charges in irrigated agriculture may be ineffective if there is a countervailing subsidy on complementary agricultural inputs. It follows that identifying the most cost-effective way of changing the decision environment requires an appraisal of the impact of all policies bearing on the decision environment.

Ensuring that the productive potential of the community asset base is preserved is also less easy to translate into effective policy. At the project level the problem is easy. Von Amsberg (1993) argues that this implies a sustainability levy, or a compensation component invested over the lifetime of a project, sufficient to yield an asset of equal value to the environmental resources used. However, where the problem is the SL of the private behaviour of members of a community, the solution is less obvious. An investment tax or charge for the environmental resources used by economic agents might generate funds adequate to undertake compensating investment, but it is not clear how such investment should be undertaken. Private investment incentives are one option, but private investment naturally favours private goods and eschews public goods. Many of the environmental resources being lost through private economic activity are in the nature of essential public goods. The private response tends to be an increase in private defensive expenditures against the loss of public environmental goods – which both influences conventional measures of income and investment (El Serafy, 1989) and fails to deal with the problem of sustainability. An appropriate investment strategy would include public cover for the depreciation of environmental public goods, or the protection of essential ecological services. In other words, while charges for environmental public goods may provide the correct signal to the immediate user, if the revenues generated are inappropriately invested, the overall strategy will be unsustainable.

Sustainable Livelihoods and Environmental Soundness 65

This chapter has argued that from an operational perspective EST should be regarded as any technology which does not threaten essential ecological services at current levels of resource use, and current methods and procedures of application. A definition such as 'cleaner' or 'less wasteful' technology is not operational, or at least is operational only in a trivial sense. The key feature of EST is that it is environmentally safe at some scale of use – that it does not threaten thresholds of ecological resilience at that scale of use – and that the appropriate policy comprises incentives to ensure that the scale of use does not exceed the safe limit. The nature of the limit and the penalties involved in exceeding the limit will be case-specific. Carbon monoxide emissions, for example, are handled through restrictions on vehicle emissions. Fisheries are protected through a combination of quotas and gear restrictions. Such restrictions have been described here in terms of a set of fixed inequality constraints on the decision problem. What is important is that such constraints should be sensitive to changes in the level of resource use under the technology as well as to changes in conditions of application.

Finally, both SL and EST have major implications for national policy, and these have been briefly discussed in this chapter. However, they also have implications for international policy. Although both concepts are intended to apply at a disaggregated level, the fact is that many of the main threats to environmental sustainability come from the degradation of global environmental public goods. The appropriate decision environment in such cases is the global environment, and the loci of compensating investment for any given project may be widespread. Since the soundness of any technology will be a function of local conditions, this does not mean that there should be uniform restrictions to protect critical thresholds. But it does mean that there should be general dissemination of information about the conditions in which technologies may threaten thresholds. This is an aspect of the technology transfer problem that has not been explored in the literature, but it is necessary if the users of technology are to evaluate its environmental costs.

In addition, the point has been made that distribution is important for two different reasons. Not only does the sustainability of the livelihoods of dependent members of the community require an appropriate distribution, but the decisions of productive economic agents are also influenced by the distribution of income and assets. The structure of pay-offs to the adoption of different technologies in different countries will depend on the distribution of income and assets between those countries. This is clear in the case of assets. The pay-off to the adoption of environmentally sound relative to environmentally unsound agricultural technologies in each of two countries, for example, may depend on the infrastructure or support services available in each; that is, the pay-off to a country from adopting an EST may be low just because it does not have the assets that would enable it to make best use of that technology. This point is partially recognized in the OECD (1992) recommendation for technology transfer to be aided by subsidizing either the technology or the finance needed to acquire it. However, it is also the case that the pay-off will depend on income. It has long been recognized that newer agricultural technologies may not be adopted in some countries because farmers cannot afford to accept the attendant risks (Lipton, 1987). In this case, the appropriate transfer may take the form of an insurance policy against such risks. The general point is that neither SL nor EST can be treated at the level of individual agents, firms, household or occupational groups, and achieving sustainability and soundness alike requires that the overall decision environment is sustainable, and that the broader asset base is protected. While it may be sensible to evaluate the sustainability of autarchic communities in isolation, the ever greater integration of the world economy makes such an approach increasingly inappropriate.

Notes

1. An adapted version of this chapter has appeared in the *International Labour Review* (Perrings, 1994). Some sections of the article are reproduced verbatim.
2. Regulatory instruments are, in reality, economic in exactly the same way as are taxes, or user charges. The economic incentive in a regulation, or a standard of some kind, lies in the cost to the individual of breaching that regulation or exceeding that standard. The economic instrument is the penalty. Hence the only difference between 'regulation' and 'economic incentives' lies in the shape of the private cost function associated with each. In the case of regulations it will tend to be discontinuous at the maximum admissible use of the resource.

Part Two
Empirical Studies

4 Technology, Environment and Employment in Third World Agriculture

A. Markandya

THE CONCEPTUAL FRAMEWORK

As one would expect, the relationships between changes in technology, the environment and employment are complex. Schematically they can be represented as shown in Figure 4.1. The choice of technology is determined by many factors, prominent among which are the relative prices of the inputs; the institutional arrangements and infrastructural provisions for the employment of factors such as labour and capital; the population pressure facing the sector and the economy at large; and the resources devoted to research and development (R & D) in this area. This choice has impacts on the environment, on employment and on incomes. The environmental impacts take place not only through the choice of technique but also through the effects of population growth and of input and output prices on incomes, particularly incomes of the poorest groups in society.

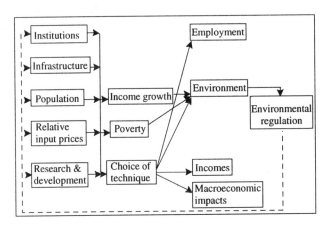

Figure 4.1 Relationships between changes in technology, the environment and employment

The net effect of all these changes on employment is difficult to assess a priori, and one is obliged to look at specific cases in detail to ascertain whether or not employment and environmental objectives conflict with each other, or can in fact be made complementary. In addition one is interested in the effects of environmental regulations on other important variables such as income, exports, agricultural growth, etc.

In the agricultural sector, on which this chapter focuses, increases in productivity have been achieved by the large-scale use of pesticides and fertilizers. It is generally recognized that this has resulted in environmental damage and that the 'price' paid in terms of the latter for the increased agricultural production has, *at the margin*, been too high. In precise terms, the costs of the marginal applications of agrochemicals has exceeded the benefit in the form of increased agricultural productivity. Hence there is a need for some form of environmental regulation on the use of these inputs. What one needs to know, however, is the impact of alternative regulations on incomes and on employment. One such policy could be to lower or eliminate subsidies on fertilizers. This would have a direct effect on the environment by reducing their use but could also reduce agricultural productivity and incomes.

The kinds of argument presented above, although complex, are in fact only partial descriptions of the linkages between environmental change and employment and development. One important route by which environmental degradation may influence the nature of development and the environment is through migration and population growth. A view that has been advanced is that soil erosion, biomass loss and unsustainable farming practices in general can accelerate migration to urban areas, thereby increasing urban employment problems. This view, however, has not been systematically examined.

At this stage we cannot say how strong are the developmental and environmental effects of technological choice. Although there is considerable piecemeal evidence on the linkages, a theoretical structure, and data to support or refute it, are lacking. What is needed therefore is more research at the theoretical level, supported by micro-level data and detailed case studies.

There are technologies that are both environmentally friendly and potentially beneficial to the poor. Examples are improved multicropping techniques, new drought–resistant and pest-resistant varieties of crops and no tillage mulching. These result in reduced run-off and soil erosion, which can, in turn, lead to higher yields and improved incomes and employment opportunities, particularly among the poorer sections of society. What must be ascertained, however, are: (a) the *incentive* struc-

Technology, Environment and Employment   71

ture under which such technologies would be adopted and developed; (b) the *institutional* changes that would facilitate the adoption of the appropriate technologies; and (c) their impacts on both employment and output 'macroeconomic' considerations such as exports, the balance of payments, etc. In case (c) there may be a trade-off between the achievement of employment, environmental and poverty objectives, on the one hand, and conventional economic development, on the other.

In this chapter, the determinants of technology in developing countries are examined and the links are evaluated between the choice of technology and the environment; technology and employment; environment and employment; and all three – technology, environment and employment. The theoretical discussion is complemented by an examination of the empirical evidence, such as is available. Particular attention is paid to the role of government policy, both in determining the choice of technology through market forces, in defining the institutional structure in which the decisions are made, and in influencing the distribution of income and access to resources, which are so vital to the impacts of technology on employment and environment.

THE CONTEXT OF THE DEVELOPING COUNTRIES

In looking at the linkages between these three concepts we need to start with clear definitions of the terms, as their usage is not consistent, and definitions used for developed countries are not always applicable for the developing countries.

Technology has been defined by Stewart (1977) as the 'hardware of production – knowledge about machines and processes', which covers 'methods used in non-marketed activities as well as marketed ones. It includes the nature and specification of what is produced – the product design – as well as how it is produced.' In many instances the North-to-South transfer of technology has not sufficiently taken into account the higher cost of capital and the lower cost of labour, and the need to import suitably skilled personnel, materials, spare parts and so on. These problems are well documented.

Employment in developed countries refers to wage employment, and unemployment thus refers to those who are not employed in wage labour. As a large proportion of the population in developing countries are not employed in wage labour but depend on traditional forms of work relationships, these definitions are not appropriate. The majority of the population in developing countries can be described as being in a situation of disguised unemployment or underemployment. This describes those people who are not actively seeking work but whose activities are unsatisfactory, because they are either not sufficiently financially rewarding or not sufficiently productive.

Environmental degradation can be viewed as a decline in the quantity and quality of the services provided by the natural resource base. This decline can be reflected in a reduced assimilative capacity of environmental 'sinks', such as air and watersheds; in a worsening of the services provided by natural resources such as forests, lakes, rivers and coastal zones; and in a deterioration in the man-made environment, in cities, towns and villages. We are mainly interested in degradation arising from human activities, as opposed to natural causes, although it is not always possible to distinguish between the two.

Developing countries' experience of technology, employment creation and environmental degradation

Recent agricultural development

In the agricultural sector, the transfer of developed country technology began in the pre-colonial era, with the production of cash crops and the establishment of plantations. Until recently, the advantages of indigenous technology were unrecognized, and improvements to agricultural production were to be achieved through the transfer of 'Northern' agricultural methods, such as mechanization, row planting, monocropping, etc.

Lipton and Longhurst (1989) have shown that the new technology associated with the Green Revolution (high-yielding varieties, fertilizers and irrigation) has made it possible to double food production per acre per person in the past 20–30 years, outpacing the growth of population, and avoiding the crisis envisaged at the end of the 1960s when it was thought that population growth would exceed food production. Initially, only the large farmers of Asia and Latin America adopted the new varieties, and early criticisms of the Green Revolution concentrated on the distributional effects of the new technology: see, for instance, Griffin (1989).

Although the production of food per person has increased dramatically since the 1960s, the number of people living in poverty has not been reduced. Sen (1981) has shown that access to food or 'entitlements' is a more important determinant of famine or hunger. Thus it is by increasing employment that the problems of poverty in the Third World will be tackled, and not simply by increasing the production of food, or even industrial goods.

Technology, Environment and Employment

Environmental degradation

Because of differences in their levels of development, the type and the extent of environmental degradation in developing countries will differ from those problems as experienced in the developed world. The greater dependence on the agricultural sector in the developing countries means that this sector will have suffered proportionately more damage than the industrial sector, although in absolute terms environmental degradation in the agricultural sectors of developed countries may have been more acute, as high levels of subsidies have encouraged the more widespread use of environmentally damaging practices. We shall now discuss six different categories of environmental degradation.

Desertification Desertification is the degradation of lands in dry areas affecting a quarter of the earth's surface (Dregne, 1983). It is not the desert expansion of popular imagination. Instead it is essentially a subtle, dispersed and continuous process which mainly occurs far away from desert fringes, with the outright conversion of fertile land into desert only taking place in extreme cases (Grainger, 1990). The cause of desertification, is poor land use – which is commonplace in the tropics and elsewhere – but the nature of the drylands means that it has more disastrous consequences in that environment. Desertification is seen as a gradual process of land degradation, with the conversion of an area into a desert being the extreme condition. In Dregne's view the process is reversible and, although there are areas that are severely eroded, 'very little land has been irreversibly desertified as a result of man's activities' (Dregne, 1983).

Soil erosion This is the commonest form of land degradation, but soil can also be degraded by high salt and alkali levels, deposits of radioactive and inorganic wastes and saturation by chemicals. Erosion affects the productive capacity of the soil, as the topsoil that is lost is usually the most productive. Erosion is difficult to detect as it occurs slowly and is visible only when the problem has become serious. It is therefore not easy to estimate the extent of the erosion problem, but broad estimates have been made for various countries. Of India's 328 million hectares, 150 million are affected by erosion, with farmers losing 6 million tons a year of nitrogen, phosphorous and potash. In Guatemala 40 per cent of the productive capacity of the land is totally lost owing to erosion, and farmers have abandoned their farms in several areas. In Nepal losses of topsoil average 35–75 tons per square kilometre per year, and in Ethiopia 2,000 tons per square kilometre are lost annually (Younis, 1987).

Salinity, alkalinity and waterlogging Increases in the area under irrigation have been one of the major means of raising agricultural production and allowing it to keep pace with population expansion. However, if irrigation is introduced into areas where there is a lack of experience and where the irrigation system is not designed to suit the local environmental conditions, irrigation can result in soil salination, alkalinization and waterlogging, which will transform irrigated areas into totally unproductive land. Throughout the drylands Grainger (1990) estimates that a total of 40 million hectares of cropland are affected by salinity, alkalinity or waterlogging. Another 500,000 hectares, or about one-eighth of the area brought under irrigation each year, becomes desertified annually. The problem is particularly severe in Pakistan, where three-quarters of all cropland is under irrigation, but 30 to 60 per cent of this irrigated land is estimated to suffer from salinity or waterlogging. Waterlogging and salination are also estimated to affect 28 per cent of all of Egypt's farmland, where average yields have fallen by 30 per cent (Speece and Wilkinson, 1982).

Chemicals Chemicals have played an important part in the huge increases in agricultural production, but the problems associated with their overuse and, more importantly, misapplication are becoming more evident. The use of chemicals to increase productivity and control pests and weeds does lead to increases in output, but run-off from nitrogen and phosphates in groundwater has resulted in dangerous levels of these chemicals that affect all species, including humans. A 1983 study estimated that in developing countries approximately 10,000 people died each year from pesticide poisoning and about 400,000 suffered acutely (Gear, 1983). On the other hand, many areas of developing countries do not use significant quantities of chemicals and are therefore not at risk. In a number of these areas an increase in productivity can be achieved through an increase in the use of fertilizers and pesticides. Moreover, although much is made of alternatives to fertilizers and pesticides, through the use of systems of integrated pest management (IPM) and organic farming, there is no doubt that such systems, *by themselves*, cannot achieve the output levels required to ensure that economic development and employment are supported.

Overgrazing Overgrazing is the result of increasing the numbers of livestock beyond the carrying capacity of the rangeland, leading to the degradation of vegetation and the compaction and erosion of soil. It therefore affects the productivity of the land, favouring less nutritional shrubs and reducing the supply of animal fodder. Overgrazing can also be a cause of deforestation and loss of crown cover when livestock are allowed to browse on trees and shrubs, thus preventing the regeneration of vegetation. This has become a severe problem in areas of the Andes, in Nepal and in Ethiopia, where the loss of vegetation increases the run-off from the higher ground, causing siltation and flooding in the lower areas.

Deforestation Deforestation results in the loss not only of tree species but also of valued plant and animal species, which rely on the forests and woodlands for their habitat. If the tropical forests were to disappear, up to 40 per cent of the world's species would be threatened (WCED, 1987). The current concern with deforestation has centred on the high rates of loss of forest cover in Amazonia and other tropical areas of the world. Landsat images indicate that deforestation in Amazonia has accelerated sharply since the mid-1970s, with the area deforested increasing from an estimated 125,000 square kilometres in 1980 to 600,000 square kilometres in 1988 (Mahar, 1989). The main causes of deforestation here are the clearing of land for agriculture and cattle ranching, road building, hydro-electric development, mining, logging and urban development. Although no detailed data exist on the share of each cause, the most important contributing factor is undoubtedly the clearing of land for agricultural and livestock production. In Amazonia, the land cleared for agriculture and pasture increased from 313,000 square kilometres in 1970 to more than 900,000 square kilometres in 1985, mostly for the purpose of cattle ranching on large-scale ranches (Mahar, 1989). Logging is also a cause of deforestation, but the extent to which logging is in itself to blame for the high rates of deforestation is unclear, as many of the trees felled for timber are cut from land cleared for agriculture.

Deforestation is also a serious problem in the open woodlands of the more arid areas of the world. Africa is the worst affected continent in terms of clearing of woodland, accounting for two-thirds of the world's woodland losses. The clearing of land for agriculture is one of the main causes of loss of woodland: in Burkina Faso an estimated 50,000 hectares of woodland are cleared each year, and 60,000 hectares are cleared in Senegal (Berry, 1984). In the Sudan 88,000 hectares of woodland are cleared each year in the Kordofan and Darfur provinces, of which an estimated 42,000 hectares become degraded and abandoned (World Bank, 1986). Trees are also cleared for livestock grazing or even used as fodder themselves. In Nepal the importance of trees for fodder outweighs their use for fuel; the area of forest needed to supply fodder to the average farm is three to five times greater than that for timber and fuelwood requirements (Wyatt-Smith, 1982; Pearce *et al.*, 1990). The use of trees as fuel is the most common reason for loss of woodlands in some arid areas, where, according to FAO estimates, the demand for fuel is outstripping the natural regeneration of indigenous woodlands. Fuelwood shortages are also common in Asia and Central America. An FAO survey found that 100 million people in 26 countries already experience acute fuelwood scarcity, defined as 'a very negative balance where the fuelwood supply level is so notoriously inadequate that even overcutting of the resources does not provide the people with sufficient supply and fuelwood consumption is, therefore, clearly below minimum requirements'. Nearly two-fifths of the population in developing countries are estimated to suffer from fuelwood deficits and are only able to meet their requirements by overcutting the remaining supplies (FAO, 1981a). There are often off-site effects of deforestation, which can be felt far from the deforested area. Loss of vegetation can lead to soil erosion and run-off, causing siltation in downstream areas. For example, the Khashem el Birba reservoir on the Atbara River in the Sudan lost half of its capacity in six years on account of siltation from the Ethiopian highlands (World Bank, 1986).

THE LINKAGES BETWEEN TECHNOLOGY TRANSFER, EMPLOYMENT CREATION AND ENVIRONMENTAL DEGRADATION

This section examines the linkages between the three processes of technology transfer, employment creation and environmental change: first by looking at the link between technology and employment creation; second by examining the link between technology and environmental degradation; and finally by analysing the linkages between the three processes and the possibilities of a vicious or virtuous cycle being set up by the interrelationships among the three.

The impact of new technology on employment creation

The introduction of new technologies can lead to increased production and increased employment, but may also result in increased production with lower levels of employment or even lower levels of production and employment. The impact of technology will depend on the appropriateness of the technology for the economic, social and environmental context into which it is introduced.

New crop varieties

In the agricultural sector, evidence on the impact of the introduction of Green Revolution technology is mixed. On the one hand, it was the large farmer who initially adopted the new plant varieties with capital-intensive methods, thus with minimal effects on employment (Lipton and Longhurst, 1989). On the other hand, it is claimed that, despite the criticisms of the Green Revolution, in the early years it did lead to substantial increases in employment (Lipton and Longhurst, 1989). Furthermore, a recent study by the International Rice Research Institute (IRRI) on the employment impact of double cropping as made possible by the adoption of new technology in the Philippines also showed favourable employment effects. First, the demand for labour increased. Even though the labour input per crop is less than with the traditional system, more harvesting and land preparation activities are involved. Second, labour demand is spread more evenly through the year. The mix of crops and staggering of production to spread labour requirements means that resources are better used in periods of lower opportunity cost. On the negative side, the study concluded that the impact of the introduction of new technology is greatly dependent on the resources of farmers. Moreover, the improved technology led mainly to increased land values and greater inequality, although all groups appeared to be better off in absolute terms.

A study of the impact of new technology on Malaysian rice production by Hart (1987) similarly found that changing from single to double cropping increased labour demand, increased the numbers of smallholders and owner operators and reduced the level of tenancy. The increased welfare for small farmers was, however, partly the result of huge government subsidies which provided them with access to irrigation systems. Small farmers also benefited from advantages such as extension services, generous price support and free fertilizers that were not available to many small producers in other countries. In this case, however, combine harvesters were subsequently introduced, leading to the massive displacement of harvesting and threshing labour, usually provided by older women. The introduction of combine harvesters also tended to further the concentration of landownership, thereby limiting access by small producers and causing the displacement of wage labourers.

Adverse environmental effects from new agricultural technologies have arisen mainly because some of the technology options designed to raise yields have been inappropriately applied. In particular, the switch to monocropping and to intensive, chemical-reliant agriculture can be damaging to the soils in marginal lands, which include most of sub-Saharan Africa and the semi-arid and arid lands, uplands, converted swamplands and forest lands of Asia and Latin America. Here some of the newest varieties, which produce high yields at low levels of fertilizer input, could prove to be an important part of a sustainable solution. However, these newer varieties also present difficulties. Some smaller farmers (e.g. those in India), having obtained sufficient returns from the new varieties using a low input approach, are now beginning to use more fertilizers and irrigation. Furthermore, new risks have been associated with the improved seeds, for the narrow genetic resources used in breeding the improved varieties have increased the risk of pests and diseases.

Unequal benefits

With only an estimated 40 per cent of farms planting new varieties (Lipton and Longhurst, 1989), not all the farmers in developing countries have benefited equally from the modern crop varieties. The benefits have been gained predominantly in Asia and Latin America, where food production has risen dramatically. Although there have been positive effects in terms of increased yields, employment and the availability, cheapness and security of food, Lipton and Longhurst argue that within countries adopting the modern varieties a number of people have become worse off. There are several reasons for the unequal distributional effects of the new technology. First, population growth has meant that, as the demand for labour has risen, so has the supply of labour, leaving real wages static. Second, new varieties have increased the demand for land, and thus the gains from new varieties have been transmitted into rising land values. Third, although new varieties may be scale-neutral, the negative effects of the new technology lie in the incentive structure into which the technology has been introduced. For example, incentives provided to the large farmers have, in some instances, resulted in labour-displacing technology such as mechanization. Large farmers have benefited disproportionately from the provision of extension advice and credit.

One criticism of the shift to the new technology is that it has been associated with an increase in temporary rather than permanent employment. A survey of the effects of the growth of the modern agricultural sector in Latin America by de Janvry *et al.* (1989) showed that there has been little improvement in the employment opportunities for permanent workers. Agricultural growth has raised land prices and motivated some small farmers to sell their land and to move to other rural areas. Although the number of farms increased by 92 per cent between 1950 and 1980, the level of landlessness has also been rising. Despite land reforms in some Latin American countries, giving access to land to the landless, the total effect has been small. The situation is worsened as the rural labour force must compete with the urban labour force in temporary labour markets in rural areas.

Other new technologies

Other new technologies associated with the Green Revolution, such as threshing machines, herbicides and milling, have tended to be labour displacing. The effect of labour displacement had not always sufficiently been taken into account as, for example, with the IRRI thresher or the research by the Centro Internacional del Mejoramiento del Maíz y del Trigo (CIMMYT) on reducing weeding requirements. The introduction of mechanization can also have positive employment effects. Examples are machinery which makes possible double cropping, the production of higher value crops or an expansion of the area planted. Mechanization can provide further benefits where off-farm employment is more profitable than returns to agriculture.

Problems with the introduction of mechanization have been that the technology, as developed in the North, is not suitable for the smaller size of farms which predominate in the South; and that it is not economically justifiable to introduce the technology when the cost of labour is as low as it is in these countries. The first problem can be overcome by changing the structure of the farming system itself as in the Russian Federation and the United Republic of Tanzania, or by enlarging landholdings as in India and Pakistan. Another way of overcoming the negative effects of mechanization has been to modify the technology to suit the factor endowments, i.e. to reduce the scale and complexity of the machinery and thereby increase the labour input. This has been achieved with the two-wheel tractor, designed in Japan and successfully used in rice-producing countries or areas such as Indonesia, the Philippines, Sri Lanka and Taiwan, China.

Despite the development of more appropriate machinery and other changes indicated above, the overall effect of introducing labour-saving machinery into surplus labour situations has tended to be negative. This is due not to the technology in itself, which can increase labour demand, but to the institutional structure into which the technology is introduced. Abercrombie (cited in Clayton, 1983) estimates that 'a total of approximately 2.5 million jobs have been displaced by tractors at present in use in Latin American agriculture' – that is, three to four jobs per tractor. In the relatively labour-scarce farming systems of Africa, the introduction of mechanization may be more appropriate. Nevertheless, problems associated with scale do still apply, in addition to the environmental risks of using heavy machinery on fragile soils. The environmental risks may be further compounded when the labour displaced by mechanization is forced to seek a livelihood by clearing forests or other undeveloped lands.

The livestock sector

New technology in the livestock sector relates to improved disease control, increased livestock productivity and reduced range degradation. The employment effects of these innovations have been fairly small, particularly when associated with attempts to commercialize livestock production through the establishment of ranches. Most of the ranches in Africa and Latin America have failed to increase productivity beyond the traditional system, and often ranches have been established to derive fiscal and other benefits. Mahar (1989) has estimated that the ranches set up in Amazonia generate very little employment, except in the early stages of land clearing, after which employment falls to about one person for every 250–300 hectares. Where ranches are established on land that was used for gathering from the wild, such as nut-growing areas in Brazil, this can lead to the displacement of considerable numbers of indigenous people.

The most worrying aspect of the livestock sector is that much of the rangeland in developing countries, particularly in Africa and the Near East, is suffering from overgrazing. In part the causes are associated with changes in technology. For instance, mechanized agriculture has been permitted to take up land previously available for grazing. In part they are attributable to government policies which unduly encourage the holding of livestock.

The environmental impact of new technology

The introduction of new technology can have both negative and positive effects on the environment. These effects operate both directly and through secondary effects of the technology at other locations, for other producers or in other sectors of the economy.

New technology may counteract population pressure; as land scarcity increases, technology can increase the productivity of land and raise incomes and living standards. Improving land productivity can reduce the degradation of the land under cultivation and also reduce pressures to move to more marginal land as, quite frequently, unsustainable farming practices are the result of poverty. Further, the development of alternative employment-generating activities through new technologies can provide opportunities for the population to earn income in less environmentally damaging activities. This does not exclude the development of new rural-based employment-generating activities or the adaptation of existing production processes through new technologies. Moreover, the development of alternative sources of and supply systems for energy, water and other inputs may allow for less damaging impacts on the environment. For example, the development of renewable energy sources can reduce the demand for fuelwood and can limit deforestation.

On the other hand, technology itself may be damaging to the environment. The cases cited above include the use of high chemical-reliant technologies on marginal soils; mechanization in inappropriate conditions; the misapplication of fertilizers and pesticides; and the industrial pollution generated by the growth of production in that sector. Also, secondary and indirect effects, related not to the technology itself but to the effects of new technology elsewhere in the economy, may have adverse environmental implications. The following negative effects have been cited above: (a) the impact of technology in displacing labour, which is then forced to colonize marginal land or to put pressure on the government to permit the exploitation of forest land for agriculture; and (b) the unsustainable exploitation of new areas, as infrastructure is added to take advantage of the introduction of the 'new' technology (e.g. the building of roads, the construction of dams).

It is important to note that, in all the cases where the adoption of a new technology was considered to have had a damaging effect on the environment, the critical factor was the *policy context*. Apart from directly influencing the choice of technique, government policies also determine which crops are grown, to what use land is put in the different ecological zones of a country, and which environmental controls are imposed on the industry, energy and transport sectors of the economy. The 'instruments' through which the policy acts are the input and output prices of agricultural products; macroeconomic variables such as the interest, tax and exchange rates; controls on land use; the use of investment appraisal rules; and the direct and indirect incentives to industry to control its negative environmental effects. The policy context of technology is discussed further below.

The environmental effects of the introduction of technologies on cropping, livestock production and forestry

It has been argued that the expansion of cash crops at the expense of food crops has, in many cases, been environmentally damaging. As Grainger (1990) points out, a common feature of all cash crops is that they require high levels of nutrients and highly productive land. If cash crops are produced on marginal land, with inappropriate practices, there is a higher chance of land degradation. The expansion of groundnut production in Niger and the Sudan has been seen as one of the major causes of desertification during the 1960s and 1970s. The area under groundnut production quadrupled between 1960 and 1972 in the Sudan and increased from 142,000 hectares in 1954 to 432,000 hectares in 1968 in Niger. As food prices rose, farmers had to increase groundnut production through the use of improved seeds which required a shorter growing period and less rain. Fallow periods were reduced and the areas planted were expanded into more marginal land. The rapid expansion in cotton production on marginal lands in the other Sahelian countries has also been a major cause in declining fertility, which has recently made it necessary to replace cash crops by subsistence crops: these are less likely to deplete the soil fertility, while the production of some staple legumes can actually contribute to soil nutrients. By contrast, Timberlake (1988) finds that the cash crops produced are adopted from temperate climates and produced in monocultures, with the physical constraints of the soils and climate not being taken into account. The introduction of continuous cultivation on sub-humid and humid tropical soils can lead to leaching and soil acidification, resulting in rapid declines in soil fertility and yields. Some cash crops, such as sugar cane, coffee, cocoa and cotton, also use less labour than traditional crops such as beans and cassava, forcing subsistence farmers on to marginal land, thereby putting additional pressure on the land.

Before population pressure increased, farmers could afford to 'mine' the land by intensive cropping because long fallow periods allowed for recovery. However, the demands on land from rising population levels and marginalization has meant that small farmers have not been able to move to new areas as fertility levels dropped. The decline in the fallow period has led to further declines in productivity and to *soil erosion* in many areas. This effect can be mitigated through the introduction of new seed varieties, which can function with low applications of fertilizers and equipment. But the lower fertilizer and pesticide requirements of the improved varieties, which result in high output and low inputs, can themselves lead to soil mining. Lipton and Longhurst (1989) argue that the ability to produce higher levels of output without fertilizer encourages farmers to abandon traditional soil and fertility conservation practices and to grow a continuous stand of the improved variety, even twice a year. The introduction of monocropping, in turn, can lead to a greater incidence of pests, diseases and weeds. The chemicals used to control these aggravate the problem by also destroying the natural predators that are usually found on agricultural land (Conway and Barbier, 1990). The spread of improved varieties to less productive areas, for example Bangladesh, south-west India and Kenya, has not yet resulted in soil exhaustion, but if these varieties were to spread to even more marginal areas, there would be clear signs of stress on more fragile soils.

The introduction of *irrigation*, particularly in areas where there is no traditional irrigation technology, can involve high levels of capital, energy, technology and skilled employees. Irrigation has tended to be less successful in Africa than in Asia and Latin America: Timberlake (1988) quotes Club du Sahel estimates of 5,000 new hectares a year coming under irrigation in the Sahel; but at the same time another 5,000 hectares were going out of production owing to waterlogging and salination.

Livestock production often constitutes the most appropriate use of arid areas, although livestock grazing was once seen as a major cause of desertification. It has been argued that traditional values led livestock producers to expand their herds, as explained by the 'cattle complex'. The current view is that livestock production is the rational approach to arid land use, and grazing areas are thought to recover from overgrazing more quickly than originally perceived. It seems that it is not livestock production in itself that is causing desertification but the introduction of new technology (e.g. the digging of wells) and the pattern of land use, namely large-scale ranching. The subsidies provided to large-scale producers do have the effect of raising incentives to overstock these ranches and reduce their long-term viability. Furthermore, the technology involved with livestock disease prevention can have harmful environmental effects, as in the use of chemicals to eradicate the tsetse fly.

The introduction of new technology in the forestry sector has resulted in several forms of environmental degradation. For instance, agroforestry practices of planting monoculture tree crops (e.g. coffee) have destroyed the habitat and resulted in soil depletion. The introduction of improved logging techniques which has increased the felling of trees for timber, particularly in south-east Asia and some western African countries, is another cause for concern. Improvements in pulping for paper production have created additional demand for tropical hardwoods, while technical improvements in the production of veneer and plywood have increased demand from Indonesia, Malaysia and the Philippines. Another concern is that, whereas most traditional arable production systems relied on clearing forest areas but allowed long fallow periods during which the forest could regenerate, the pressure on land which has often accompanied the introduction of new technologies has resulted in a reduction of the fallow period and in the ability of the remaining forests to recover, while also leading to the permanent clearing of large areas of forest.

Interrelationships among technology, environment and employment

Reinforcing effects between the three processes

The processes of technological, environmental and employment change interact in ways that are mutually reinforcing, both positively and negatively. Negative reinforcement can set up a vicious circle. This can occur in three ways.

First, the introduction of a new technology or input may adversely affect both employment and the environment at the same time. For example, the use of chemical weedkillers can reduce the demand for labour and have a negative impact on the environment. The reduction in labour demand would force workers to produce on more marginal land, thereby increasing environmental degradation. A similar tendency applies to the introduction of mechanization in certain areas. Production on marginal land will lead to declining fertility and to further reductions in employment as well as to a lack of resources to increase the productivity of the land. A vicious circle is thus established.

Second, the introduction of new technology may initially affect only employment, but the declining labour demand can have secondary effects on the environment: this lowers income further and results in further mismanagement of natural resources.

Third, new technology can initially have a negative impact on the environment, but the decline in productivity of either agricultural or industrial production as a result of environmental degradation can reduce employment in the longer term. The use of environmentally damaging technology will reduce the long-term productivity of production processes and the economy as a whole. Declining productivity will therefore reduce the demand for labour.

A most important way in which new technology affects both employment and the environment is through *distributional* effects: if only larger farmers benefit, this can marginalize small farmers and force them to produce on environmentally fragile land. The further differentiation between rich and poor that would result from the introduction of distributionally regressive technology means that the different production or land use strategies pursued by different groups, usually competitive, leads to further marginalization and degradation of the environment.

It can be argued that large producers cause more environmental damage than smaller producers, but this will depend on their incentives to manage the environment in a sustainable manner. Larger producers have tended to benefit proportionately more from tax incentives, access to infrastructure, inputs and so forth, and often have more ability to relocate elsewhere when the resource is depleted. None the less, small producers can also be seen to be involved in the process of environmental degradation. Poverty is often seen as a cause of environmental degradation. However, it is not merely poverty in itself but the associated inequality of access to new technologies and factors of production, and inappropriate incentives, that have led to production on marginal land and the misuse of natural resources. In a recent survey of the links between poverty and environmental degradation in three areas (west Java, northern Nigeria, southern Nigeria), Jagannathan found that these links were significantly influenced by the economic and social policies. 'The poor, like the non-poor, have utilized opportunities brought about by the spatial integration of economic activities, sometimes to the detriment, and at other times to the benefit, of long-term renewable resource usage' (Jagannathan, 1989).

The role of economic and social policy

From much of the discussion so far, it is becoming increasingly clear that the analysis of the relationships between technology, environment and employment is critically dependent on the economic and social policies that are in force in the countries concerned. The role of policy is summarized in Table 4.1.

The policies listed as 'General' concern the wider aspects of how technology affects the environment and employment. For instance, the amount spent on R & D will have an impact on the kind of technology that is produced and on its effects on the environment and employment: but these effects are also dependent on the macroeconomic framework, on the policies instituted to control the use of natural resources and on the rules government uses to guide its investment decisions. The following examples will illustrate the importance of the policy context.

(a) The presence of export taxes on key agricultural commodities has a major effect on the producer price and thereby on the level of production. Removing taxes, for example on coffee in Costa Rica, has been shown to increase production at the expense of annual crops with higher soil erosion characteristics (Lutz and Daly, 1990).

(b) In the debate over excessive livestock holdings in Africa, the argument has been presented that no alternative store of wealth exists which offers an equivalent return. This in turn is a function of macroeconomic policies that determine real interest rates at low or negative levels.

Table 4.1 Summary of policies that determine how technology, environment and employment interact

Technology	Economic/social policies that affect environment and employment
General	Allocations of resources to research in appropriate technology. Pricing and control of access to natural resources, such as forests, rangelands, etc. Macroeconomic variables, e.g. inflation, exchange rate, export/import taxes, criteria for investment in infrastructure.
New varieties	Land tenure, population control, subsidies on inputs, price controls on outputs.
Irrigation	Water pricing/water allocation rules.
Weedkillers/pesticides/fertilizer	Pricing of these inputs, transport subsidies of inputs.
Mechanization	Subsidies on capital equipment, development of appropriate technology, land use policy.
Import substitute industrial development	Environment protection policy, polluter-pays-principle (PPP). Protection of local industry, macro variables as under heading 'General' above, labour market policies.
Export-oriented industrial/agricultural development	Pricing of natural resources, environmental protection policy, PPP. Labour market policies. Macroeconomic policies as in 'General' trade policies of developed countries (see above).

(c) Issues of security of land tenure have been alluded to above. It is not the titling that is important in land tenure, but the security of user rights. If such rights are not guaranteed, users will take a short-term perspective and pursue unsustainable practices. On the other hand, the adoption of Western legal systems does not generally work. Instead, one has to operate within complex indigenous land tenure forms while, at the same time, ensuring that they function without maladministration (Markandya, 1990).

Although these and other examples attest to the importance of the policy context, it remains true that our understanding of the quantitative relationships between many of the key policy variables and the factors of

Technology, Environment and Employment

interest in the context of this study remain poor, or even nonexistent. This is particularly so for the macroeconomic variables: the analysis of the linkages between the environment and the macroeconomy is only in its infancy (Markandya and Richardson, 1990). Much more is known, of course, of the impact of macroeconomic decisions on employment. Yet it is difficult to see clearly whether a particular policy instituted to protect the environment is detrimental to employment or vice versa.

GOVERNMENT RESPONSES TO ENVIRONMENTAL DAMAGE, AND THE ROLE OF MULTILATERAL INSTITUTIONS

This section reviews a selection of environmental projects and programmes, undertaken either by the governments of developing and other countries or by bilateral and multilateral institutions. The review of multilateral experiences draws heavily from a recent ILO review of 70 rural development projects, 34 of which were undertaken by other multilateral institutions and 36 by the ILO itself (ILO, 1990). Attention will be focused on the concept of sustainable development, which is seen to have three dimensions: economic viability, the development of sustainable institutions and environmental sustainability. No formal definition is offered in the ILO review – and unfortunately sustainability is a concept that has many definitions (see Chapter 1) – but it is clear that three aspects are being referred to here: whether the projects are economically viable in the long run; whether the institutions required to support the programmes will endure over time; and whether the environmental improvements will endure similarly. Although all three aspects are relevant, it is specifically the last that is of interest in the context of this chapter.

Multilateral rural development projects

The ILO survey draws a number of conclusions regarding the success of the rural development projects undertaken by multilateral institutions in achieving sustainable development.

(a) In many cases multilateral projects had a too short-term horizon to ensure that economic sustainability was achieved. Projects tended to concentrate on infrastructure investments, 'with little concern for activities that revolve around the infrastructures to generate lasting benefits'. As a result, projects often died once donor funding stopped. There was inadequate *ex post* evaluation covering the project strategy as a whole. Of the 70 projects examined, the ILO review concluded that only one-third could be said to have achieved economic sustainability, the major reason for failure being a lack of counterpart funds to support the programme on departure of the donor.

(b) A lack of institutional support in the country is often a cause of project failure in the long run. Reasons include poor budgetary support within the government for the requisite institutions and a lack of grassroots involvement and beneficiary participation. The ILO survey concluded that, of its sample projects, three-fifths did not meet the criterion of institutional sustainability. A similar conclusion, in quantitative terms, was reached in a World Bank evaluation of 27 projects approved between 1961 and 1975 (World Bank, 1987a).

(c) It has been difficult to assess the environmental sustainability of the projects involved because no detailed environmental evaluation was carried out. Many projects, however, are specifically aimed at environmental improvement and are therefore compatible with environmental sustainability *ipso facto*.

The survey concludes that the success of such environmental projects is critically dependent on secure tenure and on people's participation and beneficiary involvement. Examples include labour-intensive hill terracing in Rwanda; reforestation of wasteland in West Bengal by landless women (which was successful when government schemes failed); village wood-growing projects in Burkina Faso, Mauritania and the United Republic of Tanzania; and labour-intensive rural works projects in the United Republic of Tanzania. Of the sample of projects, the survey estimates that four-fifths had been environmentally sustainable, although it is not clear how this figure was arrived at.

Specific environmental projects

Four areas of environmental concern can be selected for examination in respect of which national and international projects and programmes have been developed: soil conservation, irrigation rehabilitation, and measures to counter the effects of overgrazing and deforestation.

Soil conservation

Soil conservation techniques have been researched and practised for many years, mainly in Australia and the United States. The techniques involved are either organic, e.g. involving improved cropping practices or the planting of trees, or mechanical, e.g. involving the building of bunds and terraces to prevent water run-off and run-on. Kenya has had some success with soil conservation projects. Between 1974 and 1985, 490,000 farms were terraced out of an estimated 1.1 million farms. The benefits of the terracing programme are not easy to distinguish from other improvements in crop management and improved seeds, but Hudson (1987) argues that 'intangible benefits to the farmer may be as important as yield increases. There is evidence that conservation farming can lead to greater reliability of yields, to a better yield per unit of labour input, to better cash flow, and to better nutrition of rural farm families.'

Even when it is possible to provide the right incentives to farmers for them to carry out soil conservation works, there is also the problem, as identified in the ILO report referred to above, of ensuring that these works are maintained. One example is the USAID-funded project in Somalia, where the bunds constructed under the project in the 1960s had fallen into disrepair by the 1980s. The benefits of preventing water flows had been partially lost as the bunds were allowed to disintegrate (yields remained higher than on the unprotected land). The reasons for the lack of maintenance by the farmers were: the labour costs of maintenance were higher than the benefits of increased crop production, which was only a supplement to livestock production; the government did not continue assisting the farmers through advice and the construction of additional bunds; and the new socialist government placed greater emphasis on large-scale farming. Not only did the farmers neglect the bunds, they also cultivated the land on a continuous basis instead of regularly leaving some of the land fallow, resulting in further soil erosion.

Irrigation rehabilitation

To reverse problems of waterlogging and salinity, several countries have undertaken rehabilitation programmes and turned unproductive land back into fertile land. For example, Pakistan has implemented 42 Salinity Control and Reclamation Projects (SCARPs) to improve drainage channels and sink tubewells (Grainger, 1990). Although this programme provided substantial increases in fresh water, it has proved to be expensive and low in cost recovery. What is being considered in its place is a more appropriate pricing level, with charges based on total available land rather than on land under cultivation. The revenues are to ensure that there are adequate funds to maintain the system, thereby improving the drainage. The charges, on the other hand, are to encourage the introduction of bio-saline agriculture and crops more suited to the conditions. India has also had some success in reclaiming saline areas. Salt-tolerant trees are being planted to reduce salinity and waterlogging. The shade of the trees lessens surface evaporation, reducing the upward movement of groundwater and salts to the surface. Fallen leaves increase the amount of organic matter and improve soil structure. Cost-benefit studies in both India and Sri Lanka have shown that the benefits of increased crop yields can justify the costs of desalinization in social and economic terms (Joshi, 1983; Herath, 1985).

Overgrazing

As overgrazing is one of the major causes of desertification, the control of livestock numbers is seen as the means of preventing further desertification. This view ignores the needs of people dependent on livestock production, many of whom were severely affected by the droughts of the 1970s and 1980s. The current approach to preventing further rangeland degradation is to improve the productivity of livestock and the range, and to encourage livestock producers to increase their offtake from the range. In this way, it is hoped that livestock producers will obtain higher incomes from livestock, but without damaging the environment. The productivity of livestock is being encouraged through improved animal health, breeding and fodder programmes. Natural regeneration of rangelands has been severely impeded by overgrazing in many areas, and several countries are undertaking the seeding of rangelands with fodder grass and trees. While these technical innovations can assist in reversing the process of degradation, the socioeconomic problems of grazing control on communally managed land remains a major constraint to preventing overgrazing. Attempts to tackle the problem have met with little success. For example, the concept of group ranching has been a failure in Botswana, Kenya and the United Republic of Tanzania, and donors have lost interest in funding livestock development projects. Only 5 per cent of all development aid to the Sahel is devoted to livestock projects (Timberlake, 1988). The only promising initiative is the formation of community-based grazing associations, where the communities themselves design a grazing control system. A number of projects in Mali, Niger and Senegal are assisting pastoral groups to develop their own self-regulatory mechanisms, adapting traditional systems of control. The World Bank is optimistic about the future of the projects, but difficulties have arisen in Lesotho, where the dynamics of unpopular decision-making are hampering future progress in communal grazing control (Grainger, 1990).

Deforestation

The initial reaction of many governments and donors to the deforestation problem has been to plant large plantations of exotic species. These projects have not been very successful, as the species were not suited to the local climate and soil. The problem of low rainfall was to be overcome in some areas through irrigation, but the high costs and management requirements of this approach made it economically unviable. One of the more successful approaches has been in social forestry, notably in China and the Republic of Korea. Between 1950 and 1983, China's forest area increased by 60 per cent to 122 million hectares. In the Republic of Korea over 1 million hectares of fruit, fuel and timber trees were planted between 1973 and 1977 (Gregersen, 1982; Arnold, 1983). Social forestry has also had some success in India, particularly in Gujarat, where the State Forest Department established a separate social forestry section with staff trained to encourage the public to plant trees (Karamchandani, 1982). The concept has been less successful in Africa, mainly because the emphasis there has been on planting trees for fuel.

Agroforestry includes silvopasture, i.e. the managed utilization of woodlands and pastures for livestock production, which can include planting fodder crops where necessary. In general, the mixed approach has proved to be more productive than conventional separate arable and livestock systems (Gupta, 1983). However, agroforestry projects suffer from a failure to take account of the economic constraints under which farmers are operating, and their unwillingness to risk adopting an expensive new technology when it is unproven.

In connection with the integration of tree planting into the farming system, the assumption of the higher productivity of exotic species in comparison with indigenous species has been brought into question. In several countries higher growth rates have been recorded for indigenous species than for exotic species; annual growth rates of 4.0–4.8 cubic metres per hectare per annum were recorded for *Acacia senegal* in Chad and 0.67–2.35 in Senegal, compared with 1 cubic metre per hectare per annum for plantations of *Eucalyptus camaldulensis*. Thus, increasing the productivity of indigenous woodlands may offer a better and cheaper method of wood production than planting exotic species that are less suited to local conditions. The cost of rehabilitating indigenous woodland has been estimated at US$200 per hectare, in comparison with over US$1,000 needed to establish a plantation (Catterson *et al.*, 1985). Indigenous woodlands have the added advantage that they provide the natural habitat for local wildlife, as well as berries, fruits, nuts, seeds, gums or medicinal ingredients, which are not provided by the monoculture of exotic species.

GENERAL POLICY IMPLICATIONS

The introduction of new technologies in developing countries and the effects on employment creation and environmental degradation suggest the following conclusions.

(a) The appropriate choice of technology is a complex issue in which the sociolegal context is crucial yet is often not given enough importance. The purely physical effects of new technology on the environment will *never* be easy to determine, but there are strong arguments, based on past experience, that those responsible for the introduction of new technology should take a more risk-averse approach than in the past.

(b) A review of the different types of 'appropriate' technology has shown that it is not possible to determine whether labour-intensive technology is always more appropriate to the needs of a developing country, and that each situation needs to be judged individually. The most appropriate technology will be determined by several factors, such as the supply of land, labour and capital, the employment and environmental effects, both direct and indirect, the socioeconomic context, and the macroeconomic framework. In particular, the evaluation of alternative technologies should include the *secondary* employment and environmental effects on other activities and other sectors of the economy. New agricultural technologies may have the benefit of increased labour requirements, but may affect labour requirements in other sectors. New technology may similarly be environmentally neutral in one situation but may have downstream negative effects. It is important that these secondary effects be identified and taken into account when assessing the costs and benefits of new technology.

(c) A most important issue arising from the review of environmental programmes and projects is the involvement of local people in the development and introduction of new technologies. The inappropriate nature of some technologies has arisen from the top-down approach to planning. Thus the local conditions and constraints have not been taken into account and new technologies have failed to achieve their objectives.

(d) In increasing incomes and employment, areas of comparative advantage should be identified and developed. The over-exploitation of natural resources has led to demands for the preservation and prevention of further utilization of these resources to halt their extinction. However, it is possible to make use of products of the environment in a sustainable manner, e.g. propagation and processing of natural products, plants, honey, silk, etc., utilization of wildlife and forests, handicraft development. These products of the environment can be harvested in a sustainable manner, often by the indigenous populations of areas such as Amazonia, and this can provide a source of income to these populations which will usually be sufficient to encourage the sustainable utilization of resources and prevent over-exploitation.

(e) Developing countries should be assisted to evaluate the use of economic instruments to encourage environmentally sound technology. Developing countries can learn from the experiences of the OECD countries in the use of economic instruments to take into account the environmental costs and benefits which are normally excluded from prices. The initial process should include an examination of the impact that existing subsidies have had on the environment and whether they should be modified or replaced.

(f) Developing countries need assistance to take part in international research on environmentally sound technologies to ensure the appropriateness of technologies for their needs. This will involve strengthening national research institutions through the training of staff and collaboration with international research institutions. It is also necessary to reorient international R & D programmes towards environmental protection and employment creation, as appropriate for the traditional, informal and modern sectors of developing economies.

(g) Finally, an important handicap in devising effective technology policies which can take into consideration the employment and environmental implications of new technologies is the lack of adequate information about the linkages between technology, environment and employment.

5 Incentives for Sustainable Development in Sub-Saharan Africa

Charles Perrings

THE ECOLOGICAL AND SOCIOECONOMIC SETTING

One of the most striking coincidences of the past decade has been that between deepening poverty and accelerating environmental degradation in the arid, semi-arid and sub-humid tropical regions – the drylands – of sub-Saharan Africa.[1] In a period in which poverty has fallen in many parts of the world, sub-Saharan Africa has witnessed both declining per capita consumption and an increase in the absolute number of people in poverty (UNDP, 1990). In the same period, it has seen the progressive degradation of an environment that was already under considerable stress. By the mid-1970s the existence of a correlation between rural poverty and environmental degradation was already causing alarm. Environmental degradation was noted as 'a common factor linking virtually every region of acute poverty, and virtually every rural homeland abandoned by destitute urban squatters' (Eckholm, 1976). Nor was there any illusion about the potential longer-term consequences of the trend. The United Nations Conference on Desertification (1977) estimated that the world could lose one-third of all arable lands by the close of the century as a direct result of environmental degradation, with all that this implied for the welfare of the rural poor. In the countries of the Sudano-Sahelian region, the degradation of dryland environments was judged to have reached crisis proportions. Nor was it significantly less of a problem in Lesotho and the countries straddling the high plateau from Botswana to Kenya. Since the mid-1970s, matters have got steadily worse.

At the level of aggregates, the deepening poverty associated with environmental degradation in sub-Saharan Africa is reflected in per capita consumption. In 1985, average consumption was not only less than it had been in 1975, it was less than it had been in 1965. Average daily calorie supply as a percentage of requirements has been estimated to have fallen from 92 to 91 per cent in sub-Saharan Africa between 1964–66 and 1984–86, but the incidence of deprivation that this implies has been uneven. Two groups of countries have been worst affected. These are Chad, Ethiopia, Mauritania, Niger, Senegal, Somalia and the Sudan in the Sudano-Sahelian region, and the Central African Republic, the United Republic of Tanzania, Uganda, Zaire and Zambia in east and central Africa. In this period, daily per capita calorie supply fell by an average of nearly 20 per cent in Chad, Ethiopia and the Sudan. Nor was the burden borne evenly in rural and urban areas. With the exception of Zambia, these are predominantly rural economies, with agriculture accounting for over 70 per cent of the labour force in all but two.[2] And it is in the rural sector that poverty is most highly concentrated. While distributional data are scarce in these countries, estimates of the degree of poverty in the period 1977–87 show that the probability that a rural household was in poverty was nearly twice that for urban households. On the basis of the available data, the depth of rural poverty in this period was greater amongst the countries of central and east Africa than those of the Sahel – exceeding 80 per cent in Burundi, the Central African Republic, Malawi and Rwanda.[3]

Other indices of poverty tell a similar story. All the countries in the two groups referred to recorded zero or negative average annual per capita GNP growth between 1965 and 1988. Moreover, although measures such as life expectancy and infant mortality have not worsened in the period, the rate of improvement in these countries has been much less than elsewhere. Indeed, the populations of the Sudano-Sahelian region remain the most deprived in the world, by the UNDP's human development index (UNDP, 1990).[4] The extreme destitution of famine victims in the refugee camps of Ethiopia and the southern Sudan is only the most acute symptom of a very widespread malaise.

The root causes of the decline in these rural economies are complex, and will be discussed later. They are, at the same time, causes of the collapse of the agricultural sector, of the declining productivity associated with the degradation of the natural resource base, of the increasing level of dependence on food imports, and of the sharp fall in those non-food imports – spare parts, fuels and fertilizers – on which expansion at the intensive margin depends (United Nations, 1986). The proximate causes of the decline of the rural economy are more easy to identify. While there remains some doubt about the extent and reversibility of many of the processes of environmental degradation which lie behind declining agricultural performance, there is a high degree of consensus in the literature that such damage as has occurred is directly due to two things: increasing pressure on grazing, arable and forest resources, and an insufficiently flexible response to climatically induced changes in the sensitivity of the environment to such pressure.

Population growth is commonly believed to be the immediate source of such pressure, and it is certainly the case that high levels of population growth are associated with increasing levels of pressure on the natural resource base. But it is not helpful to see population growth as an exogenous factor driving essentially short-term processes. The strong positive correlation between the rate of population growth, resource degradation and poverty does not necessarily imply a causal relationship running in one direction only. There is some evidence that, although average fertility in sub-Saharan Africa has been falling over the period 1965–88, for example, it has been increasing in a number of those countries where environmental degradation and rural impoverishment have been most marked: Central African Republic, the United Republic of Tanzania, Uganda, Zaire and Zambia (World Resources Institute (WRI), 1990). At the very least, this indicates the need for caution in characterizing the nature of feedback effects between environmental degradation and population growth.

In the short period, it is axiomatic that population growth implies increasing demand for agricultural products. What is interesting about the sub-Saharan African experience is that for various reasons – cultural and institutional rigidities in the societies of sub-Saharan Africa, plus the effects of government fiscal, price and incomes policies – the increasing demand for agricultural products has proportionately increased the pressure on soils, vegetation and water. Resources have tended to be locked into traditional activities and traditional technologies. The result is a density of livestock in pastoral areas that is unsustainably high, given the variance in climatic conditions in the arid and semi-arid zones, fallow periods in arable areas that are unsustainably low, and an exponential increase in the demand for construction timber and fuelwood. Overgrazing, overcultivation and deforestation have, in turn, led to various processes of resource degradation that have been grouped together under the term 'desertification'. They are a reduction in the processes that involve a reduction in the productivity of desirable plants, undesirable alterations in the biomass and the diversity of micro fauna and flora, and accelerated soil deterioration: from bush encroachment at one extreme to the increasing aridity and denudation of soils at the other (Dregne, 1983). It turns out that drylands are particularly vulnerable to such processes, since they tend to be less resilient in the face of stress and shocks than other ecosystems owing to the low level of system complexity (United Nations Conference on Desertification, 1977).

The economically interesting question arising out of this is why resource users should have failed to adjust to changes in the demand in a way that is consistent with the ecological sustainability of the resource base. Eckholm (1976) ascribed this to a 'basic dilemma ... that what is essential to the survival of society often flies in the face of what is essential to the survival of the individual.' This does not question the rationality of individual resource users, but suggests that what is privately rational may not be socially rational. The 'dilemma' implies that the private and social costs and benefits of resource use are different. Put another way, the structure of incentives confronting individual resource users induces behaviour that is inconsistent with the best interests of society.

This chapter considers the incentives that currently govern resource allocation in the dryland areas of sub-Saharan Africa, paying special attention to the institutional and other factors that stand in the way of the ecologically sustainable use of resources. The next two sections consider issues that are normally treated as exogenous: climatic change and population growth. The following section summarizes the evidence for climatic change as a determinant of resource degradation in the arid and semi-arid zones, focusing on the evidence for exogenous shifts in climate. This is not to deny that the localized feedback effects discussed by Charney (1975) and Hare (1977) exist, but to isolate trends in the productivity and resilience of dryland ecosystems which are beyond the control of local resource users. The evidence turns out to be patchy and incomplete, and most projections are highly conjectural. Nor do the projections of different General Circulation Models (GCMs) always coincide. Nevertheless, there is sufficient agreement to indicate that the current global climatic trends have serious long-run implications for the use of natural resources in these zones.

The third section of this chapter considers the environmental implications of population growth. It is argued that feedback effects between resource use, income and population are an important feature of the long-run dynamics of the dryland economies, and should be accommodated in policy. This has implications for the way in which population issues are addressed. If there is positive feedback between population growth, income and resource degradation – if an increase in the rate of fertility is as much a result as a cause of resource degradation – it may not be sensible to control population growth without addressing the factors on which it depends.

The fourth section considers the linkages between resource degradation, the nature of property rights in natural resources, and the institutions which regulate the level of activity in areas of common or communal property. The section reviews the evidence on the evolution of property rights and regulatory institutions in sub-Saharan Africa, and considers the barriers to the emergence of institutions that will enable externalities to be 'internalized' or accommodated. The following section then addresses the related issues of risk and innovation, and considers the link between risk, savings and technological innovation in resource use. What is at issue here is the difference in propensity to innovate between resources users in sub-Saharan Africa and those in the countries where the Green Revolution has had an impact. The section considers the determinants of technology choice in rural sub-Saharan Africa, and the implications of technological conservatism under high rates of population growth. It addresses the apparent paradox in the extreme risk aversion of resource users revealed by the choice of technology, and the high rate of time preference revealed by the degradation of the asset base.

The remaining sections deal with the determinants of the set of relative prices confronting resource users, and the responses of resource users to changes in relative prices. The background against which these topics are addressed is the argument that government intervention in product and factor markets has deepened the wedge between private and social costs of resource use, and so has exacerbated the misallocation of resources. The sixth section reviews these arguments, and considers how far there is a systematic bias in government intervention in sub-Saharan Africa. The following section then looks at the determinants of income of resource users, and considers the evidence on the relative strengths of income effects in their responses to changes in relative input and output prices. It also considers the connection between income and the rate of discount or the rate of time preference. While this is generally treated as exogenous in analyses of resource use, there is evidence of a well-defined relationship between risk aversion, time preference and income. Poverty induces both extreme (current) risk aversion, and an unusually myopic view of the future costs of present activities (Lipton, 1968; Perrings, 1989a). This implies that increasing risk aversion and rates of time preference may be positively correlated with increasing levels of resource degradation.

The main implication of this is that a key element in any strategy for the ecologically sustainable development of the resource base in sub-Saharan Africa is the generation of employment options outside traditional agriculture in order to reduce the adverse environmental effects of poverty within traditional agriculture. What is of interest, therefore, is whether there is a set of incentives that will both stimulate the diversification of the rural economy and encourage the adoption of ecologically sustainable technologies within the traditional sector.

CLIMATE CHANGE AND ECOSYSTEM RESILIENCE

One of the less tractable aspects of the problem of resource degradation in the rural economies of sub-Saharan Africa is the role of exogenously determined climate change. At present it is not clear how far climatic shifts have already influenced the sensitivity of the natural environment to the stress imposed by economic activity, and there is a very high level of uncertainty as to the local impacts of projected long-term trends. Nevertheless, there is sufficient consensus in the literature to indicate at least the direction of change in some of the key climatic variables.

Predictions of the General Circulation Models

While the specific predictions of the different General Circulation Models vary, they are agreed that a continuation of emissions of the 'greenhouse gases' – carbon dioxide, nitrous oxide, methane and the chlorofluorocarbons – at present or higher levels will cause an increase in both mean and variance of surface temperatures. Mean temperatures are currently expected to rise at between 0.2 per cent and 0.4 per cent per decade over the next century (Intergovernmental Panel on Climate Change (IPPC), 1990). The lack of resolution in many GCMs makes the problem of predicting regional changes particularly severe, but all are agreed on two things: that mean temperatures will increase in all areas, and that the rate of increase will be greatest in the higher latitudes and least in the tropics. The models do not accommodate cloud effects which could moderate temperature changes, but it is currently conjectured that cloud effects would be unlikely to alter the direction of temperature change. A continuation of greenhouse gas emissions accordingly implies increasing rates of evapotranspiration in all areas, and this effect is argued to be crucial to the impact of climate change on agricultural productivity (Parry, 1990).

At the global level, the evidence on temperature change over the past century is not inconsistent with the broad predictions of the GCMs. Global mean temperatures have increased by between 0.3°C and 0.6°C during this period, which is within the range predicted by the GCMs given the best estimates of change in the level of CO_2 over the same period (IPCC, 1990). It is harder to assess whether changes in precipitation are similarly in line with the GCMs, for the reason that the GCMs do not agree on even the direction of change in precipitation and soil moisture at the regional level. In the case of the Sahel, some models predict an increase in both precipitation and soil moisture owing to the northerly penetration of the Intertropical Convergence Zone, while other models predict the opposite. A further possibility is that, even if there is an increase in rainfall over the Sahel, higher rates of evaporation (because of higher temperatures) may still make an overall rise in soil aridity likely (Parry, 1990).

The recent history of rainfall patterns in the Sudano-Sahelian region does not appear to support a long-term trend towards increasing precipitation. Not only has there been an unusually large number of rainfall deficit years in the region since the late 1960s, but the duration, intensity and geographical extent of droughts have been exceptional by the standards of other drought periods recorded during the twentieth century (Snijders, 1986; Grouzis, 1990). The results include the collapse of streamflow in a number of significant river systems (both the Senegal and the Niger are affected), the disruption of key lacustral systems (such as Lake Chad), and the general lowering of aquifers. Nevertheless, it is not clear if the duration and intensity of drought in the region reflects a reduction in the mean level of rainfall or an increase in climatic variability. What is known is that the variability of climate is extremely sensitive to the mean: a small increase in climatic means can be associated with a very large increase in climatic extremes. It is in fact likely that the major impact of climate change on agriculture will come not from the changes in mean temperature or precipitation but from changes in the risks of extreme events (Parry, 1990).

The importance of these trends lies in their implications for the level of stress that the ecosystems of the region can withstand without losing resilience. Resilience, as the term is used here, refers to the stability of the organization of an ecosystem, or the propensity of the system to retain its organizational structure following perturbation (Holling, 1973, 1986).[5] It is accordingly a measure of the degree to which ecosystems can withstand the shocks of extreme events and continue to function: that is, it is a measure of the degree to which they can maintain 'productivity' in the face of shocks. Typically, resilience is a decreasing function of the stress to which an ecosystem is subjected. Of the different sources of ecosystem stress, the stress resulting from economic activity is obviously of prime interest here. But resilience is also a function of environmental conditions. Many dryland ecosystems have a comparatively low initial level of resilience, and so a low level of tolerance to stress. Since an increase in soil aridity resulting from change in both mean temperature and rainfall may cause such ecosystems to lose resilience, it may also cause them to become less tolerant to stress.

There is accordingly a close link between the resilience of ecological systems and the sustainability of economic activity. If the incidence of extreme events increases with change in average climatic conditions caused by global warming, and if the resilience of those ecosystems is simultaneously reduced by that same change in average climatic conditions, it follows that the 'ecologically sustainable' level of economic activity will fall. In fact, we may assert that an economic system will be ecologically sustainable if the ecological systems on which it depends are resilient. The crucial feature of resilience is the capacity it implies to adapt to the stresses imposed on a system by its interdependence with other systems. Hence the resilience of an ecosystem used in the course of economic activity is the guarantee that it will offer the same options to future generations of resource users, and that it will not in any sense prescribe those options.

The resilience of ecosystems is itself subject to change. This may be the result of the way in which they are exploited in the course of economic activity, or it may be because of a change in environmental conditions. Although much of this chapter will consider the impact of economic activity on the resilience of the dryland ecosystems of sub-Saharan Africa, it is important to acknowledge that climate change which is largely exogenous to the region (since it contributes very little to the build-up of greenhouse gases) may independently change the resilience of those same ecosystems. While the available evidence on the impact of climatic factors on the resilience of ecosystems in the region is very sparse, there is little doubt that climatic factors are significant. Indeed, some analysts of vegetation dynamics in the region argue that climatic factors strongly dominate human activity as a source of vegetation change (see Ellis and Swift, 1988, pp. 450–9).

Climate change and agriculture in sub-Saharan Africa

The principal ecosystems of interest in the rural economies of sub-Saharan Africa are those underpinning agricultural activities. Aside from the indirect effects of greenhouse gas emissions on these ecosystems through climate change, an increase in the concentration of CO_2 is expected to have a number of direct effects on the growth rate of both economic and other plants. Available evidence suggests that increased concentrations of CO_2 will have a significant positive effect on plant growth, through the role of CO_2 in the process of photosynthesis. It has, for example, been estimated that if CO_2 were to double, the photosynthetic rate would rise by between 30 and 100 per cent (Parry, 1990). What is of particular interest for sub-Saharan Africa is that the increase in the rate of plant growth that this implies would differ from one plant group to another. The major crops in the semi-arid and sub-humid tropical zones of Africa, such as maize, millet and sorghum, would benefit much less than the crops grown in the temperate zones, such as wheat, rice and soya bean.[6] This implies that agriculture could be relatively disadvantaged in the zones in which maize, millet and sorghum are staples.

Current predictions as to the future effects of exogenous climate change in the drylands of sub-Saharan Africa have rested on a *ceteris paribus* assumption, assuming no change in the basic pattern of resource use. This may be a reasonable assumption with respect to slowly evolving ecosystems, in which the scope for a genotypic evolutionary response to climate change over the sort of time horizon under consideration is very limited. But the same is not necessarily true of human resource users. In their case, the net welfare effects of climate change will depend critically on the flexibility of their responses in respect of institutions, technology and preferences. These are considered in more detail below.

It should also be added that regional patterns of resource use may themselves have climatic effects: not all climate change in the region is necessarily exogenous. Apart from the fact that sub-Saharan Africa is contributing to the processes driving global climate change, even if in a small way, it is recognized that agricultural activities may have localized climatic effects. A common result of the overgrazing of rangelands, the extension of arable activities into marginal lands and the concentration of populations around water sources is a reduction in vegetation. It is now recognized that this may have local climatic effects. It may, for example, increase surface albedo (reflectivity of solar radiation) which causes the air to lose heat radiatively, hence to descend and to lose relative humidity (see Charney, 1975; Hare, 1977). These are not at present considered to be highly significant.

POPULATION GROWTH AND ENVIRONMENTAL DEGRADATION

Climate change is not, of course, the only exogenous factor driving resource degradation in sub-Saharan Africa. A number of other factors beyond the control of local resource users have a part to play, and are considered later in this chapter. One factor which is conventionally treated as exogenous in economic analysis is population growth. Indeed, most models of economic growth are driven by an exogenously given rate of population growth. In the development literature, this convention is reflected in the fact that a significant strand in the literature identifies exogenous population growth as a major 'cause' of environmental degradation in the low-income countries, the 'solution' to which is argued to be both institutional reform and institutional control over population levels. This line of argument is worth considering in some detail.

Population growth and resource utilization

The role of population growth in resource degradation is argued to lie in the connection between population growth and resource use. Taking population growth as a datum, a number of contributions to the literature find that population growth leads to the intensification of subsistence production: the shortening of fallow periods, the increase of stocking densities, increasing rates of timber extraction, and so on (Boserup, 1965, 1981; Darity, 1980; Pryor and Maurer, 1982). The same trend has been identified in sub-Saharan Africa (Pingali et al., 1987), and has been argued to explain the process of environmental degradation (Ruddle and Manshard, 1981). The argument is that population growth implies an expansion of aggregate demand which then forces pastoralists and cultivators to put increasing pressure on a given environment: to intensify the exploitation of land currently in use, to switch land from pastoral to arable uses, and to bring increasingly marginal land into economic use. The cultivation of land that is both less productive and more sensitive to the effects of drought leads, eventually, to the degradation of the resource through both erosion and depletion of soil nutrients. Similarly, the increase in stocking densities on remaining pastoral lands results in overgrazing, which leads to devegetation and erosion of topsoil (Dixon et al., 1989; Pearce, Barbier and Markandya, 1990). In the absence of technological change, one result of this is argued to be a Malthusian adjustment process in which famine re-establishes a rough equilibrium between the size of the population and the resource base (Eckholm, 1976; Dregne, 1983).

There are two aspects of this argument to consider. First, the presumption that population increase necessarily implies greater pressure on the environment may be warranted in many cases, but it does not have the status of a universal law. The empirical relation between resource degradation and population expansion is not a monotonic one. Kates et al. (1977), for example, argued that while there was undoubtedly a strong positive correlation between population growth and resource degradation in many instances, localized resource degradation could be a consequence of population decline as well as population growth.[7] It has subsequently been recognized that the proximate causes of resource degradation in sub-

Saharan Africa – overstocking and overcultivation – need bear no systematic relation to population increase (Repetto and Holmes, 1983; Pearce, 1988). As is argued in later sections of this chapter, change in the nature of property rights in environmental resources, taken in conjunction with the effectiveness of institutions regulating access to those resources and with the structure of economic incentives, can have devastating implications for the environment, irrespective of demographic trends.

Second, it turns out that the rate of population growth is not independent of the rate of environmental degradation, and that the feedback mechanisms between resource degradation and population growth are more complex than a simple Malthusian view would suggest. While average fertility in sub-Saharan Africa has been falling over the period 1965–88, it has been rising in a number of countries where rural impoverishment associated with the degradation of the resource base has been increasing. It is clear that current population growth rates in the countries of the Sudano-Sahelian region are not consistent with the ecological sustainability of resource use in the region, given the current technology and structure of production. Nor are those population growth rates themselves sustainable. But it is not clear that a Malthusian adjustment in the form of widespread famine will re-establish anything like a stable equilibrium population. The poverty associated with resource degradation contains within it both a tendency towards Malthusian collapse evidenced by rising mortality rates (especially infant mortality rates) in times of famine, and a contradictory tendency towards population expansion. While deepening rural poverty increases the risks of morbidity and mortality among the existing population, it also creates a powerful incentive to expand that population, partly to compensate for the increasing risk of infant mortality, and partly to compensate for the increasing risk of income failure. Children are insurance against the risks that poverty brings.

Population stability and ecosystem resilience

The key points here are, first, that the relationship between population growth in any given environment and the degradation of that environment is not necessarily monotonic. It is mediated by the institutional and economic environments within which resource users operate, and there is no systematic relationship between population growth and either institutional or economic conditions. Population growth is not always and everywhere environmentally damaging. Second, the impact of population growth on resource degradation is highly sensitive to the technology used. Although population growth has frequently been assumed to cause technology change (Boserup, 1965, 1981), population growth in much of rural sub-Saharan Africa has not had this effect. Why? Once again, the institutional and economic environments within which resource users operate are part of the answer. But a major factor is the risk management strategies that characterize decision-making at very low levels of income. Third, the impact of resource degradation on population growth is more complex than the Malthusian arguments of Eckholm (1976) and Dregne (1983) would suggest. The extreme poverty of resource users in degraded environments is itself a spur to population growth. It is not, therefore, helpful to treat population growth as if it were exogenously determined, or to seek to change the fertility rate amongst resource users without addressing the motivation for large families. What is important is to understand and address the incentives that lie behind population trends. It is certainly not helpful to assume that these trends are simply the product of social ignorance and institutional irresponsibility.

PROPERTY RIGHTS AND THE OVERUTILIZATION OF RESOURCES

A third set of factors conventionally taken to be exogenous in economic analyses of resource degradation is that of the property rights and supporting institutions under which resources are used. It is generally acknowledged that different forms of property rights in resources – open-access common property, communal property, private freehold and leasehold property – have different incentive effects for resource users. It is also generally acknowledged that the less complete are property rights, the greater is the propensity for resources to be overutilized. The more environmental services/disservices associated with a given pattern of resource use are not the subject of well-defined rights, the greater is the risk of environmental damage.

Since property rights tend to be least well defined in common and communal property regimes, which remain the dominant form of property in sub-Saharan Africa, it has been argued that Africa is in some sense predisposed to environmental degradation. Specifically, it is assumed that under traditional property rights individual resource users have enjoyed more or less unlimited access to the resource, and have based their decisions with respect to stocking densities or fallow periods on a private cost/benefit calculus that has excluded all costs carried either by other members of the same generation or by future generations. While this structure of property rights is argued to have been efficient when applied to non-scarce resources, for which the external and user costs ignored in the private cost/benefit calculus are insignificant, it is inefficient when applied to scarce resources (Ault and Rutman, 1979).

The problem, in this perspective, is that the 'natural' evolution of property rights in the face of increasing land scarcity towards more well-defined (private) property rights is argued to have been frozen. Traditional land rights are argued to have been preserved more or less intact for political or ideological reasons. South Africa at one extreme, and the United Republic of Tanzania at the other, are claimed to have preserved outmoded systems of communal property rights that have led directly to the overexploitation of land in both the South African 'homelands' and the Ujamaa villages of the United Republic of Tanzania. The implication is that the system of communal property rights should be permitted to evolve in a natural way: that is, towards a system in which the externalities created by common property are internalized, and individual resource users are encouraged to take full account of the social opportunity costs of their activities (Dorner, 1972; Ault and Rutman, 1979; Harrison, 1987).

Traditional property rights

It is beyond the scope of this chapter to review the historical development of property rights in land in sub-Saharan Africa, largely because of the very wide range of systems involved. Successive intrusions into Africa have resulted in multilayered, complex systems of rights. Nevertheless, in each of these hybrid systems it is possible to identify a substratum of property rights deriving from the customary law of 'traditional' African societies. It is this substratum of traditional rights that is argued to be incompatible with the sustainable development of the resource base under conditions of increasing scarcity. All traditional African systems of land rights shared three characteristics: land was 'owned' by the tribe, each member of the tribe had guaranteed access to land in one form or another, but the tribe retained the right to determine the level and pattern of land use (Yudelman, 1964). Absolute individual rights of ownership over land and other natural resources were 'unknown' (Caldwell, 1976). Instead, individuals held more or less limited use or usufructuary rights (Herskovits, 1940).

Usufructuaries enjoyed exclusive rights to the produce of their land, but were subject to a number of restrictions regarding its use. They could transfer their usufructs to other members of the tribe or clan, or could pass them on to their heirs. Herskovits called this 'inherited use ownership', which he distinguished from the private property enjoyed by clan or tribe members in assets other than land. The interests of the collective were served by the fact that the usufructuaries' security of tenure was subject to their observance of the laws or customs of the tribe or clan in respect the level of their activity and the technology applied. Land that was not worked in accordance with the laws or customs of the tribe or clan could be reassigned. Subsequent research on customary land law has verified this view (see James, 1971), and has confirmed the basic similarities in traditional land rights amongst all 'tribopatriarchal' societies of western Africa (Howard, 1980) and the Horn of Africa (Hoben, 1973). In societies based on hunting, gathering or nomadic pastoralism, individual usufructs typically conferred fewer rights of exclusion. No individual within the group could exclude any other individual from the land or its produce. Nevertheless, the collective continued to regulate both the general level of activity and the pattern of land use in systems based on common property (Herskovits, 1940).

The important points here are, first, that traditional tenure systems did permit the limited transfer of usufructuary rights and, second, that the usufructuary rights were collectively controlled to meet social (including environmental) goals. This makes traditional rights rather closer to the modern ideas of tradable permits or quotas than to the open-access common property model. Many traditional societies prevented usufructs from being sold outside the group, but this is effectively true in a number of modern market economies. There is no evidence for the inference that traditional systems of property rights in sub-Saharan Africa were especially insensitive to the environmental constraints they faced because they were based on communal or common property in land. Indeed, all the available evidence suggests just the opposite.

The evolution of property rights

In fact, although it is possible to identify these various elements of traditional land rights, it should not be assumed that traditional rights and patterns of resource use have been as static as Ault and Rutman (1979) claim. Recent work on land rights in sub-Saharan Africa confirms that traditional rights have evolved in response to changing institutional, economic and environmental conditions (Dyson-Hudson, 1984; Feder and Noronha, 1987; Bruce, 1988). This is consistent with the historical tendency observed elsewhere for property rights in resources to develop when the gains from so doing become manifestly larger than the costs of change. While there has been a tendency for property rights to some scarce resources to evolve in the general direction of exclusive private property hnke, 1990; Dixon et al., 1989), this has not been true of all scarce ources. This does not, however, mean that property rights to such ources have been frozen. Property rights comprise a complex collection rules governing the rights (entitlements) and obligations of resource sers, and their evolution in any given set of historical circumstances will reflect the institutional, cultural and intellectual characteristics of the society concerned. It does not follow that any particular structure of rights can be taken to be the logical end-point of evolution. A more satisfactory explanation for the failure of private property in land to develop in many parts of rural sub-Saharan Africa, despite the scarcity of the resource, is that the introduction of private property in land in sub-Saharan Africa has offered no advantages over traditional tenure systems, neither in terms of productivity gains nor in terms of the ability of the resource user to raise credit. The incentive for resource users to opt for private as opposed to some modified form of communal property in sub-Saharan Africa is not therefore obvious (Migot-Adholla et al., 1991).

Unfortunately, comparatively little empirical work has yet been done on the link between evolving property rights and the management of environmental resources in the region. Two aspects of the evolution of communal property rights are, however, worth comment. First, the institutions traditionally charged with the allocation of usufructual rights and with the regulation of the level of activity have either disappeared or been severely weakened, while replacement institutions have failed to exercise the authority vested in them. It might be expected, a priori, that this trend would exacerbate rather than alleviate the problem, and there is now considerable evidence to support that expectation. In pastoral economies, although there is still an incentive to cooperate where land is held communally (see Cousins, 1987; Wade, 1987), it remains the case that, wherever traditional institutions for regulating the use of communal lands have broken down, individual graziers have tended to add stock past the maximum sustainable point (Jamal, 1983). In Botswana, to take a particular example, traditional mechanisms for regulating grazing have degenerated. The *badisa* charged with the care of the rangelands have disappeared and range management decisions have become more individualistic and less coordinated (Abel et al., 1987). At the same time, the Land and Agricultural Resources Boards charged with the allocation of grazing land in the communal areas, the granting of water rights and the conservation of rangelands have failed to ensure that private stocking decisions are consistent with the sustainable use of the resource. The result is that livestock owners in the communal areas of Botswana currently have virtually uncontrolled grazing rights, and no individual bears the full cost of adding stock beyond the carrying capacity of the range (Arntzen and Veenendaal, 1986). Nor is the example of Botswana unique.

Security of tenure and resource degradation

The second feature of evolving property rights is that the security of individual tenure has tended to diminish. People moving on to previously unoccupied land (squatters) typically have no legal entitlement to use of the land, and occupy it only so long as they are permitted to do so by more powerful agencies. Even on lands occupied on the basis of traditional communal property rights, the security of tenure over usufructs has diminished with the weakening of institutions of collective control (Bruce, 1988). From an environmental perspective, this too has adverse incentive effects – albeit of a different kind. The collapse of regulatory institutions opens up the potential for overexploitation of the resource base by enabling resource users to ignore part of the social opportunity cost of resource use. The collapse of the security of land tenure, on the other hand, tends to increase the rate at which the future costs of resource use are discounted: that is, it lowers the incentive to manage the land on a sustainable basis. Whereas the collapse of regulatory institutions widens the scope for external effects to be ignored in resource use decisions, insecurity of tenure ensures that the future or user effects of resource use are ignored, whether or not markets for those effects exist.

One of the arguments conventionally advanced for promoting private property in dryland environments is precisely that it would increase security of tenure, and so encourage a longer-term perspective in land use decisions. This is taken to be a prerequisite for individual resource users to undertake conservation investments (Dixon *et al.*, 1989). It is not obvious that the introduction of private property in land in sub-Saharan Africa has had this effect, nor is it obvious that private property is the only tenure system consistent with security of tenure. But it is clear that the rate of discount implied by current patterns of resource use in parts of sub-Saharan Africa is inconsistent with the ecologically sustainable use of the resource base. To the extent that myopia is a function of insecurity of tenure, sustainability does imply the need to revise current tenure systems. The evidence for unsustainably high implicit discount rates in current patterns of resource use lies in the fact that present rates of resource depletion in parts of sub-Saharan Africa exceed the natural rate of regeneration of the resource base – given existing technologies. Sustainability implies either a reduction in current levels of rural economic activity (whether through a Malthusian adjustment process or through some less traumatic means of exiting the rural economy) or technological change. Once again, to the extent that the form of tenure is a factor in the choice of technology, the incentive effects of existing tenure arrangements are relevant. But the evidence to date does not support the notion that privatization in sub-Saharan Africa has led to productivity-improving investments (Migot-Adholla *et al.*, 1991).

The important point here is that, if the sustainability of resource use in the drylands of sub-Saharan Africa is not to be 'restored' (however temporarily) through a reduction in population, it implies the need for both diversification of the rural economy (Perrings *et al.*, 1989; Pearce, Barbier and Markandya, 1990) and technological change within the agricultural sector (Conway and Barbier, 1990). This makes it interesting to inquire into the structure of incentives governing the allocation of rural resources, of which the system of property rights is an important component part. It turns out, however, that the incentive effects of property rights *per se* may be less significant than previously thought, although security of tenure within any given system of property rights may still be important.

RISK MANAGEMENT AND INNOVATION IN THE RURAL ECONOMY

Traditional management of environmental risk

The main characteristics of closed agrarian economies have historically been a heavy reliance on agricultural or pastoral activities, a very marked technological conservatism, a marginal physical product of all resources at or close to zero, together with zero net savings (Schultz, 1964; Fei and Ranis, 1978). The environmental management strategies in such economies were typically also highly conservative, being designed to minimize the risks associated with the overexploitation of the resource base (Perrings, 1985). All traditional societies have sought to balance control over the population – both of humans and of herds – with control over the intensity of land use via the regulation of the frequency of cultivation or grazing. In nomadic pastoral, hunting or gathering societies the primary means of control over the frequency of land use has been, quite naturally, control over the pattern of mobility. In swidden agriculture, primary control of the frequency of land use has taken the form of group regulation of the period of fallow.

Allan (1965) identified a range of what he termed 'land use factors' in agriculture – defined as the relationship between the duration of cultivation on each of the land or soil units used in the classification and the period of subsequent rest required for the restoration of fertility, or as the number of land units required to keep a single unit in cultivation. These land use factors were the basis of collective decisions on the primary means of control – the frequency of cultivation. Very fertile areas, such as the upper Zambezi valley, Buganda or the Oalo lands of Senegal had a land use factor of less than 2, and so permitted semi-permanent cultivation. Three or four years of cultivation might be followed by one or two years when the land lay fallow – what Boserup (1965) called 'short fallows'. A much wider area had a land use factor of between 4 and 8, permitting what Allan referred to as 'recurrent cultivation', and Boserup terms 'bush fallows'. A period of between two and six years of cultivation had to be followed by anything from six to 30 years in fallow. Large areas of the Zaire basin, western Africa, the Guinea savannah, eastern and central Africa fell into this category. Finally, the semi-arid zones, including the Kalahari and much of the northern Sahel, had a land use factor of more than 10. These areas permitted only the nomadic form of agriculture that Allan termed 'shifting cultivation'.

The most remarkable feature of such land use strategies was that actual fallow/cultivation ratios were almost always significantly greater than the minimum fallow/cultivation ratios implied by the land use factor. The actual population permitted to exploit a given area under cultivation or in fallow was consistently well below the critical carrying capacity of that area (Nash, 1967; Sahlins, 1974), actual levels of activity being linked to the lowest possible rather than the expected carrying capacity. The effect of this was to augment the land required for each member of the group. Indeed, Allan (1965) argued it to be 'axiomatic' that subsistence cultivators would cultivate an area large enough to ensure the food supply in a season of poor yields as part of a risk-minimizing strategy. This is consistent with Leibenstein's (1957) model of the closed agrarian economy, which generated a 'quasi-stable' equilibrium, or a stable equilibrium conditional on the economy concerned being insulated from external influences.

The management of market risks

The opening of the agrarian economies in sub-Saharan Africa to world product markets changed these characteristics in important ways. Since certain input and output prices were taken from world markets, the relations between the productivity of the resource base and real income, and between savings/investment and environmental conditions, were both altered. Capacity utilization was no longer only a function of the distribution of rainfall or temperature, but reflected expectations about input and output prices. There were two sources of uncertainty to be accommodated – the environment and the world market. Indeed, analyses of the succession of famines that have affected the Sudano-Sahelian region in the past three decades have shown that most were induced not by the decline of food availability resulting from drought, but by the collapse of real income caused by changes in food prices (Sen, 1981; Speece, 1989).

There is little evidence that agricultural producers of sub-Saharan Africa have exploited the potential specialization gains available from international trade. Indeed, the evidence suggests exactly the opposite. Most agricultural producers have done little to change either product mix or technology (see Konczacki, 1978), and this has distinguished them from agricultural producers elsewhere. Whereas the introduction of Green Revolution technologies in Asia resulted in a 27 per cent increase in per capita food production between 1964 and 1986, Africa remained almost untouched. Indeed, per capita agricultural output in Africa showed a downward trend during this period, a fact that is only partly explained by the less favourable endowments of sub-Saharan Africa (Conway and Barbier, 1990).

Part of the explanation for the failure of producers to change the technology under which they have exploited an increasingly scarce resource base is to be found in their risk management strategies. It turns out that these strategies are largely driven by income effects, which are considered later in this chapter. What is important here is that it has long been recognized that risk aversion is a decreasing function of income, and that agricultural producers close to the poverty line tend to adopt highly risk-averse strategies (Lipton, 1968). In general, producer strategies for minimizing risk in the rural economy involve two decisions. One concerns the optimal level of activity, the other concerns the balance between production for direct consumption and production for the market (Livingstone, 1981). Neither the market risk nor the direct consumption risk of harvest failure are insurable except through the decisions made by agrarian producers in these two areas.

Producers who are near the minimum subsistence level, and are unable to take a loss below subsistence, tend to adopt what Lipton (1968) has called 'survival algorithms': environmentally conservative practices characterized by the selection of low value but robust crops or livestock suitable for both market production and direct consumption. They tend to avoid high market value but environmentally susceptible crops or livestock that are not directly consumable (Yamey, 1964, pp. 376–86).

Risk-minimizing strategies of this sort have biased production and consumption decisions in sub-Saharan Africa in favour of tried practices and traditional products, and against technological innovation. This is particularly the case where technological change is itself a source of risk, as with Green Revolution technology (see Schultz, 1964). It follows that the costs of such technological conservatism, in terms not only of the specialization losses of trade but also of the degradation of the natural environment, are implicitly accepted as part of the premium to be paid for the risk of participating in the world market.

A similar set of considerations informs the savings behaviour of agricultural producers. Savings in the open agrarian economy may notionally be made in the form of either financial or real assets: in money, bank deposits, etc., or in natural assets, grain and livestock. Partly because saving in financial assets tends to be limited by the availability of financial institutions, and partly because of the risks attached to saving in financial assets, the safety-oriented survival algorithms employed in sub-Saharan Africa have favoured saving in real assets. Savings in real assets, particularly grain or livestock, provide both a means of direct consumption in the case of drought and a degree of protection against the volatility of product prices (see, for example, Speece, 1989). However, they also tend to reinforce the technological conservatism of agricultural producers, since they make it very difficult to find substitutes for existing capital assets.

In terms of its environmental effects, this technological conservatism of agricultural producers has been an important source of stress on the ecological systems of the natural environment. Given a breakdown of controls over the level of activity, the intensification of agriculture within the existing technology has not surprisingly led to diminishing returns caused, in large part, by the degradation of the resource base. This is reflected in the fact that, as the level of agricultural activity has intensified, so has the sensitivity of output to climatic variation. Even minor variations in rainfall or temperature have been associated with major output losses. That is, the resilience of the ecosystems involved has fallen to the point where normal perturbations in rainfall lead to system failure. Moreover, where expansion has taken place at the extensive margin, it has involved increasingly unsuitable vegetational or topographical conditions, with similar implications for the sensitivity of output to climatic variation (Barbier, 1988; Pearce et al., 1988a; Mosley and Smith, 1989).

Water risks in dryland environments

By comparison with the technological conservatism of private risk management strategies, the collective management of water supply risks in the dryland environments of sub-Saharan Africa has been recklessly innovative. A common element in the rural development strategies adopted in all high-risk areas of sub-Saharan Africa in the past three decades has been the attempt to control fluctuations in the natural supply of water through the provision of dams or, more commonly, wells or boreholes, an immediate consequence of which has been the disassociation of stocking or cultivation levels and the minimum expected level of rainfall.

The results of this are now well documented. It is generally recognized that the most stressed localities of the Sahel and similar arid zones are water sources. The concentration of livestock in these areas means that, whatever average stock levels may be, these particular spots are subject to excessive pressure. Since the number of available water sources tends to diminish with a reduction in rainfall, it follows that the problem is exacerbated in dry years. Put another way, rural development programmes which have revolved around the construction of wells have created at least one of the necessary conditions for desertification to occur (Bernus, 1980; Dregne, 1983).

What has changed is that one element in traditional collective risk management – the restriction of stocking levels to the minimum expected carrying capacity of the range – has been discouraged in order to increase aggregate agricultural output. But this has not affected the risk management strategies of individual producers. Hence, herds that have been built up in wet years have not been reduced when the rains have failed, as livestock owners have sought to protect themselves from both the anticipated scarcity of food supplies and expected rises in prices. Where wells provide temporary water sources in the early stages of drought, the decision to destock tends to be deferred for even longer, leading to additional pressure on drought-stressed pastures (Hare, 1977). In addition, water sources have become foci for human as well as livestock concentration, and this has given rise to an increase in the rate of vegetation destruction in the neighbourhood of such sources in a variety of ways. Traffic, cultivation and above all woodcutting, both for arable land and for fuel, have all proved to be destructive of the vegetational cover. In the Sahel the 'sedentarization' of nomadic pastoralists around water sources is reported to have resulted in dramatic environmental degradation in the neighbourhood of settlements. More generally, the rate of woodland clearance in these areas far exceeds the natural rate of regeneration, and yet still leaves a current fuelwood 'deficit' (Berry, 1984; Food and Agriculture Organization (FAO), 1985; World Bank, 1986; Markandya, 1991).

The general point here is that the risk management strategies adopted by low-income agricultural producers have been built around a technologically conservative pattern of resource use designed to guarantee output in the short run. Risky innovations in terms of both input and output mix have been avoided. Moreover, the lower the private return to agriculture, the more risk averse has the behaviour of agricultural producers become. At the same time, however, traditional collective risk management strategies have been largely abandoned as successive governments have promoted an increase in aggregate agricultural output as a major plank of their rural development or food self-sufficiency programmes, and as population growth has stimulated the demand for food. In the absence of collective control, agricultural activity has been intensified, resulting in stocking densities or fallow periods that are well beyond the capacity of the resource base to sustain. While a change in technology appears to be a prerequisite for the sustainable development of the rural economies of sub-Saharan Africa, the private incentives at present militate against the kind of risk-taking behaviour that is needed for farmers to consider either diversification or innovation.

PRICE INCENTIVES

The agricultural price system

Historically, the opening of the agrarian economies of sub-Saharan Africa to wider markets created at least the potential for gains from trade, but it also introduced both new sources of uncertainty and new constraints on the behaviour of producers. Prices for the main cash and food crops, and for the main animal products produced in sub-Saharan Africa, have historically been extremely volatile. To the extent that independent domestic markets have existed for agricultural products, this volatility has reflected the considerable variance in domestic supply resulting from fluctuating climatic conditions. But much of the volatility in product prices has reflected conditions that are exogenous to the region. Partly as a result of this, most governments have long invoked measures to stabilize producer prices, usually based on a system of administered prices under monopsonistic marketing arrangements. The economic environment within which agricultural producers have made their decisions has accordingly one or both of two elements: a set of competitive 'parallel' markets in which prices have been driven by local supply conditions, and a set of near-monopsonistic markets dominated by centralized marketing boards, in which prices have been set independently of domestic supply conditions.

One very influential view of the environmental crisis in sub-Saharan Africa is that it is predominantly an agricultural crisis, and that it is driven by the contradictory signals coming out of these marketing arrangements

(Pearce et al., 1988a). Certainly, considerable attention has recently been paid to the microeconomic decisions of resource users in sub-Saharan Africa, and to the role of economic incentives in such decisions. The main finding in these studies is that the existing system of administered prices in African agricultural markets has both driven a wedge between the private and the social opportunity cost of the use of the resource base, and reduced rural incomes.

It is widely believed that administered prices have driven the cost of tradable inputs below their social opportunity cost, thereby encouraging their overutilization. For example, destumping subsidies in agriculture and stumpage fees or royalties in forestry have encouraged deforestation at excessive rates, both in terms of rates of felling in timber concessions and in terms of the clearance of ever more marginal land for agricultural purposes (Berry, 1984; World Bank, 1986; Warford, 1987; Repetto, 1989; Perrings et al., 1989), with significant implications for the subsequent rate of soil loss (Younis, 1987). Subsidies designed to promote cash cropping as a means of increasing export revenue have been argued to contribute to leaching, soil acidification, loss of soil nutrients and the reduction in the resilience of key ecosystems (Timberlake, 1988; Grainger, 1990). The result is that the price system has been unable to guide the allocation of resources in a socially efficient way. Resource users have faced a positive incentive to misallocate resources.

The second effect reflects the control of output prices under monopsonistic marketing arrangements. The marketing boards set up in many countries in sub-Saharan Africa to handle the purchase, grading and marketing of a variety of agricultural products are argued to have depressed producer prices below world prices (taken to define the social opportunity cost of the resource). The significance of this, at the risk of oversimplification, is generally taken to be the fact that it has reduced producer income and so discouraged investment in land conservation (see Warford, 1987; Ghai and Smith, 1987). It has already been argued that the impoverishment of resource users encourages an extreme degree of (current) risk aversion which inhibits technological innovation. It turns out that impoverishment also encourages a myopic approach to the intertemporal allocation of agricultural resources, since the future costs of resource use must be subordinated to the need for immediate survival. In addition, it ensures that the income effects of changing relative prices will tend to offset the substitution effects, so dampening the response of agricultural producers to relative price changes (Perrings, 1989a).

The main conclusion of the literature on the incentive effects of the existing agricultural price system in sub-Saharan Africa is that the liberalization of agricultural markets is a necessary condition for the conservation of the resource base (Bond, 1983; Cleaver, 1985; Repetto, 1986, 1989; Barbier, 1988, 1989; Warford, 1989). Indeed, the liberalization of the agricultural sector has become a key component of the conditionality attached to structural adjustment lending by both the World Bank and the International Monetary Fund (Johnson, 1989). The near-unanimity of this conclusion is compelling, but care needs to be taken in interpreting it. The literature is not nearly so convincing on the consequences of price reform as it is on the need for it.

The price implications of liberalization

The liberalization of agricultural markets has implications for both the level and the stability of prices. It is expected that average producer prices will rise, and although little attention is paid to the question of stability in the literature, it must also be expected that producer prices will become more variable. If we take the issue of price stability first, even though this is seldom addressed explicitly in these recent analyses, it is implied that monopsonistic marketing arrangements have imposed welfare losses that significantly outweigh any welfare gains from stabilization. This may be true, given the level of administered producer prices; but it leaves open the possibility that stabilization close to the expected value of world prices for tradables would be preferred to the kind of price volatility experienced in many world agricultural markets. Indeed, this may be particularly important in sub-Saharan Africa, where the dominant cause of famine has been the failure of real purchasing power rather than a decline in the availability of food (Sen, 1981). All that can be said at present is that the price stability implications of the liberalization of agricultural markets in sub-Saharan Africa have yet to be analysed satisfactorily.

The implications of liberalization for the price level are easier to identify. In most cases, the average price of food crops has been held well below the expected world price in the interests of controlling consumer costs, and so urban wages (Markandya, 1991). There has been an anti-agricultural bias in policy, reflected in the distortion of tradable input and output prices resulting from the monopsonistic pricing of outputs, the monopolistic pricing of inputs and the income-depressing effects of taxes and tariffs. Indeed, one of the dominant motivations for liberalization is the belief that it will result in an increase in producer prices. This, it is claimed, will lead to the expansion of agricultural supply, and will provide both the incentives and the resources to conserve the agricultural resource base.

Incentives for Development, Sub-Saharan Africa

In the short run, it seems clear that the liberalization of agricultural markets should result in a rise in producer prices. But what the long-run effects will be is not nearly so clear. The long-term trend for real non-oil commodity prices has been a declining one. In respect of agricultural products, the index of world prices of non-food agricultural products has declined much more sharply than that for foods. But as Figure 5.1 shows, both have fallen dramatically. Removing the wedge between administered and market prices may not therefore be sufficient to ensure a sustained increase in the real income of agricultural producers. The efficiency arguments in favour of liberalization remain, but it cannot be assumed that it will necessarily have a positive impact on long-run producer prices.

Incentive effects

Evidence on the impact to date of agricultural price liberalization on the allocation of resources in the agricultural sector has been very mixed. A number of countries have introduced the principle of export parity pricing for major cash crops – but there has been no continent-wide trend towards

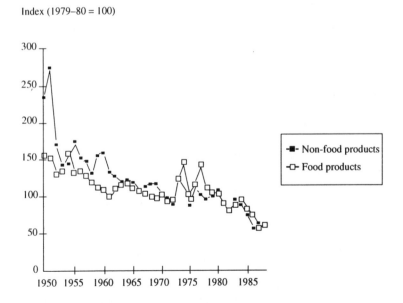

Figure 5.1 Real agricultural commodity prices, 1950–86 (*Sources*: World Bank, 1987; WRI, 1991).

this objective (Mosley and Smith, 1989, pp. 321–5). Indeed, in most countries agriculture continues to be 'penalized' by the protection of the industrial sector, either directly or through the overvaluation of the exchange rate. But where price regimes have been liberalized, the supply responsiveness of farmers has been muted.

The short-run price elasticity of supply of individual cash crops has been shown to be positive and sometimes high, but findings on long-run supply elasticities are mixed. One recent survey found that the long-run supply elasticities of all crops tested was positive (and generally identical to the short-run supply elasticity) (Gammage, 1990).[8] However, other studies have found generally low, and in some cases even negative, long-run elasticities (Green, 1989; Perrings, 1989b; Rao, 1989).

While no attempt has been made to disentangle the income and substitution effects of the changes in relative prices that have followed liberalization, a number of explanations have been offered for the lack of a sharper response. The first is that price responsiveness is more tightly constrained by institutional factors in sub-Saharan Africa than elsewhere (Delgado and Mellor, 1984; Lipton, 1987; Junankar, 1989). Certainly, most of the evidence on own- and cross-price elasticities of supply and demand that is used to support the use of economic incentives in agriculture is drawn from non-African data (Markandya, 1991). A second explanation is that, since a substantial proportion of goods and services are not traded, farmers are less sensitive to price changes than elsewhere (Ghai and Smith, 1987; Beynon, 1989). A third is that constraints on the supply of basic natural resources limit the capacity of those who do participate in the market to respond to price incentives. Raising the return on productive assets will not help if people do not have access to those assets. In many parts of sub-Saharan Africa, for example, it has been argued that poorer farmers cannot increase productivity at the extensive margin simply because they do not have access to land (Feder and Noronha, 1987). Moreover, even where farmers do have access to productive assets, the lack of physical and financial infrastructure may limit their capacity to respond to the incentives offered by changing real returns on those assets (Addison and Demery, 1989; Beynon, 1989).

The problem of common property environmental resources

It is not easy to distinguish between the various effects of price liberalization in practice. The 'anti-agricultural bias' in the system of administered prices implies that average agricultural income is below the level that would obtain under the same distribution of the agricultural product at socially efficient prices – which are no less than export and import parity prices in the case of tradables, and are equal to the marginal social cost of resource use in the case of non-tradables. In other words, an anti-agricultural bias implies non-neutral intervention in respect of both input and output prices. Price liberalization in the case of the markets for tradables may be a necessary condition for the alignment of private and social costs, but it is not a sufficient condition. If non-tradable inputs are priced below their marginal social cost, and if there are no other restrictions on the use of such resources, price liberalization may increase, not decrease, the level of environmental stress.

A particular feature of the price structure of agriculture in much of sub-Saharan Africa has been the implicit subsidy on agricultural land offered by traditional (common or communal) land tenure systems. Land has been almost universally priced below its marginal social cost, and in most cases has been 'free'. Yet the cost of arable and range land to the user is not at issue in the liberalization of agricultural markets. The general implication of this is that an increase in rural incomes will not, by itself, affect the gap between the private and social costs of environmental resources.

To be more specific about this last point, there are three elements in the cost of committing an environmental resource to some economic use: (a) current and future direct costs of use, including the costs of extraction or harvest of potentially depletable resources, the site costs of resource-using processes and the costs of waste disposal; (b) current external costs of use comprising intersectoral external costs resulting from the interdependence of activities reliant on a common environment (damage done to third parties which is not compensated within the existing structure of property rights, such as the cost to 'downstream' activities of the loss of watershed protection from deforestation); and (c) future indirect external costs of use comprising 'user costs' or future opportunities forgone from committing a resource to a current use (costs imposed on future generations of committing a resource in a way that deprives future users of the benefit of its use, such as the loss to future generations everywhere caused by the extinction of a species in one location now). Price liberalization in respect of the markets for tradable inputs and outputs in agriculture leaves the last two of these untouched.

At present there is a strong presumption in the literature that market prices are better environmental indicators than either administered prices or market prices net of taxes. This implies that government intervention is systematically biased against the ecological sustainability of resource use. While it is possible to determine a systematic bias in the administered prices of tradables, since these have tended to be well below the market prices which give the lower bound of the social opportunity cost of resource use, it is not possible to identify any systematic bias in respect of other forms of government intervention. It is certainly the case that many macroeconomic policies which do impact on resource prices and interest rates tend to be targeted at short- or medium-term objectives. Nevertheless, the presumption of a systematic bias in government intervention seems untenable.

INCOME AND ASSET DISTRIBUTION

One factor that recurs again and again in the analysis of both causes and effects of environmental degradation in sub-Saharan Africa is the level and distribution of income and assets. While rural poverty is very clearly an effect of rural resource degradation, it is also a cause. Following the famines of the Sahel and the Horn of Africa in the 1970s, Sen (1981) made the point that the basic cause of famine was not the decline in the availability of food caused by drought, nor was it the economic dislocation caused by war. It was the collapse of the purchasing power of agricultural producers: what Sen called the failure of trade, direct and transfer entitlements. While something similar can be said of the recurrence of famine in these areas, it is now clear that the decline in the availability of food is a large part of the problem. However, this decline results not from drought but from the degradation of the resource base. The resilience of the ecological systems on which agriculture depends has been reduced to the point where even modest fluctuations in rainfall and temperature have severe effects on output, and this is the direct result of land use practices that are largely driven by poverty. Poverty distorts attitudes to risk and innovation, just as it distorts savings and investment decisions and the weight given to the future effects of current activity. Poverty also dominates the responses of resource users to changes in relative agricultural input and output prices, in taxes and in transfers. It generates a vicious circle of responses in which the options available to successive generations are steadily eroded. It is in this sense that the sustainability of resource use implies the need to tackle the poverty of current generations (see World Commission on Environment and Development (WCED), 1987).

Rural poverty in sub-Saharan Africa

As was remarked earlier, the evidence for the increasing impoverishment of people in many of the dryland regions of sub-Saharan Africa is overwhelming. This is reflected both in the number of people judged to be in need by some absolute measure and in the distribution of assets and income. Measures of the depth of absolute poverty include the proportion of the population judged to be in need of emergency relief, or to have access to goods or services judged to be basic needs, such as drinking water or sanitation services. By both sorts of measures, absolute poverty is increasing in parts of rural sub-Saharan Africa. The first measure tends to focus on short-term poverty, the second to capture trends. But the two measures often coincide. In the Sudan, to take one example, the percentage of the population judged to be in need of emergency relief is no higher, and the percentage of the population without access to safe drinking water is lower, than in 1980 (WRI, 1990).

At the same time, there has been a marked and continuing tendency for the distribution of both assets and income to widen over time. In part, this reflects the erosion of traditional rights of access to the resource base, which has given rise to a long-term trend involving widening disparities in the farm and herd sizes of arable and pastoral economies (Ghai and Radwan, 1983). Nor has this trend been interrupted by the decolonization process. Colclough and Fallon (1983), for example, have reported that the distribution of cattle ownership in Botswana has become increasingly skewed following independence.[9] A study of agriculture in northern Hausaland in the middle 1970s found that, although land was communally owned, 50 per cent of the farmers worked only 22 per cent of the available land (Norman, 1977). Moreover, it would seem that the Gini coefficient for land distribution in the 'traditional' agricultural sectors of most African countries in the same period lay between 0.42 and 0.55 (Ghai and Radwan, 1983), which is very different from the egalitarian distribution expected by Ault and Rutman (1979) under the archetypal traditional tenure system. It turns out that gender is an important factor in this trend. Female-headed households typically have access to a much smaller asset base, whether in terms of land or livestock, than male-headed households, and it is not coincidental that relative poverty in the sense of relative deprivation is reckoned to bear most heavily on women (UNDP, 1990).

Poverty and the rate of discount

The main problem with poverty in the context of the intertemporal allocation of environmental resources is the effect it has on the individual resource user's rate of time preference. This has two main implications. First, those in poverty will tend to discount the future costs of current activities at a much higher rate than others. The higher the discount rate, the less relevant are the very important user costs of environmental resources. Second, since the rate of discount determines the optimal rate of extraction of potentially exhaustible resources, those in poverty will tend to deplete such resources at a much higher rate than others. It can be shown that, if the price of a natural resource is constant over time, it will be optimal to deplete that resource so long as the discount rate exceeds its natural rate of regeneration. And if the price is not constant, it will be optimal to deplete the resource so long as the discount rate exceeds its natural rate of regeneration plus the rate of change in its price. The so-called Hotelling rule for exhaustible resources is a special case of this (see Clark, 1976; Dasgupta and Heal, 1979).

Poverty in the sense in which it is used here is 'subjective' poverty. That is, it is defined by the household's perception of what constitutes a minimum acceptable level of consumption. Such perceptions vary both from household to household, and within households over time. An exogenous shock which necessitates a downward revision of consumption levels will be accepted with a lag which depends on the speed of adjustment of consumption expectations. For those at the subjective poverty point this implies some period of dissaving. Indeed, the zero savings level of income for any household may accordingly be regarded as the 'subjective' poverty point for that household, while dissaving may be taken to indicate that the household is in 'subjective' poverty (Drewnowski, 1977).

The recorded dissaving that has occurred in many sub-Saharan African countries during the past decade is consistent with this view (see World Bank, 1990). What is more important from an environmental point of view is that there has been additional dissaving in terms of real assets that is not accounted for in the standard system of accounts. Because the natural resource base is not valued, neither the flow of environmental services nor the degradation/depreciation of the resource base is recorded in the accounts, and there is no available measure of such dissaving (Perrings *et al.*, 1989; Markandya and Perrings, 1991). Nevertheless, the depletion of ground water, soil nutrients or vegetation in order to maintain current consumption represents dissaving just as much as the drawing down of financial assets for the same purpose. Moreover, the lower the initial level of consumption, and the longer the history of poverty, the more this dissaving involves degrading the basic agricultural assets.

ENVIRONMENTAL DEGRADATION, TECHNOLOGY AND EMPLOYMENT

Exogenous factors

While the existence of a vicious circle of poverty makes the problem of environmental degradation in sub-Saharan Africa extremely intractable, recent work gives at least some purchase on it. There is still a paucity of empirical data on the problem, but enough indicative work has been done to identify at least some of the necessary conditions for the ecologically sustainable growth of rural income and employment in the region, and to distinguish between necessary and sufficient conditions. In drawing out the policy implications of this work, however, one needs to bear in mind that there are a number of global trends which affect the options open to resource users in the region, but which they cannot influence. Global climate change is likely to impact on the sub-Saharan African physical environment in the longer term in a number of ways, but these impacts remain highly uncertain at present. Mean temperatures are conjectured to rise, albeit by less than in higher latitudes. Movement northward of the intertropical convergence zone is conjectured to increase rainfall in the Sudano-Sahelian region, but soils are still conjectured to become more arid owing to higher rates of evapotranspiration. The variance of temperature (and rates of evapotranspiration) is conjectured to increase, bringing greater risk of extreme events. In sum, the sub-Saharan African resource base is conjectured to be more at risk of degradation from economic activity in the future than it has been in the past.

Similarly, there are international political and economic trends that will impact on the economic environment within which rural resources are allocated, but these are also highly uncertain. There is a strong trend towards the liberalization of international financial markets, but it remains unclear how far this trend will be extended to other markets. Currently, it is possible to identify a much weaker trend towards the liberalization of product markets, but agricultural product markets remain highly protected, and subject to a considerable price intervention. Indeed, since the mid-1980s, the United States and the European Union have both increased the level of export subsidies on agricultural products, while maintaining farm income and domestic price support schemes. At the same time, protective measures against agricultural products from the low-income countries have increased (Conway and Barbier, 1990). Taken together with the instability of exchange rates, these trends suggest that the longer term is likely to see continuing pressure on agricultural incomes in sub-Saharan Africa. Yet the options open to agricultural producers are very limited. Entry and exit to most national labour markets remains strictly controlled, and such liberalization as has taken place – within the European labour market, for example – has paradoxically raised the barriers to entry from regions such as sub-Saharan Africa. The prospects for agricultural producers choosing to exit agricultural markets in the region tend to be limited either to the domestic formal and informal labour markets, or to a very restricted set of foreign markets, such as the market for mine and farm labour in South Africa.

In the longer term, it is trends in this last set of markets that are likely to be most important for the ecological sustainability of resource use not only in sub-Saharan Africa, but globally. Unless there is free mobility across national boundaries, Malthusian crisis may be the only means of reducing intolerable levels of stress on the environment in the very low-income countries. For the present, however, mobility between national labour markets remains highly restricted, and there is little indication of any impending change. This implies that domestic policy initiatives are going to have to carry the burden of reducing stress on the resource base, whilst simultaneously generating sufficient new employment opportunities to cover the expansion of the domestic labour force resulting from population growth.

Labour force growth and absorption

It is estimated that the average annual growth in the labour force in sub-Saharan Africa will be 2.7 per cent between 1985 and 2000, and will be over 3 per cent for many of the countries in the dryland areas of the Sahel, the Horn of Africa, east and southern Africa (World Bank, 1987b). The scope for absorbing increases of this magnitude in the urban industrial sector is limited. The industrial sector was estimated to be able to provide only 2.5 per cent of new employment in sub-Saharan Africa in the late 1980s (van Ginneken, 1988), and the rate of increase in manufacturing employment remains well below that in Latin America and Asia (van Ginneken and van der Hoeven, 1989). The service sector has been growing at a faster rate, but it too remains small. This implies that in the absence of further structural change the majority of new entrants to the labour force may be expected to be absorbed by the agricultural sector, or to be unemployed.

Yet on all the available evidence, the agricultural sector will not be able to absorb the projected increase in the labour force in the region without massively diminishing returns. Per capita physical production of both 'all agricultural products' and 'food products alone' is currently declining, reflecting both the use of increasingly marginal lands and the degradation of existing arable and range lands. There is certainly some potential for increasing both productivity and employment in the agricultural sector. Indeed, various commentators have argued that there exists considerable scope for productivity improvements in agriculture, even on the most marginal land. Conway and Barbier (1990), for example, cite a number of examples to show that, while levels of productivity on marginal lands in sub-Saharan Africa will always be below that of 'favoured' lands elsewhere, they can be significantly greater than levels of productivity now being attained. Given the use of the right farming system techniques, along with adequate research and extension, plus an appropriate set of economic incentives, they argue that it is possible to achieve sustainable results in the 'most difficult agricultural conditions'. Nevertheless, it is clear that the agricultural sector alone will be unable to expand at a rate equal to the rate of increase of the labour force into the indefinite future.

In the past, the reaction to food deficits caused by increasing population pressure in the rural economies of sub-Saharan Africa has been the introduction of policies to expand agricultural output in order to achieve food self-sufficiency (see Rukuni and Eicher, 1987). Indeed, these policies have been argued to be one of the proximate causes of resource degradation in the region, in that they sought to expand output in the short to medium term often without regard for longer-term environmental consequences (Perrings et al., 1989). More recently, the focus has been switched to a policy of promoting food security through income security, which directs attention away from food production and towards the conditions necessary to assure a level and distribution of income that will guarantee the means to acquire food. This suggests that the appropriate response to an expanding labour force (where there is limited scope for out-migration) is to generate sustainable employment opportunities in whatever sectors offer a comparative advantage.

Categories of incentive

Conway and Barbier (1990) suggest that there exists a set of economic incentives that will induce agricultural producers to choose techniques of production that are consistent with the ecologically sustainable development of agricultural resources. We need to consider whether there also exists a set of incentives that will encourage diversification of the rural economy into non-agricultural activities. They distinguish two classes of incentive: 'variable' incentives, and 'user-enabling' incentives. Variable incentives consist of subsidies, user charges which are used to bring the private and social costs of resource use into line with one another. User-enabling incentives, on the other hand, address the institutional environment within which resources are allocated.

To see how each class of incentive is argued to operate, consider the recommended liberalization of agricultural markets reported earlier (Bond, 1983; Cleaver, 1985; Repetto, 1986, 1989; Barbier, 1988, 1989; Warford, 1989). This is an example of a policy recommendation directed at Conway and Barbier's variable incentives. Liberalization affects traded goods and services, and works on the relative prices of those goods and services. The problem of the misallocation of resources is, however, less tractable in the case of non-traded resources in common property – land, water and vegetation – which have been adversely affected by the collapse of systems of collective control over the level of activity. This is what the 'user-enabling' incentives referred to by Conway and Barbier are designed to address. Such user-enabling incentives may involve the allocation of individual tenure, or the introduction of new and more secure forms of usufructual tenure, or the strengthening of institutions charged with regulating the level of activity (such as grazing associations), or any combination of these things.

Income effects

The idea behind the allocation of property rights is that these are a precondition for the emergence of markets in the resources concerned, and so for the internalization of the environmental externalities associated with the existing communal tenure arrangements. However, this begs a very important question. From the standpoint of efficiency, the allocation of property rights may be sufficient to internalize environmental externalities irrespective of equity considerations (at least in the small numbers, low cost of negotiations case). But if the distribution of income is as important in the allocation of natural resources in rural sub-Saharan Africa as I have suggested earlier, resources may be allocated efficiently, but still not sustainably.

In the case of both tradable and non-tradable resources, the key assumption behind the recommendations of researchers such as Conway and Barbier (1990) is that the set of relative prices associated with the existing system of administered pricing has been a major factor in the choice of technology (Markandya, 1991). By implication, the choice of 'traditional' input and output combinations is driven by the relative prices of traditional versus non-traditional inputs and outputs. This is undoubtedly part of the story, but it is only part. The choice of technique is also influenced both by the degree of risk aversion and by the time preference of resource users. Since both these are partly a function of income, it turns out that choice of technology is also driven by income. To be sure, the income of resource users is affected by relative resource prices, and this is one motivation behind the recommended liberalization of agricultural markets; but income also depends on both the distribution of assets in society (endowments) and the system of transfers. If one ignores transfers for the moment, it appears that if endowments are such that resource users are impoverished, traditional techniques may still be optimal even if these are inconsistent with the ecologically sustainable use of resources. A collapse in the income of agricultural producers (for whatever reason) may provoke an increase in the level of intensity with which the land is exploited, irrespective of the costs in terms of reduced future productivity. In the extreme case, where the ecological system is both locally and globally unstable, such behaviour can be fatal but still optimal (Perrings, 1989a).

The same general point is true for the diversification of the rural economy. While a revision of tenure arrangements may be necessary for diversification, it is not sufficient, since the distribution of assets also has a role to play. The absence of well-defined transferable property in either land or the rights to use land discourages the realization of assets, so inhibiting the mobility of land users. Similarly, traditional usufructual rights that specify the nature as well as the level of land use discourage the reallocation of land from one use to another. However, while a redefinition of rights (an appropriate set of user-enabling incentives, in Conway and Barbier's terms) may overcome these obstacles to diversification, the endowments of resource users may be such as to make the risks of diversification unacceptable.

In both cases, the system of transfers becomes extremely important. The use of tied transfers as a means of stimulating technological change in sub-Saharan Africa is well established. Indeed, the destumping, herbicide, pesticide, fertilizer, seed and draft power grants and subsidies that have been blamed for much of the recent land degradation in the region are examples of such tied transfers and have been extensively investigated for their effects on technology choice. But transfers may also have a crucial role to play in influencing risk-taking behaviour. This specific issue has so far been neglected in the literature on incentives, although it has been considered in the main body of this chapter. The general point, however, is that where the extreme risk aversion of impoverished resource users has prevented the innovation that would enable them to realize the specialization gains from international trade, the compensation principle suggests that there exists a set of transfers redistributing the gains from trade which leaves everyone better off than they would be under autarky (Perrings, 1989a).

CONCLUDING REMARKS

The thrust of much of the recent work reviewed in this survey is that an important component of an ecologically sustainable development strategy should be the removal of those barriers to adjustment in the economic system which have locked resource users into environmentally damaging practices. The arguments for the liberalization of agricultural markets and for the rationalization of property rights are based on the assumption that the economic system will adjust to environmental imbalances if it is permitted to do so. While the assumption is well founded, it is clear that a *laissez-faire* approach will not be sufficient to avert the continued environmental degradation of regions such as sub-Saharan Africa. Aside from the well-understood problems of the public good nature of environmental services, and the impossibility of assigning property rights to all economically interesting environmental effects, there is evidence that the implicit rate of discount of resource users, together with their degree of (current) risk aversion, are both incompatible with the ecologically sustainable use of the resource base. Even if all resources are allocated on the basis of their social opportunity cost (even if they are correctly valued), it may still be optimal for the very poor to degrade the environment. That is, the poverty of resource users is just as much a barrier to adjustment as the monopsonization of agricultural product markets, or the artificial preservation of inappropriate land tenure arrangements.

An equally important component of an ecologically sustainable development strategy is therefore the employment generation that will sustain an increase in both the earned and the transfer income of resource users. What is needed is a structure of incentives that will not only weight the environmental effects of economic activity appropriately, but will stimulate both the adoption of ecologically sustainable technologies in agriculture and the diversification of the rural economy. Moreover, where change in the relative prices of resources is not sufficient, what is needed is a structure of incentives which addresses the distribution of both assets and income. The difficulty here, as Markandya (1991) has recently pointed out, is that it is not yet clear what are the distributional or employment implications of different incentive structures. He describes research on the cross price elasticities of demand for labour and the range of inputs targeted for their effect on the environment as 'sporadic', and claims that the traditional partial equilibrium analysis of environmentally innovative activities is inadequate to identify the longer-term feedback effects on employment. It is not, for example, obvious that labour-intensive technology is necessarily more environmentally sound than capital-intensive technology. What this means is that the revision of the system of incentives, through the liberalization of either markets or institutions, is necessarily in the nature of an experiment in which the potential economic benefits and environmental costs are largely unknown. The application of a precautionary principle in these circumstances suggests a conservative approach, informed as far as possible by the analysis of the general equilibrium longer-term implications of change.

Notes

1. The term 'drylands' comprises both arid lands with insufficient rainfall to sustain crop production but capable of supporting pastoralism (< 250 mm rainfall per annum), and semi-arid lands capable of supporting some drought-resistant crops (< 600 mm rainfall per annum). In Africa this includes not only the Sahara and Kalahari deserts but all of the Sudano-Sahelian region and much of east, central and southern Africa.
2. Ghana at 59 per cent and the Sudan at 65 per cent. The proportion of the labour force in Zambia in agriculture was reported to be 37.9 per cent in 1985–87 (UNDP, 1990).
3. No data are available for Burkina Faso, Mauritania, Mozambique, Senegal, the United Republic of Tanzania, Uganda and Zambia in this period, all of which have very high levels of rural poverty.
4. The human development index (HDI) is equal to 1 minus an average deprivation index, which is a simple average of three indicators: life expectancy, literacy and GDP. On a scale in which Japan has an HDI of 0.996, 20 countries have an HDI of less than 0.3. All but three of these are in sub-Saharan Africa. The eight most deprived countries in terms of this index are: Burkina Faso, Chad, Guinea, Mali, Mauritania, Niger, Sierra Leone and Somalia.
5. Note that resilience does not imply the stability of the species populations of that system, i.e. their propensity to return to an equilibrium value (whether stationary or cyclic) following perturbation. That is, it does not necessarily follow that, if the relative sizes of the populations of different species changes, a system will lose resilience. An important qualification to this is that an ecosystem will tend to be more resilient, the greater the 'interconnectedness' of species within the system (interconnectedness being a measure of the interactions between species populations). The greater the interdependence

of species in an ecosystem, and the more species it involves, the greater its potential resilience. Hence biodiversity loss, which implies a reduction in the 'interconnectedness' of an ecosystem, does imply a reduction in system resilience.

6. The first group of plants are known as C4 plants, since the first product in the growth sequence has four carbon atoms; the second group are known as C3 plants. The latter respond much more positively to increased concentrations of CO_2. See, for example, Akita and Moss (1973).

7. Where, for example, the emigration of labour results in environmental maintenance tasks being neglected, land degradation and a declining population may go together. Evidence from elsewhere supports the general proposition that population changes can work both ways. In the dryland environments of Mexico, for example, Garcia-Barrios and Garcia-Barrios (1990) argue that emigration is one of the main factors in the weakening of institutions charged with regulating access to resources held in communal property.

8. While the short- and long-run supply elasticities for cocoa, coffee and sisal were reported to be significantly different, those for cotton, groundnuts, palm kernels and palm oil were reported to be identical.

9. The trend was started by a colonial policy to combat the drought of the early 1960s, when exclusive rights to the water issuing from new boreholes were allocated to larger cattle owners – typically those with traditional positions of authority – giving them '*de facto* exclusive rights to extensive areas of new grazing land'. Despite the fact that this violated customary law with respect to both grazing land and natural resources (aquifers), the policy was maintained for at least a decade after independence. The result was that poorer herdsmen were moved on to progressively more marginal land, and Colclough and Fallon estimated that by 1976 39 per cent of all rural households owned no cattle at all, while the distribution of cattle amongst the remaining 61 per cent was highly unequal (Colclough and Fallon, 1983, pp. 144–5).

6 Quantification of the Trade-Offs in Zambia between Environment, Employment, Income and Food Security

Dodo J. Thampapillai, Phiri T. Maleka and John T. Milimo

AIMS AND SCOPE

The aim of this chapter is to formulate an environmental policy for an area in the Middle Zambezi Valley on the shore of Lake Kariba in Zambia (see Figure 6.1). The study area lies somewhat less than 200 kilometres to the south of Zambia's capital city, Lusaka. It is bounded by Lake Kariba and Zimbabwe to the south and the escarpment to the north, and consists of the administrative districts of Siavonga, Gwembe and Sinazongwe.

The policy issues that arise in this area are representative of many that are rapidly becoming common in Africa. The need for an environmental policy stems from the conflicts and public concerns caused by several development project proposals in the area for the clearing of approximately 162,000 hectares of medium-density forest for agricultural development with irrigation. Proponents of these projects claim that irrigated agriculture could raise the levels of income and consumption, and ensure stable supplies of food during periods of drought. It is indeed true that droughts and famines are common in the area, which moreover is one of the poorest in Zambia in terms of per capita income (approximately 150 kwacha (K) per year), and where nutritional intakes are claimed to be well below recommended levels (Zambia, National Commission for Development Planning, 1986, 1988). However, the concern regarding these projects is due to the environmental costs, which are likely to be high. The clearing of the forests would lead to the loss or displacement of flora and fauna, some of which are fast becoming scarce in Africa, as well as contribute to climatic changes that would lead to desertification.

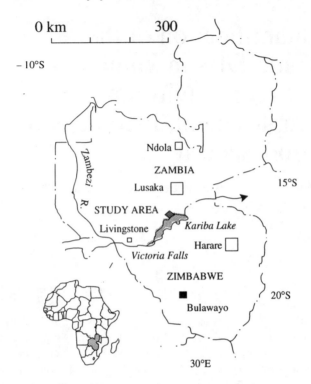

Figure 6.1 The location of the study area

Zambia's present economic difficulties also lie behind the moves for agricultural development. Zambia's dependence on a single commodity, namely copper, and declining commodity prices have led to falls in export earnings, which in turn have triggered off a host of other difficulties. These include a contraction in GDP, large deficits in the balance of payments, a shortage of foreign exchange and the expansion of the debt burden. As indicated in Table 6.1, the Zambian economy has displayed real losses in GDP since the early 1980s. Further, the economy cannot depend on a single commodity alone. Hence agricultural development has been seen as a viable avenue for diversification. The calls for agricultural development also stem from the fact that, although nearly 60 per cent of the population are engaged in agriculture, food imports have grown rapidly over the years.

This chapter focuses on the proposed schemes for irrigation in the context of four social objectives: income maximization (IM), employment

Table 6.1 Zambia's economic performance: Trends in selected indicators

Indicator	1980	1982	1984	1986	1988
Per capita GDP (1980 kwachas)	547	526	470	465	409
Current account balance (millions of 1980 kwachas)	−848	−449	−226	−549	−
Terms of trade index	60	28	26	23	−
Total foreign debt (millions of US$)	−	−	3,080	4,785	5,557
Inflation (%)	−	6.2	19.5	82.3	32.5
Labour force (millions of persons)	2.7	2.9	3.1	3.6	3.9

(EM), environmental quality (EQ) and food security (FS). All four objectives are pertinent as well as important. The extremely low incomes, the level of unemployment and the frequent droughts underlie the need for the IM, EM and FS objectives.

The concerns regarding a range of environmental costs and the evidence of the adverse environmental effects of the attempts at accelerated agricultural development elsewhere in Africa (for example, Burkina Faso, the Sudan) justify the EQ objective. The inquiry seeks answers to a set of questions, which include: (a) What is the contribution of irrigated agriculture to growth in incomes and employment? (b) What is the nature of the trade-offs between IM from agriculture and EQ? (c) What are the impacts on IM and EQ of imposing the requirement of full employment? (d) What are the impacts of imposing the requirement of stable food supplies? (e) What is the value of environmental resources and services, and are these worth sacrificing for additional income and employment? and (f) Can incomes and employment be improved without sacrificing environmental resources?

THE STUDY AREA

The Kariba dam

The dam was constructed in 1959 to generate hydroelectricity for Zambia and Zimbabwe. However, in recent years there have been proposals to use surplus water from the lake for irrigation purposes. Some of these proposals involve the access to vast tracts of land by multinational enterprises which have the investment capacity to clear large areas of forests very quickly and to initiate agricultural production.

The Zambezi River Authority, the organization charged with overseeing the operation of the Lake Kariba dam, has indicated that the lakewater level should be at least 475 metres deep for hydroelectricity generation, and that this level has been well above this critical point over the past 25 years. Further, according to Clayton (1985) the electricity generated by the dam far exceeds the power requirements for these two countries. For example, the combined demand for electricity from Zambia and Zimbabwe is approximately 14,092 gW, while the capacity of the Kariba dam is approximately 19,425 gW. Hence, it is plausible to assume that irrigation and hydroelectricity would not be in conflict, at least in the short to medium term, as long as the lakewater was 475 metres deep. Although irrigation and electricity generation may not be in conflict, however, irrigation *is* in conflict with the maintenance of environmental quality, as we illustrate below.

Soils, ecology and rainfall

The soils of Lake Kariba, as described by Scudder (1962), are of pre-karoo and alluvial formation. Whereas the pre-karoo soils are at the foot of the escarpment, the karoo and alluvial soils are found between Lake Kariba and the escarpment and around the shores of Lake Kariba, respectively. From the soil scientists' point of view, the alluvial soils are highly suitable for cropping activities, whilst the clay soils may be cultivated only with irrigation (Maclean, 1969). However, these soils have a high sodium content, and hence irrigated agriculture could lead to problems of salinity as experienced at several locations elsewhere.

The proposals for agricultural development have been confined to an area well inland from the lake, comprising medium-density forest with approximately 1,500 trees per hectare, mainly *Acacia*, *Agansonia* and *Combretrum*. The fauna are predominantly wide-ranging species of birds and herbivores such as monkeys (including baboons), deer, zebras and giraffes. The adjoining hilly area is a dense stand of forest with larger fauna and flora. While large herbivores such as elephants and buffaloes are found, most of the animals are carnivores such as lions, leopards and rhinos. One significant implication of clearing the land area intended for agriculture (i.e. the medium-density forest) is that many of the herbivores would be displaced into the adjoining dense forest area where they are likely to fall prey to the carnivores. Accordingly, the costs of agricultural development include not only the destruction of the habitat and its species but also disturbance to the prevailing ecological balance between the medium- and high-density forest areas.

The average rainfall during the wet season (November to April) is 721 millimetres. Rain during the dry season is exceptional: the mean rainfall of 21 millimetres during this season makes it practically impossible for crops to be grown without irrigation.

Population and economic activity

Some 12,000 families live in the medium-density forest area, in small sets of communities engaged in subsistence agriculture. At present approximately 14,000 hectares of land are cultivated as smallholdings, that is approximately 1 hectare per family. The main crops grown are sorghum, maize and sunflower, plus some cotton. The average annual per capita income is extremely low, namely K150 at 1985 prices. This low income coincides with a high level of unemployment and underemployment. The precise extent of labour utilization is difficult to assess owing to the subsistence nature of production. Surveys also suggest that the calorific intakes are well below recommended levels, and the frequent droughts no doubt exacerbate these problems. Existence under these circumstances places considerable pressure on the environmental resources of the area, since families are prompted to use the forest resources for food and fuel as well as to attempt to enhance their incomes by supporting the poaching of wildlife resources, some of which are indeed scarce. Hence, whilst environmental considerations are undoubtedly important, they cannot easily be isolated from considerations pertaining to income.

Although subsistence agriculture is confined to sorghum, maize, sunflower and cotton, the climatic features of the region also permit the cultivation of other crops, such as rice, soya and wheat. The development proposals include these crops. The income derived from current agricultural practices is shown in Table 6.2.

As expected, the pattern of income is unstable, with the fluctuations in income being associated with fluctuations in rainfall. Clearly, the maximization of income, and the minimization of output instabilities resulting from the irregular rainfall in the study area, are important social objectives for the region. The implementation of irrigation programmes could serve to achieve these objectives. However, it would be less than prudent to confine the costs of developing irrigated agriculture to the traditional resources of land, labour and capital. The natural environment and its

Table 6.2 Income from current agricultural practices in the study area
(millions of kwachas)

Year	Maize	Sunflower	Cotton	Sorghum	Total	Income/hectare
1976	2,527	735	4,502	0	7,764	92
1977	1,838	1,257	8	0	3,103	37
1978	1,892	369	2,713	0	4,974	59
1979	2,051	1,748	3,505	0	7,304	87
1980	1,166	440	3,524	0	5,130	61
1981	1,537	178	1,511	0	3,226	38
1982	410	111	1,032	9	1,562	19
1983	689	141	588	17	1,435	17
1984	882	373	362	597	2,214	26

Source: Zambia, Department of Agriculture, and Central Statistics Office, Lusaka.

components are also vital resources that are either impaired or lost in the development of irrigated agriculture. Hence, in this chapter, we evaluate the viability of irrigated agriculture by incorporating the natural environment as an essential ingredient. The models that permit this evaluation are described in the next section.

METHOD OF ANALYSIS: FORMULATION OF MODELS

Nature of the models

The literature on agricultural development in drought-prone areas is explicit on the need to take into account the risks and uncertainties that are associated with drought (Hazell and Norton, 1986). Further, the literature also favours the use of mathematical programming models for the determination of land use decisions that involve the spatial allocation of resources because of the ability of such models to evaluate an infinite set of resource allocation patterns from a set description of the prevailing resource endowments. However, this advantage inevitably depends on whether or not the mathematical descriptions of the endowments and their utilizations are an adequate representation of reality, especially in terms of the processes of utilization (McMillan, 1975).

Problems involving risks and spatial patterns of resource allocation are perhaps best resolved by quadratic programming models with a quadratic objective function and linear constraints. In their application to agricultural land use decisions, these models are formulated to maximize the expected revenue from various land uses minus the cost of risk taking, subject to the usual resource endowment constraints of land, labour and capital. The quadratic form is due to the costs of risk taking being defined in terms of variance of returns from land use. Thus, risk is defined as dispersions around the mean (namely, expected income) and the model is formulated to choose patterns of resource allocations that minimize these dispersions. Although quadratic programming is theoretically adequate, its applicability is limited because suitable algorithms are not readily accessible and because of difficulties associated with computation.

However, a simplified and more readily applicable transformation of quadratic programming models was presented by Hazell (1971), where the variance term was replaced by mean absolute deviations. This replacement enables the model to be linear instead of quadratic. Since these models are designed for the 'minimization of total absolute deviations' from a prescribed objective, they are usually referred to by the acronym MOTAD. Further, they are also more readily applicable owing to the availability of a variety of linear programming algorithms. Although the usefulness of linear programming could be questioned because of its assumptions (i.e. linearity, additivity and convexity), empirical findings, for example in Johnson (1966), suggest that the restrictiveness of these assumptions would not be significant, especially in the context of agricultural land use decisions. Hence, linear programming has found widespread application in agriculture: see, for example, Hardaker and Troncoso (1979), Hazell and Norton (1986), and McCamley and Kliebenstein (1987).

In this chapter, a variant of Hazell's MOTAD model is used, namely Target-MOTAD, which was introduced by Tauer (1983). Whilst the original MOTAD models measure risk in terms of absolute deviation of income from mean income, Target-MOTAD models measure risk as absolute deviations from prespecified values of target income. The relevance of Target-MOTAD derives mainly from the fact that mean income need not necessarily satisfy basic needs, whilst target incomes do. Hence, it is possible to define the risks that are prompted by droughts in terms of the failure to meet the basic human needs.

Given that there are two distinct seasons, wet and dry, Target-MOTAD models are formulated for each of these seasons. Further, the model formulation is also performed in the context of two options: 'with irrigation' and 'without irrigation'. Given that rainfall is virtually nonexistent during the dry season, the 'without irrigation' option is confined to the wet season. In each of the models, environmental quality constraints are explicitly introduced in order to facilitate the derivation of trade-offs between IM and EQ.

The models

The main components of the model for each option are: (a) a rainfall simulator; and (b) a Target-MOTAD model. The simulation of rainfall was performed with the rainfall records for the period 1952/53 to 1985/86. Following Sharma and Nyumbu (1985), the simulated value of rainfall was reduced by 10.5 per cent to allow for run-off losses, and was then used to determine expectations of income, soil moisture levels and the risk of drought and crop failure.

In the Target-MOTAD model for the 'with irrigation' option the model solution for a given season reveals: (a) the mix of crops to be grown; (b) the amount of lakewater to be used for irrigation; and (c) the amount of credit to be given to the farmers. The aim is to maximize net income, represented by the expected value of income from cropping less the cost of irrigation less the cost of credit, and subject to the constraints of land, labour, soil moisture, capital and credit, limit on irrigation, deviation from target income and risk aversion levels, and targets on the EQ objective.

The model for the 'without irrigation' option differs from the model described above by the absence of the variables pertaining to irrigation and its application to the wet season only. The inclusion of credit as a decision variable was deemed relevant, given that the literature on development economics has recognized the lack of access to institutional credit by the poor as an important source of persistent poverty. The issues surrounding limited access to credit are detailed elsewhere (Anderson and Thampapillai, 1990). The evidence of strong associations between poverty and the lack of access to institutional credit has often been used as the argument for a relaxation of the requirements for credit access. Hence, the incorporation of credit as a variable in this model would make it possible to see whether ready access to credit has an effect on agricultural production and thereby on income levels: that is, we assume that the crop production and irrigation activities involve credit.

The objective function of the Target-MOTAD model concerns the maximization of expected income from seven crops: maize, cotton, sunflower, soya, sorghum, rice and wheat. It is assumed that these seven crops have to be produced, within the limitations of the constraints, in five zones of the study area. Following information that was furnished by farmers and agricultural officers in the area, the target income was nominated as K100 million for the 'with irrigation' option, and as K20 million for the 'without irrigation' option. Further, in quantifying the risk-taking behaviour of farmers in the region, we assumed that they were willing to tolerate losses (deviations) of up to K20 million with irrigation, and up to K4 million without irrigation. The rationalization of these values is based on the premise that the people in the study area generally tolerate losses of up to 20 per cent of expected income.

Ideally, the constraints pertaining to EQ targets need to be defined in terms of the various species of fauna and flora and of the impacts that agricultural production decisions have on the environment. As indicated above, the area intended for agricultural development is home to a diverse range of species. However, since the numbers of the various species are unknown, we have assumed that the EQ target can be defined as the number of hectares of land preserved. By systematically varying this target, i.e. by varying the amount of land that is preserved or denied for development, we can derive a trade-off function that reveals the relationship between IM and EQ. As shown below, these trade-off functions can be also derived by imposing the requirements of full employment and maintaining at least the minimum requirements of food production to avert the adverse consequences of drought.

The models were applied across 30 scenarios that were generated from values of simulated rainfall. Each scenario had six sets of rainfall values, where each set consisted of two rainfall values for a given year: that is, one for the wet season and another for the dry season. Each set was hence used as a state of nature in the Target-MOTAD model for quantifying the risks, i.e. estimating the deviations from target income. Since each state of nature is a year and we consider six states of nature, the solution of the model application can be interpreted as resource allocation strategies for a six-year period.

APPLICATION OF THE MODELS AND EVALUATION OF THE OPTIONS

As indicated in the opening section, the main aim of this chapter is to evaluate the viability of the irrigation option whilst recognizing environmental resources as essential ingredients of development. However, the discussion below opens with an examination of the impacts of irrigation development without any environmental considerations being taken into account. The changes introduced by the environmental quality objective are examined later. These changes reflect on the objectives of full employment and food security as well as of income maximization.

Income impacts of agricultural development

The comparison of the two options in terms of specific variables is presented in Table 6.3. Clearly, the implementation of irrigation is by far a superior option in terms of income. Net income with irrigation is almost five times the amount without irrigation. Further, the model solution for the option without irrigation produces an income that is almost twice as much as at present. With both options, full employment is feasible. This indicates that, even without irrigation, income can be improved by modifying the current patterns of allocation of resources. However, these increases in income, and full employment, are possible only when the entire land area is cleared for agricultural production.

Further, the information presented in Table 6.3 does not pertain to model solutions that are restricted by minimum food supply requirements. When such restrictions are not imposed, the irrigation option would

Table 6.3 Comparisons of the effects of the two irrigation options (in millions of kwachas)

Item	With irrigation		Without irrigation
	Wet season	Dry season	Wet season
Net income	81.92	69.34	33.04
Amount of credit	5.81	5.81	3.68
Net income without credit	25.04	14.44	9.06
Increase in income relative to present income (with credit)	78.01	65.84	29.54
Increase in income relative to present income (without credit)	21.54	10.94	5.56
Irrigation cost	5.20	17.80	–
Crops grown	Maize, cotton	Maize, cotton	Sorghum, rice, wheat, soya
Extent of land used (hectares)	162,000	161,401	162,000
Unemployment	Nil	Nil	Nil

prompt a shift in resources away from food crops to cash crops, such as cotton, which require more water: that is, in the absence of irrigation, resource allocation is confined to the cultivation of food crops which need less water.

It should also be noted that the patterns of resource allocation in the context of both options (i.e. with or without irrigation) are based on the assumption of access to institutional credit. The net incomes reported in Table 6.3 are of course net of interest charges. The importance of credit was assessed by solving the models without the credit variables. In the case of the 'with irrigation' option, the value of net income fell from K82 million in the wet season to K25 million, a drop of nearly K60 million. In the case of the 'without irrigation' option, this fall was about K24 million, i.e. a drop from K33 million to K9 million. Hence, the viability of both options in terms of generating additional income is dependent on providing access to institutional credit.

As a proportion of GDP at 1985 prices, the 'with irrigation' option contributes approximately 5 per cent, which is significant given the relatively small size of the project. The direct per capita gains from irrigation per year to those living in the study area amount to nearly K1,800. Given that existing per capita incomes in the area are around K150, the income increase is indeed significant. When secondary effects are included by using a spending multiplier of 3.33, a value that was used in other studies – for example, Maleka (1989) – the income benefits from irrigation amount to some K500 million, or approximately 15 per cent of GDP at 1985 prices. The contribution of the 'with irrigation' option to per capita national income is K62.5. This is undoubtedly a sizeable contribution.

Although both the 'with' and 'without' irrigation options generate increases in income, these increases are depicted by model solutions that involve clearing the entire 162,000 hectares of forest land. Given that the scarcity of ecosystems and wildlife has become an issue of global significance, the impacts of preserving the land area (that is, prohibiting any form of development) need to be assessed against the monetary gains that can be generated by irrigation. This is the subject of the following subsection.

Impacts of introducing the environmental objective

As indicated, owing to the paucity of data the environmental objective was quantified in terms of the extent of land preserved. The assumption is that, when a patch of land is preserved from agricultural development, all items that are carried by this patch of land are maintained. From an ecological

Table 6.4 The impacts of EQ on the employment IM objectives

Extent of EQ (%)	Value of the IM objective (millions of kwachas)		Extent of employment (%)	
	With irrigation	Without irrigation	With irrigation	Without irrigation
0	151	33	100	100
10	137	32	98	100
20	122	30	96	100
30	108	28	94	100
40	94	25	92	84
50	80	22	86	72
60	63	18	69	60
70	48	14	52	50
80	32	9	35	40
90	15	5	17	29
100	0	0	0	0

stance, this assumption is indeed weak, since the compartmentalization of land into parcels of hectares must invariably disrupt ecological systems. Nevertheless, the assumption is retained for reasons of empirical convenience, and is used for the derivation of the EQ–IM trade-off functions.

The derivation of the trade-off function is performed by systematically reducing the amount of land that is made available for agriculture, and maximizing the IM objective for each stage of reduced land availability. The value of the IM objective for various extents of preservation is presented in Table 6.4 for both the 'with' and the 'without' irrigation options. The trade-offs between IM and EQ are presented in Figure 6.2. As expected, the trade-off function for the 'without irrigation' option is well within that of the 'with irrigation' option. These functions provide useful information in the sense that decision-makers are made aware of the amount of income benefits that have to be sacrificed in order to preserve the environment. As indicated in Table 6.4, the pursuit of the EQ objective also involves sacrifices in terms of employment benefits. It was therefore possible to demonstrate the trade-offs between employment and environmental quality objectives as portrayed in Figure 6.3. It is evident that with both options of irrigation the impact on employment is hardly noticeable until EQ exceeds 40 per cent preservation. The effects of recognizing employment are considered in more detail below.

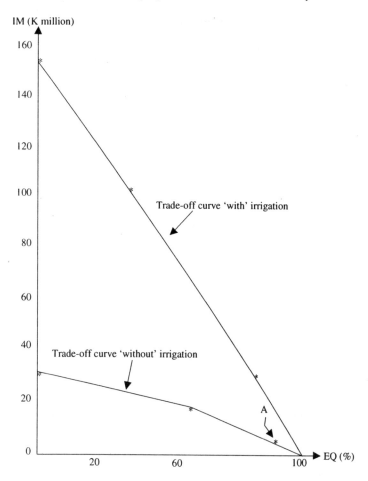

Figure 6.2 The IM–EQ trade-off curves for both irrigation options

The trade-off curves that have been shown here are not efficiency frontiers in terms of all resources. Labour and capital resources become redundant as the extent of land preserved increases. Instead, these curves are efficiency frontiers in terms of land (and its environmental resources), since at any point on the trade-off curves these resources are exhausted between the two objectives.

The IM–EQ trade-off curve for the 'without irrigation' option makes it possible to assess the efficiency of the present management of land use in

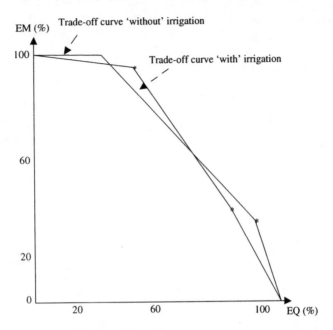

Figure 6.3 The employment–EQ trade-off curves

the study area. As indicated above, the trade-off curves shown here are economic efficiency frontiers in terms of the allocation of land between EQ and IM, and the models used also recognize the perception of risk and uncertainty by the residents of the study area. The surveys carried out in the study area reveal that the annual regional income is nearly K3.35 million and that approximately 16,000 hectares of land are being cropped. This places the present management strategies on the frontier at point A in Figure 6.2, noting that 90 per cent preservation corresponds to the utilization of 16,000 hectares of land. The result indicates that the residents of the study area are most probably internalizing all risks and uncertainties that surround their existence and are using their land resources efficiently.

There are two further issues that deserve attention: (a) the employment impacts of pursuing the EQ objective warrant further consideration in view of the high degree of unemployment and underemployment of labour in the study area; and (b) given the frequent incidence of drought, targets for food production to ensure stable supplies of food become important. The first of these issues stems from the concern that rural unemployment invariably leads to rural–urban migration and places undue pressures on urban resources. The targets for food production are also warranted on the grounds that the nutritional intake of the residents of Lake Kariba is well below recommended levels. The impacts of these other social goals are considered next.

Impacts of other social goals

Given that countering the process of rural–urban migration is recognized as a matter of great importance, the employment goal was defined as full employment. This merely involved the redefinition of the labour constraint in the models to ensure that all available labour in the region is utilized. The food security goal was incorporated by adding two more constraints to each model, to state that the production of maize and of sorghum should each be at least equal to a minimum threshold. This threshold was determined on the basis of the nutritional requirements of the area's residents and was defined as the allocation of at least 30,000 hectares each for maize and sorghum. The impacts of the employment (EM) and food security (FS) goals are summarized in Table 6.5.

The models were solved for each of these policy goals separately as well as together. It was not possible to generate a trade-off function

Table 6.5 Impacts of employment and food security goals

	Value in millions of kwachas of the IM objective when:			
EQ (%)	EM and FS absent	Only EM present	Only FS present	EM and FS present
0	151	142	128	116
10	137	136	123	115
20	122	122	110	109
30	108	108	97	97
40	94	93	82	0
50	80	0	66	0
60	63	0	51	0
70	48	0	0	0
80	32	0	0	0
90	15	0	0	0
100	0	0	0	0

between EQ and IM in the context of the 'without irrigation' option when either or both of the EM and FS goals were included. This was not the case in the context of the 'with irrigation' option. The implication of this result is as follows:

> If irrigation is not made available, and the goals of minimum food supply and full employment are to be satisfied, the entire land area has to be cleared: that is, the EQ objective is not compatible with the other objectives.

The trade-off functions that incorporate the impacts of EM and FS goals are shown in Figure 6.4. The impacts of these other policy goals in terms of the maximum area of land that can be preserved and the value of pre-

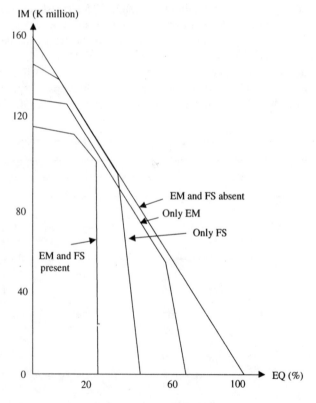

Figure 6.4 Comparison of the IM–EQ trade-off curves for the 'with irrigation' option: Presence and absence of the EM and FS goals

Table 6.6 The opportunity cost value of preservation and the maximum feasible extents of preservation for different policy scenarios

Policy scenario	Maximum possible extent of preservation (hectares)	Value of maximum preservation as an opportunity cost (millions of kwachas)
EM and FS absent	162,000 (100% EQ)	151
Only EM present	78,000 (48% EQ)	142
Only FS present	101,000 (62% EQ)	128
EM and FS present	57,000 (35% EQ)	116

servation (in terms of opportunity costs) are summarized in Table 6.6. For example, if only the food supply goal is introduced, it is possible to preserve nearly 62 per cent of the land area, whilst the maximum extent of land that can be preserved with the full employment goal is 48 per cent of the total land area. When both food supply and employment goals are introduced, the maximum amount of land that can be preserved falls to 35 per cent of the land area. The opportunity cost value of preservation reveals the amount of income benefits that have to be sacrificed to achieve the maximum feasible extent of preservation under each policy scenario, and is equal to the highest attainable value of IM in each scenario.

If the prevailing set of policy objectives is the maximization of income, the maintenance of environmental quality, stable food supply and full employment, the search for a policy option is confined to the innermost trade-off curve in Figure 6.4. This curve is reproduced in Figure 6.5. The type of information that can be gained from Figure 6.5 is as follows: given the recognition of stable food supply and full employment, an income target of K100 million and an EQ target of preserving 35 per cent of the land area are inconsistent. However, under similar circumstances, an income of target of K100 million and an EQ target of 28 per cent are both attainable. Likewise, an income target of K50 million and 32 per cent preservation for EQ are also attainable.

The net present value (NPV) of income gains from a selected set of policy scenarios is summarized in Table 6.7. The policy scenarios are all centred on the irrigation option. The NPVs were computed by assuming a 5 per cent discount rate and a 25-year time horizon, and by assuming the net incomes generated by the model solution to be annuities over this time horizon. Even if these NPVs may appear to be significant, especially in the

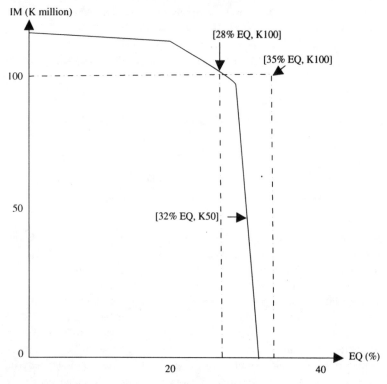

Figure 6.5 The IM–EQ trade-off curve with EM and FS.

context of a community that is faced with poverty, there is an explicit need to attempt the valuation of environmental quality by more direct means (that is, by methods other than those involving opportunity costs of preservation). This is because of the vast extent of land that has to be cleared for agricultural development. In other words, the estimation of the true social value of the environment needs to be assessed. We attempt this in the next section.

EVALUATION OF ENVIRONMENTAL QUALITY

It should be noted that, given the sparseness of ecological data and limitations of time, the methods to evaluate environmental quality are subject to a number of the deficiencies commonly cited in the relevant literature. Three approaches were used to attain a monetary assessment of environmental quality: (a) estimation of the Zambian community's demand for

Table 6.7 NPV of the IM objective for selected policy scenarios (with irrigation)

Extent of EQ (%)	NPV in millions of kwachas of the IM objective when the policy scenario is:			
	EM and FS present	Only EM present	Only FS present	EM and FS absent
0	1,600	1,959	1,766	2,084
10	1,587	1,877	1,697	1,890
20	1,504	1,683	1,518	1,683
30	1,338	1,490	1,338	1,490
40	0	1,283	1,131	1,283
50	0	0	911	1,104
60	0	0	704	869
70	0	0	0	662
80	0	0	0	442
90	0	0	0	207
100	0	0	0	0

preserving the study area; (b) estimation of the demand of those outside Zambia for preserving the study area; and (c) estimation of replacement costs. The value of preservation in terms of demand was approximated by adding the values that were derived from the first two approaches.

The Zambian demand for preservation

The Zambian community's demand for preservation was estimated by applying the method of contingent valuation to a selected number of individuals in Lusaka. This method involved specifying a particular monetary value and finding out whether the subject was willing to pay either more or less to have the study area preserved. If, for example, the subject were willing to pay more, the value is systematically bid up until the subject is not willing to pay any more. Alternatively, if the subject had been willing to pay less, the value is systematically bid down until the subject is not willing to pay less. The average annual willingness to pay (WTP) elicited by this method was approximately K250 per person. However, all members of the sample who were subjects in the valuation process had annual incomes of over K12,000. Further, an examination of past income distribution data suggests that those earning over K12,000 per year amount to only about 10 per cent of the population. Hence, the benefit of preserving the study area by this method was estimated as:

(The average WTP)(0.1)(Population of Zambia) = K(250)(0.1)(7.2) million
= K180 million

Whilst some may argue that the value of preserving the environment by this method could be an overestimate, it should be noted that this value is conservatively based only on the WTP of 10 per cent of the population whose preferences are known. That is, if the WTP of the entire population had been considered, the value of preservation would probably have been much higher. Hence the value of K180 million can be regarded as an estimate at the lower end of the spectrum of values for the preservation of the study area.

However, there are some difficulties in comparing the values of WTP that have been ascertained by bidding games with those that are determined by prevailing prices (despite the premise that prices are in fact a manifestation of WTP). For example, the gains of irrigation development were valued by prevailing prices. These difficulties arise because most of the respondents in the contingent valuation method were, on average, willing to pay only between 0.2 and 0.5 per cent of their income towards preservation. In the meantime, their WTP for, say, steady employment or stable supplies of food is likely to be higher. However, some of these discrepancies may also arise from the public good characteristics of environmental services and resources.

Foreign demand for preservation

Two methods were applied to estimate this demand. In the first method, the demand of those outside Zambia for preserving the study area was equated to the expenditure incurred by tourists who visit the area. Tourism data on Zambia reveal that the number of foreign tourists steadily increased from around 87,000 in 1980 to about 189,000 in 1990. This data source also reveals that tourism revenue (in 1985 prices) rose from K25 million in 1980 to K137 million in 1990. Using projections in the Fourth National Development Plan, we find that the average annual value of tourism amounts to K123 million. Further, it is believed that at least 50 per cent of the tourists who visit Zambia would visit the study area.

Hence, by way of the expenditure method, the WTP for preservation by those outside Zambia was equated to K61.5 million.

The second method of valuation was also related to tourism. A questionnaire that sought a subject's response to changes in the cost of a visit to the study area from his or her place of origin was distributed to a sample of tourists in Australia as well as in Lusaka. The number of visits had to be specified in the context of a six-year period. (The choice of a six-year period was governed by the Target-MOTAD model being also set in that context.) The cost of a visit ranged from US$500 to US$10,000. So, for example, a sample question reads as follows:

> If the cost of a visit to the Lake Kariba area of Zambia were US$1,000 (including return airfares, accommodation and meals), how many trips would you make during a six-year period?

As expected, more people opted for the US$500 visit, whilst virtually no one was willing to make even one visit at US$10,000. The information from each individual was used to derive individual demand curves, which were then horizontally aggregated. The area below the aggregate demand curve becomes a measure of the value of the benefits of preservation, which amounted to only US$0.8 million (K1.6 million in 1985 prices).

In fact, the second method follows the principles adopted by Hotelling (1949) in deriving the demand for recreation by the method of travel costs. In the context of the study area, this method has probably yielded an underestimate, since the sample surveyed is not likely to be truly representative of those visiting the study area. Alternatively, the tourism revenue derived from actual expenditures by tourists, and the assumption that 50 per cent of the tourists will visit the study area, are not entirely unrealistic. This is because Lake Kariba is one of the biggest lakes in the world, and its surroundings are full of varied wildlife. Hence the value of preservation in terms of demand may be approximated by the sum of the value derived by contingent valuation, namely K180 million, and that based on tourism revenue, namely K61.5 million. Consequently, the value of preservation by way of WTP is K241.5 million.

Method based on replacement

The method of replacement cost is based on the premise that the value of an item is equal to the cost of replacing it. Hence, if irrigation development is deemed essential, the wildlife and plant species, especially those that face the threat of extinction, should be rehabilitated into a similar environment elsewhere. These rehabilitation costs are equivalent to the costs of replacement. The cost estimates as provided by the Department of Wild Life and National Parks ranged from US$800 million upwards (i.e. values starting from K1,600 million in 1985 prices). If such a method of valuation is regarded as relevant, any development that involves the disruption of the ecological system becomes non-viable.

Past agricultural developments have treated environmental losses as inevitable and, by implication, justified by the profits of agriculture. Whereas this justification has often relied on pointing out that large expanses of the natural habitat still remain, this is no longer true.

An evaluation of EQ values

The values derived from the application of the various methods are summarized in Table 6.8. A comparison of these values with the net incomes generated by agricultural development indicates that preservation is far more valuable than agricultural development involving the clearing of the entire land area.

Table 6.7 showed that the highest NPV for the IM objective was K2,084 million. As shown in Table 6.8, the lowest present value for the preservation of the environment (at least within the confines of this study) is K4,830 million. Further, the valuation of the environment by replacement costs does not permit any consideration whatsoever of development. Hence it appears that irrigation development is not a viable option when the criterion of decision-making is the maximization of net social benefits, i.e. benefits accruing from all social goals.

Table 6.8 Summary of the monetary values of preserving the environment

		Method of valuation	Monetary value (millions of kwachas)	Present value (millions of kwachas)
1.	(a)	Demand within Zambia	180	3,600
	(b)	Tourism expenditure	61.5	1,230
	(c)	Total WTP	241.5	4,830
2.		Replacement cost	1,600	32,000

Zambia: Environment, Employment, Income, Food Security

However, let us suppose that the valuation of the environment by replacement costs is rejected on the grounds that the biological species that are lost could be found elsewhere in Zambia. The validity of the valuation methods based on the demand for the environment then becomes crucial in the choice of an appropriate policy. Further, given the extent of poverty and unemployment in Zambia, income considerations do inevitably dominate the choice of policy decisions. For these reasons the value of the environment might be relegated to the commercial component of the demand for the environment. This commercial or tangible component is in fact tourism revenue, the present value of which is K1,230 million. Thus, decision-makers are likely to acknowledge tourism revenue from overseas visitors as a measure of value, and disregard the local Zambians' expressed WTP, which does not enter into any actual transaction.

For reasons of simplicity, we assume that EQ benefits are uniformly distributed across the study area, and that the value of preservation is directly proportional to the extent of preservation. If such an assumption is acceptable, it is possible to search for a policy on the basis of the 'tangible value of policies'. This value can be defined as:

(The present value of the IM objective) + (The present value of the tourism revenue component of the value of EQ)

This tangible value was computed for the maximum feasible extent of preservation in the context of the four different policy scenarios that were presented in Table 6.6, namely absence of EM and FS, recognition of EM only, recognition of FS only, recognition of both EM and FS. These tangible values are presented in Table 6.9.

Interpreting Table 6.9, the policy 'only EM', for example, imposes the requirement of full employment, and the maximum extent of land that can be preserved in this context is 48 per cent. The estimate of tourism revenue in this instance is assumed to be 48 per cent of the tourism revenue that is earned when the entire land area is preserved: that is, K(0.48)(1,230) million = K590 million. As indicated, 'tangible value' is maximized when a policy of preserving nearly half the extent of the study area is pursued along with full employment. Hence, in terms of whether Zambia should proceed with agricultural development or, alternatively, preserve the Middle Zambezi Valley in its natural state, the analysis thus far would favour the latter option or an option that permits partial development. However, should pecuniary considerations dominate, a partial

Table 6.9 The tangible value of policies

Policy	Maximum possible extent of preservation (%)	Tourism revenue (millions of kwachas)	Value of IM (millions of kwachas)	Tangible value (millions of kwachas)
EM and FS absent	100	1,230	0	1,230
Only EM present	48	590	1,269	1,859
Only FS present	62	762	690	1,452
EM and FS present	35	430	1,325	1,755

development to satisfy the goal of employment appears to be more relevant. To a different degree, either solution permits Zambia to protect and develop its comparative advantage in tourism by virtue of its wildlife and environmental endowments.

CONCLUSIONS

The results of this study indicate that income, employment and environmental goals could be complementary in terms of preserving and enhancing the quality of environmental resources for tourism development. Evidence from elsewhere in Africa, especially Kenya and Zimbabwe, supports this type of complementarity. Hence, it may be prudent for Zambia to exploit its comparative advantage in terms of the endowments of wildlife and other natural amenities to generate income by way, for instance, of wildlife park development. Instead of clearing the 162,000 hectares of land for agricultural development, it may be more desirable to increase tree densities by afforestation and thereby permit the growth of wildlife populations. Such action, apart from its beneficial environmental effects, may generate more income in the long run, subject to the development of tourism. Furthermore, environmental protection and maintenance itself constitutes a source of employment. Accordingly, an important question to be analysed is the extent to which the quality, as well as the quantity, of environmental resources could be enhanced while optimizing income and employment benefits. This aspect, however, has not been analysed because of resource limitations. None the less, the analysis presented is sufficiently robust to recommend against the large-scale clearing of the environment for irrigation development. Instead, the preferred policy is one of partial development. While satisfying the goal of full employment, under partial development nearly half the study area would be left in its preserved state.

It is possible that decision-makers in the study area as well as elsewhere in Zambia would support an option of partial development instead of extensive deforestation for development. The reason lies with the emergence of an environmental awareness that has pervaded all levels of Zambian society. This awareness is reflected in the fact that Zambians portray their image overseas in terms of wildlife and natural resource endowments. There are some 19 protected areas in Zambia, spanning approximately 6 million hectares, which is nearly 9 per cent of the national land area and well above the African average of 3.5 per cent (World Resources Institute (WRI), 1990b).

The value of this study is not confined to the specific policy options addressed. The study establishes a model that can be applied to a variety of decisions that involve conflicting purposes. In view of this, it seems appropriate to conclude with a note concerning the need for further study.

Further research is important since the problems addressed in the Middle Zambezi Valley are representative of the environmental problems found across most of Africa. Extensions to this study would necessarily involve a more detailed definition of the environment and thereby illustrate, as far as possible, precise changes that could occur in the 'economy–environment' relationships. These changes could be demonstrated in the context of pursuing traditional growth-oriented activities, such as agriculture, as well as environment-oriented activities, such as afforestation and preservation. A possible way of achieving this is to formulate a systems framework that consists of an environment simulator as well as an economy model.

The basis for such an extension has already been established in this study, since the framework we used consisted of a rainfall simulator and a stochastic land-use model for agricultural development. The rainfall simulator can be replaced by the environment simulator, the components of which could include predictors of key environmental variables such as climate changes, soil loss, salinity levels of the groundwater table, and the population sizes of biological species. The economy model would replace the land-use model and could include income-generating activity variables as well as traditional fiscal and monetary policy instruments. By permitting interactions between the two models within the systems framework, policy instruments that stabilize the environment as well as sustain the economy can be more adequately defined. For example, the interactions could be performed in the context of a recursive two-stage analysis. The first stage might involve the determination of the status of environmental variables, and the second stage the allocation of resources, including those environmental resources that were determined in the first stage to maximize income. The imposition of criteria such as renewability in the second stage would constrain the extent to which income can be maximized, and the decisions taken in the second stage would in turn influence the status of the environmental variables. A sustainable state of equilibrium between the two stages can be defined as one in which the resource allocation decisions in the second stage retain the status of the environmental variables of the first stage in terms of criteria such as renewability.

7 Soil Conservation and Sustainable Development in the Sahel: A Study of Two Senegalese Villages

Elise H. Golan[1]

THE ANALYTICAL FRAMEWORK

The Sahel and Sahelo-Sudan zones of Africa are experiencing widespread soil degradation and desertification. As the vast majority of the population of these regions depends on agriculture for its livelihood, the need for governments in the area to provide policy to address the problem of environmental degradation is critical. The welfare of the rural population in the Sahel depends on the ability of Sahelian governments to provide policy that will create incentives for soil-preserving land management practices. If this is not possible, the welfare of these peoples depends on the ability of Sahelian governments to provide policy which will create employment opportunities outside the farming sector.

The aim of this chapter is to examine some of the policy options available to Sahelian governments and to evaluate, mainly through simulation experiments conducted with a Social Accounting Matrix (SAM), the socioeconomic impact of such policy on two sample villages in Senegal's Peanut Basin. The two village SAMs were constructed so as to provide information on income, employment and land use patterns for large, medium and small compounds, and for the individual classes of compound members. Because of the attention paid to the complexity of land management and landownership in the sample villages, the two SAMs presented here are equipped to examine the employment and distributional effects of various conservation options.

In order to situate and focus the policy discussion, it is necessary to examine those factors which have combined to accelerate and entrench the desertification process in the western African Sahel and Sahelo-Sudan zones.

First, a principal contributing factor of desertification in the Sahelo-Sudan zones is the fact that soils in this area are structurally fragile, with low humus content and low water retention capacity. They are particularly vulnerable to hydromorphy, hard clay pan, laterization, and wind and water erosion. In addition, the low fertility and fragility of the soil acts as a constraint on plant productivity, thereby reducing the potential of plant cover in protecting the soil from erosion (Gorse, 1987).

Second, there has been a continuous and at times rapid destruction of tree varieties in the area resulting from domestic and international trade in charcoal and gum arabic. The devastating effects of deforestation are well documented in the literature (Gritzner (1988) gives a good overview).

Third, the rapid expansion of and continued dependence on peanut cultivation has been doubly detrimental to the environment. The introduction of peanut cultivation resulted in a reduction in fallow land and a shift in livestock grazing patterns. For food security reasons, peanut crop production did not displace staple crop cultivation, but instead took place on the fallow areas around and between villages. Because peanut crop residue offers only about one-third of the dry-matter fodder of traditional subsistence crops, the expansion of peanut production not only eroded traditional buffer zones but also forced nomads and other herders north, on to the fragile desert fringe (Franke and Chasin, 1981). Both effects have served to reinforce the desertification process.

Fourth, population pressures in the heartland of the Sahelian and Sudanian zones, as opposed to the desert fringe, make these areas particularly vulnerable to desertification. In particular, the Senegalese Peanut Basin, the location of this study, has been identified as an area where, by 1980, actual rural population already greatly exceeded the carrying capacity of the land, in terms of both agricultural and livestock production and fuelwood consumption.

Fifth, to date, the results of research into alternative, soil-friendly agricultural techniques for the Sahelo-Sudan zones have been disappointing, and it has been argued that, without a technological breakthrough, significant change in carrying capacity in the Sahelo-Sudan heartland is not possible (Gorse, 1987; Perrings, 1991). Because rapid population growth and increased agricultural production have not been accompanied by technological innovation, stress on the environment has not been matched by technologically induced changes in carrying capacity. Overcultivation, overgrazing and deforestation, the major contributors to desertification, have been the result.

Sixth, since 1968 a period of drought has affected the Sahel region of western Africa. In spite of expectations to the contrary, this drought has not abated to date. The history of the drought has been long, painful and even horrific. In its first stages, from 1968 to 1972, the Club du Sahel estimated that between 50,000 and 100,000 people died in the region as a result of the drought. The FAO estimated that 3.5 million head of cattle (25 per cent of the total) died in the Sahel in 1972–73 alone. Rainfall increased between 1973 and 1975 but has since remained below the average for this century. Average rainfall between 1977 and 1981 was no higher than that for the period from 1968 to 1973, and 1983 and 1984 were even drier than 1972. Drought continued during the second half of the 1980s, and, despite heavy rainfall and excellent crop yields in 1988, there is still no evidence that the drought has subsided. The continuation of the drought has led to speculation that the region is experiencing a long-term warming trend. Possible explanations for this trend include the albedo effect (i.e. the proportion of light reflected by a surface, a result of overcultivation, overgrazing and deforestation in the region), El Niño, and the 'greenhouse' effect. (Grainger (1990) gives an excellent overview of the history and possible explanations for the drought in the Sahel.)

Even though the drought itself may be a result of the desertification process, it is also serving as a catalyst to the process. With the continuation of the drought, streams dry up, water-tables drop, salination occurs, and in many cases irreversible damage is done not only to the environment but to human and animal populations as well. In addition, the possibility of long-term climatic change in the area raises the issue of climatic variability. It has been argued that even a small rise in climatic mean could result (and maybe already *has* resulted) in large increases in climatic extremes, and that the shock of more extreme climatic events may lead to the destruction of the resiliency and sustainability of an ecosystem (Parry, 1990; Perrings, 1991).

The six factors listed above have combined to accelerate the process of soil degradation and desertification in the Sahel and Sahelo-Sudan zones. Conservation policy designed for the region must be considered against the background of these issues.

THE PEANUT BASIN

The bulk of Senegal's population lives and farms in the Peanut Basin. This region extends southward from Louga to Kaolack and eastward from Thies to Tiaf and broadly follows the historical and current distribution of peanut production in Senegal. Like the whole of Senegal, the Peanut Basin has a late summer rainy season (about four months) and an extended dry season (about eight months). Rain levels in the Basin range from approximately 800 mm in the south to approximately 475 mm in the north, with rains becoming more variable as one moves north. The most widely distributed soil types in the Basin are desaturated ferruginous tropical soils, and the natural vegetation ranges from wooded savannah in the south to sparsely wooded, shrubby steppe in the north. Giant baobab trees are prevalent throughout the area.

In the Basin, the Wolof and Serer ethnic groups predominate, with Peul, Lebou, Malinke, Toucouleur, Nouminda and Bambara scattered throughout the region. In 1980, population densities in the Basin ranged from 30 to 40 people per square kilometre in the north and east to approximately 100 people per square kilometre in the south-centre districts. These numbers represent high population densities for a desert region such as the Sahel.

The farming system and land tenure

Historically, land in the Basin was claimed by the first settlers by right of their having cleared it by fire. These men became known as the 'masters of fire' or the *borom daye* (Wolof). They usually claimed vast areas of land with up to six days of burning. Being unable to cultivate the totality of their holdings themselves, these men accorded use rights or 'rights of hatchet' to men who could cultivate the land. Once given use rights, the 'master of hatchet' or *borom n'gadio* (Wolof) had incontestable, irrevocable rights to that land as long as he paid a yearly homage to the master of fire. Usually, this annual payment was symbolic (an ear of millet for example), but in different areas and at different times in the Basin's history the homage payment became a substantial portion of the year's harvest.

Rights of fire and hatchet were (and still are) passed from father to son. In the area of the Basin where this study was conducted, farmers reported that the right of fire had died out during the period of French rule and that at present only the right of hatchet remains.

Farm production is organized at the compound level. The compound is composed of any number of households. The nucleus of the compound is typically composed of one male, who has the right of hatchet, and his household (wives, children, older parents, aunts, sisters, unmarried male relatives, etc.). The male with rights of hatchet is not only the head of his household but also the head of the compound. Other households in the compound are typically headed by male relatives of the compound head (brothers, sons or cousins). These secondary households are either independent or dependent households, where the primary distinction is that independent households prepare their own meals and are responsible for meeting their own millet needs. The position of head of the compound, along with the right of hatchet, is passed from father to eldest son. If the eldest son is unable or unwilling to assume control, responsibility is passed to the most appropriate male (or occasionally female) relative.

In the Basin, peanuts and millet are the primary cash crop and primary staple crop, respectively. The compound head is responsible for distributing compound land between millet and peanut crops. He oversees the compound's millet fields and has the ultimate responsibility for assuring the food needs of the compound. If there is an independent household in the compound, the head of this household will also oversee a millet field in order to supply his household's grain needs. The millet flow between the compound and independent member households seems to be fluid and varies, with transfers taking place in both directions.

After allocating enough land to millet production, the compound head distributes the remaining land among the various compound members for cultivation. Occasionally, land is set aside for manioc, vegetables and condiments. Wives, unmarried compound members (called *sourga*), older male children, heads of households, brothers, cousins, aunts, uncles, etc., can all be allocated land to cultivate for their personal benefit. Peanuts is usually the crop of choice. Peanut and millet fields are customarily rotated on a yearly basis so that from one year to the next many compound members do not know which fields they will be allocated for their personal peanut crops. The compound head also cultivates a peanut field for his own cash needs. All compound members donate labour to the compound's millet fields, but assuring enough labour for the peanut fields is usually the responsibility of each field manager and labour swaps are arranged on an individual basis.

Two types of hired labour are common in the Peanut Basin. One type is the *firdou*. The *firdou* travel throughout the area supplying supplemental labour as needed. They are usually paid in cash and are given food and lodging during the duration of the labour contract. *Navetanes* are the second common type of hired labour. These men hire out their labour in return for the loan of peanut seeds and a parcel of land. Traditionally, *navetanes* work on the compound fields in the mornings and on their personal peanut field in the afternoon. At the end of the season, they repay the peanut seed loan with interest. *Navetanes* live and eat with their host compounds during the growing season and then return home during the dry season; they usually return year after year to the same compound.

Inputs such as seeds, pesticides and fertilizer are acquired by the compound in a number of ways. Up to and including 1986, the year of this study, farmers had access to government-provided peanut seed (which included pesticide) that was distributed on credit through the farmers' cooperative. With this system, farmers received the peanut seed and pesticide at the beginning of the season, and after the harvest they were responsible for reimbursement with interest. As heads of compounds and sometimes independent heads of households are usually the only compound members with access to peanut seed credit, it is their responsibility to determine the allocation of this peanut seed and pesticide among the other compound members who wish to plant peanut fields. Peanut seed was, and is, also available for sale on the open market and, of course, some farmers reserve seed from year to year (though this practice cannot be continued over a long period owing to the eventual deterioration of seed quality). Pesticide and fertilizer are both sold on the open market, though fertilizer is extremely expensive and relatively difficult to obtain.

Tool use in the Basin is mainly restricted to horse- or donkey-pulled ploughs and small hand-held implements. The *iler*, a metal arrow-shaped piece attached to a long stick, appears to be the most popular tool in the area. It is principally used for weeding but has many other uses, from field preparation to seeding and harvesting.

The legal land tenure system in Senegal is complex. A Law of National Domain was passed in 1964, which granted the state ownership rights to previously unregistered land. As a result, approximately 98 per cent of all Senegalese land became part of the National Domain and subject to state control. However, owing to the incomplete and uneven application of the Law of National Domain, usufruct or ownership rights over most agricultural land are still determined through traditional practice.

DATA COLLECTION

Every region in Senegal is divided into administrative units called village sections, each of which is governed by a democratically elected rural council. Depending on the population density of an area, these sections are composed of one or more villages. At the village section level, land disputes and inheritance are decided by the rural council.

Two village sections in the Peanut Basin, Keur Marie and Keur Magaye,[2] were chosen for the study sample. They were chosen mainly on the basis of logistic criteria. Each section is less than an hour's drive from the huge daily market in the regional capital of Kaolack, and is within walking distance of a large weekly market. The sections were chosen so that a number of marketing opportunities would be readily available to the farmers and also to reflect the variation in population density in the Peanut

Basin. Keur Marie is located to the west of Kaolack where population densities are highest, while Keur Magaye lies to the east where population densities are at their lowest.

A list of compound heads for each section was acquired from the sections' extension agents. Compounds were then chosen randomly from the list with approximately one-third of each section's compounds being selected for the study. In Keur Marie, 22 compounds were interviewed and 26 in Keur Magaye. Interviewing was conducted from January to May 1987. The interview period corresponds to a part of the year of low agricultural activity in the Peanut Basin. Data were collected as soon after harvest as was feasible, given the time constraints of the compound and the availability of the enumerators. The figures thus obtained rely mainly on the 1986 agricultural season.

The compilation of data formed part of a study on African tenure and land management conducted by the United States Agency for International Development (USAID), the World Bank and the Land Tenure Centre at the University of Wisconsin (Hardy, 1989; Golan, 1990). Findings have permitted the construction of a SAM for each of the two villages. These SAMs provide a methodological framework for the analysis of land management, land tenure and incentives, which, in turn, provide a key to the policy issues examined in the present study.

Village surveys

Although Keur Marie is situated in a densely populated area and Keur Magaye is not, the basic economic structure of the two villages is similar. Both have single cash crop and single staple crop economies that revolve around peanuts and millet. The flow pattern is uncomplicated. Most purchased inputs and consumption items are imported from outside the village, and most cash crops are exported to outside the village. There is relatively little interaction between compounds within each village, and neither village has a store or market.

Examination of the flows into and out of the villages makes clear the cash crop/staple crop dichotomy which exists in the two villages. Peanut seeds are mainly purchased from the government cooperative, as is pesticide. The peanut crop is then sold to the cooperative. Millet seeds are almost exclusively retained from previous harvests, and most millet production is consumed in the compound. Tables 7.1 and 7.2 illustrate the flow pattern between the village and the rest of Senegal. Table 7.1 gives the percentage breakdowns of the amount of seed that village compounds retained for cultivation in 1986. As shown, almost all the compounds'

Table 7.1 Retained seeds: total seeds planted in 1986 (%)

Seed	Keur Marie	Keur Magaye
Peanuts	34	31
Millet	93	89
Other crops	39	29

Table 7.2 Harvest retained for compound consumption (%)

Seed	Keur Marie	Keur Magaye
Peanuts	7	2
Millet	99	98
Other crops	51	50

millet seed needs are met by retained stocks in both villages. The small amount of peanut seed retained by the compounds from one year to the next can be explained by three facts: (a) peanut seed is supplied by the government cooperative on credit, an important consideration for an area with a 'hungry season'; (b) peanut seed storage can prove quite costly: whereas, on average, it requires only 3 to 4 kg of seed to plant 1 hectare of millet, it requires about 90 to 95 kg of peanut seed to plant a corresponding area; and (c) peanut seed quality deteriorates from one year to the next, and farmers must periodically supplement retained stocks with higher-quality seeds.

Table 7.1 also shows that in both villages the largest portion of other crop 'seed' is purchased, either from village neighbours, the weekly market or elsewhere in Senegal. This reflects the fact that in Keur Magaye manioc is a booming crop, with farmers currently planting first-time plantations, and in Keur Marie the vegetable and melon market is organized by Lebanese merchants from Dakar who supply seeds, fertilizer and pesticide on credit.

It is interesting to note that in both villages the amount of peanut seed retained for cultivation in 1987 exceeds that in 1986. In Keur Marie 50 per cent more seeds were saved for use in 1987, and in Keur Magaye 14 per cent more. The large increase in retained seeds in Keur Marie probably

Soil Conservation and Sustainable Development: the Sahel 167

reflects the unease that Keur Marie farmers are experiencing over the expected change in the government's peanut seed distribution policy. The relatively small increase in Keur Magaye probably reflects the confidence that the farmers in Keur Magaye have in the ability of their cooperative president to assure them of a continued supply of government peanut seed.

Neither village substantially increased its retained millet seed: 0 per cent for Keur Marie and 7 per cent for Keur Magaye. The increase in Keur Magaye is too small for us to state with confidence that farmers in this section anticipated expanding millet production in 1987. But the increase in other crop seeds retained in Keur Magaye reflects a definite trend; farmers are allocating more and more land to manioc production. From 1986 to 1987, 511 per cent more seeds were retained for this purpose. Indeed, conversations with farmers make it clear that a manioc craze is sweeping this village section. No similar craze has hit the vegetable and melon farmers in Keur Marie, and manioc is not popular in this area.

If we now turn to consumption, we find in Table 7.2 the percentage of harvest retained for compound consumption. Again, the distinction between the cash crop and the staple is clearly drawn; very few peanuts are kept for compound consumption and very little millet is sold. In both villages, the other crop, primarily melons and vegetables in Keur Marie and manioc in Keur Magaye, is almost equally sold and consumed by the compounds. It should be noted that not one farmer in either village listed a neighbour as the purchaser of peanuts, millet or other crops and that conversely, in the consumption questionnaire, not one farmer stated that he or she had purchased any of these crops from a neighbour for consumption. This implies that a village market for consumption crops does not exist and that the consumption of these goods by village compounds represents a compound's own consumption exclusively.

In both villages, the primary village-produced consumption good is each compound's own millet. Almost all other consumption goods are imported – clothes, foodstuffs (sugar, tea, coffee, condiments, fish, etc.) and small manufactured goods (matches, pots, utensils, etc.).

The amount and type of village-produced goods and services exported outside the village also reflects the millet/peanut dichotomy: peanuts are sold outside the village and millet is for its own consumption. The organization of the peanut market shows some difference in the two villages. Keur Marie farmers sold their peanuts exclusively to the government-sponsored cooperative, while the farmers in Keur Magaye sold 27 per cent of their marketed surplus to peanut merchants from the rest of Senegal. As for millet, the farmers in Keur Marie did not sell any of their harvest to the outside world (or to anyone), while in Keur Magaye a paltry 1 per cent

Table 7.3 Factor value-added as a percentage of farm income

Village/factor	Large compounds	Medium compounds	Small compounds
Keur Marie			
Manager labour	6	7	15.4
Household labour	15	10	22
Compound labour	2	6	0.2
Village labour	2	2	5
Secure fields	44	46	26
Moderately secure fields	8	2	0
Insecure fields	5	14	0
Borrowed fields	11	4	20
Animals	7	9	11
Keur Magaye			
Manager labour	3.4	5	13
Household labour	6	6	15
Compound labour	3	2	1
Village labour	0.6	0.2	1
Secure fields	38	53.5	9
Moderately secure fields	19	0.3	12
Insecure fields	17	23	11
Borrowed fields	9	6	32
Animals	4	4	6

was sold at the weekly market or to merchants from the rest of Senegal. In Keur Magaye the manioc crop was sold at the weekly market and to other points in Senegal, while in Keur Marie all the vegetables and melons were sold to the Lebanese merchants from Dakar who were responsible for introducing these crops in Keur Marie. In neither village section were any services sold to the rest of Senegal, but more than 75 per cent of the commercial activities (76 per cent in Keur Marie and 87 per cent in Keur Magaye) were completed in locations other than the village.

The main economic interaction within the villages takes place on the input side: there is fairly extensive borrowing and lending of fields among compounds, and compounds use not only household and compound labour but also labour from other compounds ('village labour') and from other villages ('imported labour'). In fact, the complexity of these SAMs results from detailing intra-compound interactions and stratifying village landowners. The importance of inter-village and inter- and intra-

Soil Conservation and Sustainable Development: the Sahel 169

compound interaction is illustrated in the factor value-added calculations. Here the percentage of farm income generated by labour and land transactions is clearly shown.

Table 7.3 shows that labour value-added percentages across labour types and compound types are comparable, with all compounds deriving most labour value-added from household, manager and compound labour, in descending order. There is comparatively little labour interaction between compounds in the same village, and little imported labour from outside the village. The distribution across labour types in both villages is similar, with small compounds deriving almost twice as much value-added from labour as medium or large compounds do. This means that, relative to the income gained, small compounds expended about twice as much labour time on agriculture as large and medium compounds did. Animal use across compounds and across villages is not strikingly different. Secure fields account for most of the land value-added, though in Keur Magaye the smaller compounds relied most heavily on borrowed land.

Though the farming system and tenure patterns in the two villages are similar, the size of landholdings is quite different. This observation is not surprising, given the fact that Keur Marie is located in a more densely populated area. The 26 compounds in Keur Magaye controlled almost three times the amount of land controlled by the 22 compounds in Keur Marie. They also had twice as much land in fallow as did the Keur Marie sample. The statistics on size and number of parcels are presented in Table 7.4.

Table 7.4 Parcels and hectares by village: Keur Marie and Keur Magaye

Parcel	Number of parcels		Hectares		% of hectares	
	Marie	Magaye	Marie	Magaye	Marie	Magaye
Total	138	213	190.6	543.6	100	100
Owned and operated	98	150	152.0	412.1	80	76
Borrowed	25	24	25.2	40.6	13	7
Lent	15	39	13.4	90.8	7	17
Borrowed and lent	0	1	0	0.1		
Fallow	34	74	35.2	209.0	18	39
Fallow:						
Owned and operated	30	71	31.4	203.6		
Borrowed	3	3	3.2	5.4		
Lent	1	0	0.6	0		

Owned land area per compound in Keur Magaye ranged from 0.1 to 59.7 hectares, while that in Keur Marie ranged from 0 to 18.1. The average owned land area per compound in Keur Marie was 7.5 hectares (standard deviation (sd) = 5.0), as compared with 19.3 hectares (sd = 15.8) in Keur Magaye. The average amount of land to which each compound had access during the previous year (i.e. owned (including fallow) + borrowed − lent) was 8.1 hectares for Keur Marie (sd = 5.1) and 17.4 hectares for Keur Magaye (sd = 13.0). The average amount of operated land per compound (i.e. owned land + borrowed − lent − fallow) was 9.4 hectares for Keur Magaye and 6.5 hectares for Keur Marie. The gap in the ratios of land per compound in the two villages narrows as soon as reallocation resulting from borrowing and lending is considered, and narrows further when fallow land (particularly the huge tracts in Keur Magaye) is removed from the calculations.

The average distance between a compound and one of its parcels was approximately 1 km (sd = 1,312 m). Compound members usually walk this distance, though for the longer treks, which range up to 6 km, horse or donkey carts are sometimes employed. For the most distant parcels, small huts are built in the parcel for overnight stays by compound members. Table 7.5 shows the average distances between parcel and compound for each village and for each type of parcel. Given the size of the holdings in Keur Magaye, it is not surprising to find that on average they lie almost four times further from the compound than those in Keur Marie.

POLICY ANALYSIS

The following policy options for land conservation will be analysed, either through experimentation with the SAM model or through examination of

Table 7.5 Average distance between compound and parcel (metres)

Parcel	Keur Marie	Keur Magaye
All parcels	441	1,632
Owned and operated	489	1,465
Lent	272	2,163
Borrowed	356	1,879
Fallow	584	1,679

the SAMs themselves: taxes and subsidy cuts to reduce peanut cultivation, migration and diversification, reforestation through subsidies and usufruct reform, and land tenure reform.

Financial incentives to reduce peanut production

Three assumptions lie behind the argument supporting the use of taxes and subsidies in the control of desertification. First is the assumption that the social cost of land use is higher than the private cost: the farmer or herder overutilizes land resources because he or she does not consider the true cost to society today or in the future. The taxation of marginal land use or tree harvesting or grazing represents an effort to internalize a negative externality. By levying a tax on the producer, private and social costs can be reconciled (and a high discount rate compensated for). With the tax, the individual producer will factor the true cost of expanded cultivation, grazing or wood gathering into his or her utility-maximizing calculus. Likewise, a subsidy on environmental conservation will help to tip utility-maximizing calculations in favour of conservation practices.

The second assumption is the existence of an alternative environment-friendly agricultural technology or an alternative extra-agricultural income-generating option. The objective of taxes and subsidies is to change the cost-benefit calculations of the producer in such a way that a soil-saving technology or conservation scheme or income option that was once unprofitable, or at least not the most profitable, becomes the profit-maximizing choice.

The third assumption is that the appropriate decision-maker can be identified and influenced. Any effort to use financial incentives to influence the profit-maximizing calculus of the producer crucially depends on determining who the producer is and tailoring policy to his or her incentive structure. Close attention must be paid to the complexity of the compound farming system if the compound is taxed or subsidized in order to enhance conservation practices in the region.

The financial incentives to curtail peanut production will be examined in light of these three considerations, as will the financial incentives examined later in the chapter.

Tax on peanut production

In the first SAM experiment, a simple tax to curtail peanut cultivation is imposed. Because expanded peanut production has been identified as a driving-force in overcultivation, overgrazing and deforestation in the

Sahel (Franke and Chasin, 1981), this is a logical tax to consider. For the purpose of the experiment, it will be assumed that the tax is collected as a value-added tax by the government cooperative at the time of sale of the peanut crop (it is assumed that the private market peanut price will fall to meet the government 'price'). The purpose of the tax is simply to reduce the amount of land currently under peanut production. A value-added tax has four advantages: (a) it is easily collected and monitored; (b) villagers retain control of land management decisions; (c) a value-added tax should not disrupt informal tenure arrangements in the villages (if a marginal land-use tax were imposed instead, large landholders might just repossess, and then leave fallow, land that had been previously lent to land-poor compounds. If this were the case, the land-poor compounds would bear the entire burden of increased land conservation. A value-added tax is also regressive because large producers and small producers face the same percentage charge, but this could be avoided by devising a tax schedule dependent on total production); and (d) this tax would be applicable to every land manager in the compound, not just the compound head.

In this experiment, a tax of 20 per cent is applied to peanut production. The price elasticity of hectarage of peanut production in the Basin is calculated at approximately 0.4 (Beghin, 1988). This means that the hectarage response to a 1 per cent change in producer prices is calculated at approximately 0.4 per cent. A tax of 20 per cent would lead to a change in hectarage of 8 per cent. Calculated at this elasticity, peanut production in Keur Marie would fall by 2.24 hectares to a new total of 28 hectares, and in Keur Magaye 8.44 hectares would be taken out of peanut production for a new total of 105.5 hectares.[3] Average peanut production in Keur Marie is calculated at approximately 870 kg per hectare, so that post-tax production amounts would equal 22,183 kg. Multiplying by 80 CFA francs[4] (price per kg inclusive of the tax) gives a new exogenous peanut revenue of 1,774,640 CFA francs for Keur Marie, a net difference of −638,573 CFA francs. In Keur Magaye average peanut production is calculated at approximately 1,100 kg per hectare, meaning that post-tax production would equal 106,880 kg. Multiplying by 80 CFA francs gives a new exogenous peanut revenue of 8,550,400 CFA francs for Keur Magaye, a net difference of −3,066,096 CFA francs.

The impacts that these changes in peanut revenue would have on the village economies can be calculated with the relevant multiplier derived from the SAM of each village and are shown in Table 7.6. For changes in exogenous peanut revenue, the overall multiplier for compound income is 1.5 for Keur Marie and 1.2 for Keur Magaye. The multiplier for the whole economy is 5.0 for Keur Marie and 3.9 for Keur Magaye.

Table 7.6 Income effects of: (a) a tax on peanut production (b) a cut in seed subsidy; and (c) migration

	Keur Marie			Keur Magaye		
	Large	Medium	Small	Large	Medium	Small
Number of compounds	6	9	7	8	9	9
Average number of people per compound	17	11	7	14	10	9
Base income[1] per compound	2,342	955	406	2,889	1,650	1,062
Base income per capita	138	87	58	206	165	118
1. Tax on peanut production:						
Income change	−196	−140	−65	−656	−388	−252
Percentage change	−8	−15	−16	−23	−23	−24
2. Cut in peanut seed subsidy:						
Income change	−588	−421	−195	−612	−362	−235
Percentage change	−25	−44	−48	−21	−22	−22
3. Migration experiments:						
Income change	+140	+126	+108	+116	+111	+104
Percentage change	+6	+13	+26	+4	+7	+10

1. In 1987 US dollars, converted at a rate of 333 CFA francs per US dollar.

Reduced government programme for peanut seed allocation

Another way to restrict peanut production, at least in the medium run, would be to reduce the government credit programme for peanut seed. At the time of the study, there was very little private supply of peanut seed in Senegal, implying that the government effectively had the ability (at least in the short run) to regulate peanut production through regulating the peanut seed supply. This hypothesis is supported by the observation that farmers in the two sample villages stated that the primary constraint to expanded production was the lack of peanut seed. Not only are private traders failing to fill the gap left by the government programme, but they are also unable to offer the low credit terms inherent in the programme. The government programme distributes seed at the beginning of the season and seeks reimbursement at harvest time. The subsidy represented by this programme is reinforced by the fact that peanut seed debt has often been written off by the government. In the following experiment it is assumed that, if peanut seed supply were constrained, peanut production in both villages would decrease.

For this experiment, peanut seed supply to each village is cut by enough to reduce each compound's peanut cultivation theoretically by 1 hectare. In Keur Marie the 22 compounds would contract peanut production by 22 hectares and experience a decrease in peanut income of 1,914,000 CFA francs. The 26 compounds in Keur Magaye would contract peanut production by 26 hectares and suffer a decrease in peanut income of 2,860,000 CFA francs. The absolute and percentage effects on compound income for both villages are shown in Table 7.6.

Social and environmental impacts

In the two experiments conducted above, it was assumed that farmers in the Basin are overutilizing the environment through the production of peanuts. The question is, if additional pressure were applied via the peanut production tax or the peanut seed cut (in addition to the pressure already imposed by an increasingly hostile environment), what type of modification in the farming system would emerge? The tax and subsidy cut should succeed in removing land from peanut cultivation, but would they also result in the adoption of an environmentally sustainable agriculture? In order to deal with this question, the probable response of the rural community to the reduction in the tax and seed programme must be analysed.

Given the fact that to date there are no viable technological alternatives for the region, the most probable response of the rural community would

Soil Conservation and Sustainable Development: the Sahel 175

be diversification and migration. The added pressure placed on the system via the tax and seed cut would not be likely to persuade farmers to consider new technologies or crops, because few alternatives exist at present. Nor, because of the lack of markets, would farmers expand millet production. The probable result of these policy changes would be a shift away from agricultural production. In addition, there would probably be a shift in cultivation patterns and income distribution within individual compounds.

In fact, diversification and redistribution are what have been observed from the Peanut Basin data. In the village of Keur Marie a number of factors have combined to put additional stress on the agricultural system: (a) the higher population density and longer cultivation of the Keur Marie area seem to have resulted in greater soil degradation than in Keur Magaye: this hypothesis is supported by the fact that the land value-added calculations for Keur Marie are lower than those for Keur Magaye; (b) owing to both greater population pressure and a wider greater application of the Law of National Domain, the farmers in Keur Marie are less secure in their landholdings than those in Keur Magaye (Golan, 1990); and (c) because of the structure of the government cooperative in Keur Marie, farmers in that village section are less secure in their continued access to government-supplied peanut seeds (Golan, 1990).

Contrary to predictions (Boserup, 1965; Pingali *et al.*, 1987), the technological response to these stresses has been nil. It had been hypothesized that, in response to population growth coupled with land degradation, compounds might adopt technological changes such as the introduction of manuring, better seed or chemical pesticides, or might respond with institutional changes such as consolidating tenure rights, reducing hired labour and increasing migration from the village (Adelman *et al.*, 1990). Of these hypothesized responses, it appears that only the institutional changes are beginning to take place in the Peanut Basin sample.

An examination of the farming practices in the two villages reveals that there are no differences in capital improvements in the land, agricultural innovation or fallow and rotation schedules. Neither village has undertaken aggressive measures to prevent soil degradation. Very few improvements have been made to land in either village (no bunding, no contour planting, etc.), and fallow land is not being managed. For the most part, rotation schedules are adhered to, but, as we shall see, a few cases of disregard for this simple, inexpensive soil-saving technology have been observed. Tables 7.7 to 7.9 give details on land stewardship.

The most common type of improvement made in both villages is tree planting. Yet the real investment in land maintenance found in this respect

Table 7.7 Land improvements

Nature of improvement	Keur Marie	Keur Magaye
No improvements	75	133
Trees	66	70
Fences	8	14
Pasturage	2	1
Manure	0	1
Wells	8	0
No response	0	3
Total number of parcels	138	213

Note: No borrowed parcels are included.

Table 7.8 Reasons for leaving parcel fallow

Reason for fallow land	Keur Marie	%	Keur Magaye	%
Lack of peanut seed	22	65	49	52
Lack of seeds other than peanut	0	0	2	2
Lack of labour	1	3	2	2
Give land a rest	8	2	16	17
Too far from compound	0	0	1	1
Poor quality land (insect holes, etc.)	1	3	23	24
No response	2	6	2	2
Total number of parcels	34	100	74	100

is probably inflated, as scrubby bushes were often defined as trees. The fences in both villages were found mainly around the small fields of vegetables, melon and manioc. The wells in Keur Marie provide water for the vegetable and melon fields. No wells were found in Keur Magaye because the water table is too low. The information presented in Table 7.3 above corresponds to 80 fields with one or more improvements in Keur Magaye, and 63 fields with one or more improvements in Keur Marie. This translates to 38 per cent of Keur Magaye's 213 parcels and 46 per cent of Keur

Marie's 138 parcels. The quality and quantity of land improvements in both villages are disappointing.

Fallow land is also not being actively managed for soil conservation. What emerges from the Peanut Basin study is almost invariably that compounds had more land in fallow than they would have preferred and that, apart from deciding which parcels to choose, very little strategy went into determining the amount of fallow land, i.e. compounds simply left land in fallow owing to one or more resource constraints. Of the 108 parcels left fallow in the two village sections, 76 were so because of a lack of seeds or labour, six because they were too far from the compound or were of exceptionally low quality (insect holes or perpetually harassed by animals); no answer was given for two others. Only 24 of the 108 were left fallow 'in order to give the land a rest'. The proportion of land left fallow for productive reasons is almost exactly the same for the two villages: 23 per cent for Keur Marie and 22 per cent for Keur Magaye.

The responses by village are presented in Table 7.8. What becomes evident through these responses is that any other constraint to efficient land management (such as tenure insecurity or labour shortages or other input shortages) is overridden by the inability of farmers in the Basin to acquire enough peanut seed. Because the quality of peanut seed stock cannot be maintained with a farmer's own reserves, farmers are dependent on the government or traders for their supplies. Up to and including the 1985–86 season, the amount of government-supplied peanut seed was being curtailed, and private traders had not yet taken up the slack. As seen from the responses above, the ability of a farmer to acquire this scarce input was the primary factor in determining how much land was left fallow.

Table 7.9 Crop rotation schedules

Crop	Keur Marie	% of fields	Keur Magaye	% of fields
No rotation for three years:				
Peanuts	1	1	0	0
Millet	2	2	11	7
No rotation for two years:				
Peanuts	2	2	2	1
Millet	12	12	8	5

Table 7.9 presents statistics on rotation schedules. Data were collected for the three-year period from 1984 to 1986. The sample size was 103 fields for Keur Marie and 151 fields for Keur Magaye. For the most part, an every-other-year rotation schedule between peanuts and millet is closely maintained throughout the sample area. Farmers recognize the benefits to soil quality and hence to productivity. As most compounds within the sample have enough parcels to rotate them between millet and peanuts, the decision not to rotate is surprising. It does appear that, on average, farmers in Keur Marie are not so meticulous about following rotation schedules as those in Keur Magaye. In particular, they are more likely to designate a field as the compound's millet field and cultivate millet there for sequential years.

Contrary to the lack of evidence concerning technical responses to population pressure and soil degradation, there is evidence to support the hypothesis that compounds in the more populated village of Keur Marie are making institutional changes. In response to population pressure and related stress on the environment and the tenure system, the villagers in Keur Marie have begun to express strong reservations about lending land outside the compound, and labour use patterns in the two villages seem to be diverging. What is particularly important is that remittance, salary and extra-agricultural income play a much more important role in Keur Marie than in Keur Magaye. The 22 compounds in Keur Marie earned 2,669,520 CFA francs (31 per cent of income) from remittances or salaries originating in 'Other Senegal' while the 26 compounds in Keur Magaye earned only 539,300 CFA francs (3 per cent of income) from the same sources. These numbers are not quite so significant as they appear, because a large portion (1,800,000 CFA francs) of the amount earned by Keur Marie households was in fact earned by one compound head who worked as a trained bookkeeper in Kaolack. None the less, even if his salary is excluded from the total, remittance and salary income in Keur Marie was still 869,520 CFA francs, or 10 per cent of total income as compared with 3 per cent in Keur Magaye. As for village-based extra-agricultural activities, in Keur Marie 818,000 CFA francs of compound income (or 10 per cent) was gained through service or commercial activities, while in Keur Magaye this sum is 817,725 CFA francs (5 per cent of total income). Income diversification is one possible reaction to a diminished resource base and dwindling agricultural income, and it appears this reaction is taken up by the farmers in Keur Marie.

Intra-compound redistributional responses to the added pressure in Keur Marie also seem to be occurring. The amount of land allocated to insecure managers (who possess merely the right of access to land) is much lower in Keur Marie (10 per cent of village land) than in Keur Magaye (32 per cent). It appears that these insecure managers must also content themselves with the smallest tenurial bundles. In difficult times, for example when compound land becomes scarce or peanut seed becomes difficult to acquire, insecure managers, the majority of whom are women, are usually the first to lose their access to land and seeds. If the compound is poor, women are usually the poorest members; if the compound is rich, women are usually allowed access to money-making projects, notably the growing of peanuts.

It is hypothesized that the probable response of the village communities to the additional pressure of a tax on peanut production or a cut in peanut seed supply would be similar to that observed in the sample data. Because farmers in the Peanut Basin would be constrained in their reactions to the tax or subsidy cut by the lack of technological options, their reaction would probably be to diversify and reduce the agricultural workforce. If opportunities were created in the urban centres or in the rural areas themselves, the results of the forced migration and diversification, as spurred by the tax and peanut seed cut, need not be the impoverishment of the rural populations. But, with no alternative income-generating options, the cost of conservation would be a substantial reduction in compound employment and income. For Keur Marie, a 16 per cent decrease in income is forecast as a result of the value-added tax. This would bring farmers' per capita income down to US$49 a year. Likewise, in the absence of replacement employment and income, a 48 per cent drop in income (to US$30) is predicted for the small farmers in Keur Magaye as a result of the cut in peanut seed supply.

An overall reduction in employment and income is expected to involve a redistribution of intra-compound income, with women and other insecure managers being the first to lose access to agricultural income opportunities. Wives and lesser compound members' access to sustenance, shelter, clothing and income is always difficult to specify and safeguard. Nevertheless, it should not be overlooked that, with the two policies examined above, particularly with the reduction in peanut seed allocation, the economic well-being of insecure managers could be doubly threatened.

The environmental improvement purchased by the tax and subsidy cut would probably be small in the short to medium term. In particular, any change in the albedo effect would be slow to materialize. In addition, though the tax results in a maximum reduction in peanut production of 2.24 hectares in Keur Marie and 8.44 hectares in Keur Magaye, and the seed cut in a reduction of 22 hectares in Keur Marie and 26 hectares in Keur Magaye, the environmental benefits might be only partially realized,

Table 7.10 Input-output calculations for peanut and millet cultivation

Item	Peanuts		Millet	
	Keur Marie	Keur Magaye	Keur Marie	Keur Magaye
Hectares	33.7704	106.9873	83.1033	92.2788
Output per hectare (kg)	871	1,183	343	348
Seed use per hectare (CFA francs)	167	193	3.2	3.6
Pesticide use per hectare (CFA francs)	209	297	1.5	41.4
Fertilizer use per hectare (CFA francs)	88	448	0	454
Animal use per hectare (CFA francs)	6,470	3,505	2,623	2,024
Workdays per hectare	36	26	26	21
Output per workday (kg)	24	45	13[1]	16[1]

1. The millet output in both villages, but particularly Keur Magaye, was inordinately low owing to locust infestation.

because migration and diversification would be the probable response of the rural community. Stress on the environment would be reduced through curtailing peanut production, but this reduction would not be accompanied by ecology-minded upkeep of the land. There is evidence that the regeneration process in the Sahel will not occur without active human assistance (Gritzner, 1988). It will not be enough to abandon the land to natural forces. In this case, any policy which leads to rural out-migration would not necessarily lead to a slowing of the desertification process. A reduction in cultivation would need to be reinforced by conservation measures in order to be environmentally beneficial.

Even where difficulties of irreversibility and causation are overcome, and measurements linking farming intensity and soil degradation are pursued, these measurements are nevertheless of questionable value when analysing a reduction in agricultural activity. For the Peanut Basin, the differences in population densities and cultivated number of hectares per person in the two sample villages suggest that the farming intensity in the two villages might be different (Pingali *et al.*, 1987). Calculating input use per hectare, output per hectare and the value of land services further suggests that this intensification has resulted in soil degradation (see Tables 7.10 and 7.11).

The tables map out a process of intensification similar to the one hypothesized by Boserup (1965). Intensification in the highly populated area of Keur Marie has resulted in greater labour use per hectare and a lower productivity of labour. But, contrary to predictions (Boserup, 1965; Berry and Cline, 1979; Pingali *et al.*, 1987), this has not translated into higher productivity per hectare. It can be argued that the process of degradation is at such a stage that even high levels of land preparation, weeding, etc., cannot compensate for the loss in soil quality. Before this claim is made, though, the figures on fertilizer, pesticide and seed use must be examined. At first glance it appears that the much greater use of fertilizer in Keur Magaye could account for the difference in productivity: but an

Table 7.11 Value of land services per hectare (residual calculations)

Village	Secure fields	Moderately secure fields	Insecure	Borrowed	All fields (average)
Keur Marie	25,880	33,558	70,851	22,719	27,913
Keur Magaye	54,598	65,401	56,743	56,743	56,404

examination of the primary data reveals that the large quantity of fertilizer for Keur Magaye includes a 'gift' of fertilizer to the village chief. The bulk of the 90,000 CFA francs-worth of fertilizer used in Keur Magaye was spread on three fields (two peanut, one millet), and probably could not have skewed productivity results to the extent exhibited above. Pesticide use in Keur Magaye was also higher than in Keur Marie. This result is explained by two observations: (a) the insect problem is greater in Keur Magaye than in Keur Marie and therefore for the same amount of protection more pesticide would have to be used in Keur Magaye; and (b) the government cooperative responsible for pesticide distribution is weaker in Keur Marie than in Keur Magaye. It could be that the managers of the Keur Marie cooperative were unable to secure enough pesticide for their members. This weakness and the inability of the government cooperative to secure enough peanut seed could also help to explain the lower seed ratio in Keur Marie. Again, though, the difference in pesticide and seed use in the two villages is probably not significant enough to translate into such a marked difference in productivity. The difference of 88 CFA francs in pesticide use corresponds to a difference of 35 g per hectare, and the difference in seed use of 26 CFA francs corresponds to about 100 g of shelled peanuts per hectare.

If, given the evidence, the case can be made that intensification in Keur Marie has resulted in soil degradation, the usefulness of such a statement is still nebulous. It has been suggested in the literature that the elasticity of farming intensity on soil degradation is approximately 0.38 (Adelman et al., 1990). Rough calculations using the peanut cultivation data collected from the two sample villages (millet harvest amounts are not useful because of locust infestation) suggest that in the Peanut Basin this number could be as high as 0.49. But, even if the data are stretched to this conclusion, the applicability of such a calculation is questionable. In giving up 19 per cent of their income, farmers would not be repaid with a 49 per cent increase in productivity. Nor would the country. Eventually, considering questions of irreversibility, environmental forces in the area might regenerate themselves to the extent that a 49 per cent increase in productivity could be achieved, but the condition for such an increase would be that the land be left uncultivated. Without an alternative agricultural technique which could capture the increased productivity of the soil, the benefits of environmental regeneration would need to be measured in terms of existence value, option value or discounted possible future use.

The need for diversification out of agriculture is evidenced by the social and economic cost that the intensification of the farming system and soil degradation are imposing on the farmers of the Sahel. In addition to reducing peanut cultivation, the probable result of a tax on agriculture would be to speed up the process of diversification out of agriculture. If the government had established the infrastructure to absorb agricultural labour, the result of taxation need not be a reduction in the rural population's income. Without diversification opportunities, however, the result of taxation would be the further impoverishment of the rural poor and a shift in intra-compound income.

Migration

In the next set of experiments, it is assumed that opportunities are available for rural labour outside agriculture. This assumption permits an investigation of the effects of increased migration on the village economies. Many experts on desertification in the Sahel recommend regional diversification and migration (both national and international) as a necessary step in combating soil degradation and income erosion (e.g. Perrings, 1991).

Migration of two compound members

As a first step in the migration experiment, it is assumed that two people from each compound migrate to Dakar. This number was chosen because on average there are two single or childless adults attached to each compound. With 300 workdays calculated at the agricultural salary of 500 CFA francs per day, these migrants would each make 150,000 CFA francs a year. If they remit 10 per cent (remember that they are marginal compound members), the increase in remittance income to each compound would be 30,000 CFA francs or US$90.

The first entry in Table 7.12 shows that the boost to the village economy of an increase in exogenous large-compound income in Keur Marie is 2.8 times greater than the original injection. In Keur Marie an exogenous transfer to small compounds has 53 per cent more impact on the village

Table 7.12 Overall multipliers for injections of exogenous income

Compound group	Keur Marie	Keur Magaye
Large	2.8	2.0
Medium	3.6	2.7
Small	5.2	2.2

Table 7.13 Compound income multipliers for injections of exogenous income

Compound group	Keur Marie	Keur Magaye
Large	1.6	1.3
Medium	1.8	1.5
Small	2.3	1.3

economy than a comparable transfer to large compounds, while in Keur Magaye medium-sized compounds have the largest overall multiplier. Table 7.13 shows that the same ranking holds for the multipliers for compound income.

In both villages and for every compound size, village production (and hence consumption) of millet has the largest multipliers. As millet is currently the largest village-produced consumption good, it is not surprising that the SAM model predicts that increased income will translate into higher consumption and production of this good. This problem arises because the SAM imposes unitary income elasticity. As millet is not exported by either village in substantial amounts, it is not reasonable to expect that, as compounds in the village get richer, a ceiling on millet production and consumption would not be reached. Compounds would not be interested in growing and consuming ever-increasing amounts of millet. In order to correct for this difficulty, marginal millet consumption values (estimated), as opposed to average consumption values, were inserted in the normalized SAMs and new multiplier matrices were calculated. In calculating the new multipliers, it was assumed that larger percentages of new compound income were used for purchases at the weekly market or in Kaolack. The new compound income multipliers for injections of exogenous income are shown in Table 7.14.

The lower numbers in Table 7.14, when compared with those in Table 7.13, reflect the fact that in the recalculated model, as income grows and millet consumption needs are met, increasing amounts of income are channelled into consumables which are purchased outside the village. The net effect of the increase in remittance income on compound income is shown in item 3 in Table 7.6 above.

In the first step of the migration experiment, it is assumed that neither millet nor peanut hectarage is reduced. Compounds typically strive to supply migrants with the staple crop (if at all possible), meaning that millet production would not fall. In addition, because labour is not a constraint to peanut production (see the discussion of fallow land above), it is

Soil Conservation and Sustainable Development: the Sahel 185

Table 7.14 New compound income multipliers for injections of exogenous income

Compound group	Keur Marie	Keur Magaye
Large	1.2	1.2
Medium	1.3	1.3
Small	1.5	1.2

assumed that, with no change in the supply of peanut seed, peanut production would not fall in response to the migration of two compound members. Instead, the compounds would just increase or reallocate compound labour (or village or imported labour) to any peanut fields that the migrants might have vacated. With this scenario, the net impact of migration would be a reduction in the number of labourers per hectare, thereby further reducing the possibility of labour being expended on intensive agricultural procedures such as weeding, manuring, bunding, etc. Environmentally speaking, the net result of the first step of the migration experiment would probably not be positive.

In the next step of the migration experiment, two more adults migrate from the compound to the capital city of Dakar. This time, each migrant remits 43 per cent of his income to the compound. This means that each compound experiences an increase of 130,000 CFA francs in remittance income. By now, though, it must be assumed that some kind of labour constraint is being reached and that compounds must begin decreasing peanut production. For this stage of the experiment, each compound reduces peanut production by 1 hectare.[5] The net impact of this reduction plus the increase in remittance income is exactly the same as the first stage of the migration experiment as shown in item 3 in Table 7.6. In this case, each compound was forced to reduce peanut production by 1 hectare. This is unlike the case where total village peanut revenue was restricted and individual compound response was determined by the model. For this reason, the peanut revenue reduction was forced through on a per compound basis. The slight adjustments in inter-compound labour and animal exchanges, which would be expected in response to decreased peanut cultivation, are lost in this process.

Even though land is taken out of peanut production, the net effect of further migration on the environment might still be negative. Not only are there fewer labourers per hectare, but the production system is gravitating towards a monocrop system. Millet is becoming the single major crop, thus reducing the possibilities for crop rotation. Successful rotation between millet and fallow is extremely labour intensive, requiring the regular clearing of overgrown fallow fields. In addition, in areas such as the Peanut Basin, where usufruct rights are established only if land has been *mise en valeur*, extensive fallow periods could lead to compound land being repossessed and reallocated by the regional council. Repossession is particularly worrying if the compound population is decreasing, as in the migration experiments. A logical consequence of this fact would be extensively but sparsely planted millet fields. Land would be *mise en valeur* as evidenced by the millet crop, and hence tenure rights would be secured. Accordingly, the net effects of migration and tenure insecurity could prove harmful to the environment.

Moreover, the social cost of the continued migration could prove very high. If two more adults leave each compound, single-parent households (probably headed by women) would be the result. This is a scenario which has been played out in many regions of Africa. The men migrate to job opportunities sometimes hundreds of miles from home, and the women and children remain behind. The increased income generated by the men is vital to the household, and the homestead cannot be abandoned, nor can the staple crop that it produces.

Note that the experiments conducted above have tacitly assumed a zero or very small rate of population growth. With a high population growth rate, the rate of out-migration would have to be even higher to result in reduced cultivation levels.

Regional development

Regional development which leads to the employment of compound members at a nearby regional centre, such as Kaolack in the Peanut Basin, would lead to the same results as detailed in the migration experiment, except that the social costs would not be so high. Nuclear families would not be broken up if compound members could commute to regional jobs. In addition, the 'weekend' hours of commuters could be spent in land maintenance projects such as weeding, bunding, manuring, etc., thereby raising the possibility that regional development, as opposed to migration, could be made environment-friendly.

Reforestation

The establishment or re-establishment of tree stands, shelter belts and forest cover throughout the Sahel is essential for the retardation of the desertification process (Gritzner, 1988; Stiles, 1988). In 1985, 15 of the 20 anti-desertification programmes operating in Senegal involved reforestation or were related to agroforestry. Meanwhile, there continues to be a serious lack of fuelwood and building materials in the Peanut Basin and other densely populated areas (United Nations Environment Programme (UNEP), 1985). The results of plantation schemes have often proved disappointing, and further examination of the policy tools used to encourage tree cultivation is warranted.

Subsidies on tree planting

The complexity of the compound farming system suggests that the targeting of subsidies could prove a difficult task. The compound incentive structure is often splintered and rarely obvious. Inattention to the incentives of individual classes of land managers could serve to dampen the usefulness of subsidies for conservation. For example, suppose that tree planting is found not only to reduce wind erosion and run-off but also to increase yields within ten years of construction. Further, suppose that the government grants a subsidy on seedlings to all compounds in the Peanut Basin. Even then, there is reason to suspect that the incentive structure of the compound is such that tree planting may not be adopted on many compound fields.

The rationale for adopting the planting of trees is that the costs of planting and maintenance are outweighed by the benefits, but if the distribution of costs and benefits is splintered, so that decision-makers share unequally, efficient investment might not be made. In the Peanut Basin, where land is habitually rotated between millet and peanuts, many land managers cultivate different fields each year and are often unsure as to which fields they will cultivate from year to year. Those field managers with insecure or moderately secure tenure rights, who tend to work a different field every year, could have little incentive other than to get as many peanuts as possible out of the land during the year that they manage it. This could particularly be the case of a peanut field that is passed from one wife to the next in successive years. Even where a strict rotation between millet and peanuts is maintained, the incentives of the compound head as millet overseer to make improvements in the land could be diminished by knowing that, in the off-years, a nephew or wife will be using the field for personal peanut cultivation.[6]

It seems that land managers disregard not only the cost of inefficient land management to society in general but also the cost of inefficient land use to their fellow compound members. If this is the case, even a full subsidy on seedlings might be insufficient to convince an insecure field manager to adopt the technology, since labour requirements alone might dissuade him or her from planting trees. Monimart (1989) also raises this point.

We estimated that 71 per cent of village land in Keur Marie and 55 per cent of village land in Keur Magaye was managed by secure field managers. This essentially means that management practices on at least 29 per cent of the land in Keur Marie and 45 per cent of the land in Keur Magaye are unsusceptible to change via subsidies, whether they be subsidies on seedlings, walling material, contouring equipment, or whatever. The same could be said for taxes if there were any dichotomy between the payer of the tax (usually the compound head) and the actual land manager and if taxes were not, or not easily, passed on. It could be that, owing to the complexity of the compound system, a large percentage of land managers in the Sahel are impervious to many fiscal policy instruments.

Usufruct reform

With the 1964 Law of National Domain, close to 98 per cent of all the agricultural land in Senegal became national property. Farmers retain usufruct rights to land as long as they actively cultivate it, i.e. if the land is *mise en valeur*. If, as a policy to encourage tree cultivation, the status of *mise en valeur* were explicitly changed to include the cultivation of specific soil-saving tree species, farmers seeking to solidify their tenure rights would have every incentive to plant trees in fallow fields or even interspersed in cultivated fields. In many traditions, tree planting is a symbol of ownership (Bruce and Fortmann, 1985). Recognition of this fact by the legal system, though not a difficult or costly policy change, could well result in extensive tree planting.

As evidenced through discussions with the farmers in Keur Marie (the more populated of the two sample villages), the introduction of the Law of National Domain has added an element of insecurity to the traditional tenure system. Farmers in this village are nervous about lending land outside the compound for fear of permanent reallocation through the Law. These farmers are looking for opportunities to consolidate their rights over land. In areas with laws similar to Senegal's Law of National Domain, where land registration is not an option, tree planting represents one concrete method of establishing ownership, or at least usufruct rights. If trees such as *Acacia senegal, Acacia mellifera, Commiphora myrrha,* etc., were planted, tenure security could be accomplished, plus any additional income gained from gum or oil extraction, or fruit, nut or bark collection.

If tree cultivation were encouraged through land tenure laws, difficulties anticipated with other policies because of the splintered incentive structure of the compound might be avoided. In order to retain control of compound land, the compound head would have a strong incentive to see that trees were planted on compound land. Other compound members would likewise have an incentive to help to retain compound land. Furthermore, if root rights could be established (along similar lines as those found in some coffee-producing areas), the benefits of harvest (gums, nuts, fruits, barks, leaves) could be consolidated in the actual planter of the tree. This would also encourage the cultivation of trees.

Land tenure reform

Much of the literature on desertification in the Sahel pays attention to the issue of property rights and efficient land or resource use. Three reasons are given to support the argument that, if resource rights could be firmly established and then registered, resource use would become more efficient: (a) it is hypothesized that individualized, registered rights have the property of securely internalizing in a single user the costs and benefits of resource use, with efficient resource management as the result; (b) individualized resource ownership can transform a resource into a commodity, and where resources have been 'commoditized', allocative efficiency will result: the bidder willing to pay the highest price is assumed to put the land to its highest valued use; and (c) private ownership of land (or possibly any resource) could entail mortgage rights, thus permitting farmers to acquire credit, which otherwise might be impossible to secure. Theory states that, when land can be mortgaged, credit will result in productive improvements in the land and a greater use of purchased inputs, the net result being higher long-run yields.

For these reasons, many theorists advocate the privatization of resource rights. None the less, the evidence examined suggests caution in advocating and justifying the expense of any land or resource registration scheme. To date, land tenure studies conducted in Africa (World Bank et al., 1989), including the study from the Peanut Basin (Golan, 1990), suggest that the benefits accredited to private registered land ownership as contrasted with traditional ownership have yet to manifest themselves. It appears that other constraints to efficient land management (environmental, infrastructural, market, etc.) are currently more binding than possible constraints that traditional tenure might pose.

In addition, owing to the complexity of the compound system, caution must be exercised in devising a land or resource privatization scheme. Just as land tenure rights can be depicted as sticks in a bundle, so can rights over any resource. And although the compound head is usually accorded the majority of rights, his rights are nevertheless conditioned and constrained by the rights possessed by other compound members. The mere existence of all the land managers in a compound is an important factor when any change in resource rights is contemplated.

To use land tenure as an example: if an individualized tenure system is contemplated, the question immediately arises: under whose name should the compound land be registered? The most obvious choice is the compound head; his is the name that appears on all administrative lists, from village to cooperative. But, if all compound land is registered in the compound head's name, the primary logic for registration breaks down. The compound head becomes more secure in his control over compound land, but the other field managers are dispossessed of their tenurial rights. The hypothesized link between tenure security and efficient resource management is broken. The compound head would be vested with the legal authority to sell, rent or mortgage land, and the tenurial status of the other compound members would become very insecure. For example, in Senegal, in five of the 11 cases of registered land, the land was repossessed or is subject to court proceedings because of failure to reimburse mortgage loans. In these cases, registration clearly presents a threat to the tenure rights of wives, brothers, sons, etc. Managers who once had secure rights of access to particular fields would lose that security. Even those managers with the general right of access to land would lose their small bundle of tenurial rights. This corresponds to 85 fields that would be less efficiently managed as a result of registration in the name of the compound head.

Another registration option would be to register land in the name of the manager with the most secure rights to that particular piece of land. This would amount to compound heads, with dependent and independent household heads having their traditional rights of access to a field legally recognized. Again, wives and younger compound members would not be legally assured of continued right of access to land, and their incentive structure would thus be weakened. In addition, this detailed registration strategy would pose a number of difficulties. Three of them are listed here.

First, the bureaucratic machinery needed to handle the complicated registration of not just compound heads but other compound members as well may prove difficult to maintain. The Senegalese bureaucracy is still processing registration claims that were filed in the two-year grace period granted by the 1964 Law of National Domain a generation ago. The number of requests for registration was staggering – almost 13,000 – but many of these had already had cadastral surveys, and a generation is a long time.

Second, within the compound, the boundary where one member's tenurial bundle stops and another's begins is often vague. The legal and social battles that could result from a registration effort could be overwhelming. Registration schemes offer a potential of triggering off violent boundary disputes between neighbours and a major source of friction within and between compounds.

Third, to date, Senegal has not experienced land fragmentation problems like those experienced in some countries of eastern Africa. In Senegal, land that remains within the compound can be aggregately managed. Fields can be joined together to support overlapping crops, or fields can be redistributed among field managers to achieve proper crop rotation. Though tenure bundles are individualized within the compound, land use strategies can be organized at the compound level. Private registration could weaken the compound structure and result in a greater degree of land fragmentation.

CONCLUSIONS

The paucity of technical packages that would allow for the continuation of agriculture in the Sahel, while at the same time reversing or at least arresting the desertification process, has led many experts to conclude that migration and regional diversification are necessary to combat soil degradation and income erosion in the Sahel. Even policy which provides incentives for agricultural change would probably result in migration and diversification out of agriculture. Because of the lack of viable agricultural alternatives, it was shown that a tax on peanut production or a cut in the peanut seed supply would probably serve as catalysts to the process of diversification. But, whether pushed via taxes on agriculture or pulled via government schemes to absorb agricultural labour, reduced cultivation caused by migration and diversification need not lead to improvements in the environment unless it is accompanied by agroforestry innovations or other conservation projects. Reductions in the agricultural labour force could actually reduce the possibility for labour-intensive soil conservation (weeding, land preparation, contouring, walling, intercropping, etc.), increase the possibility for millet monocropping, and increase land tenure insecurity because of current usufruct laws. Without complementary conservation programmes, it was concluded that a tax on peanut production and a reduction in the peanut seed subsidy would have minimal environmental benefits but a significant social impact. Without extra-agricultural opportunities, the results of the tax and seed cut would be the further impoverishment of the rural poor and a shift in intra-compound income distribution. With migration and diversification opportunities, the fall in rural income caused by reduced peanut cultivation could be compensated for; but if migration is pursued, the cost to the family structure could be high.

It was also demonstrated that efforts to establish complementary conservation practices and foster the active participation of the rural community are complicated by the complexity of the compound farming system. Because of the diversity in the security of the position of field managers in the compound and the possible splintering of incentive structures, subsidies on conservation projects such as tree planting, walling, contouring, etc., could be dampened by as much as 45 per cent. For this reason, subsidies and taxes must be tailored to the incentives of the individual field managers for maximum success. As an alternative, it may prove more effective to encourage tree cultivation (or other conservation practices) through changes in the current usufruct laws.

The final policy examined concerns the individualization of tenure. Evidence drawn from this study and elsewhere suggests that the benefits accredited to private registered landownership, as against traditional ownership, have yet to manifest themselves. It appears that environmental, infrastructural and market constraints to efficient land management are currently more binding than the constraints that traditional tenure might pose.

The above findings require further qualification. A difficulty with the taxation of peanut production or seed use lies in the mismatch between lever and effect. The environmental predicament addressed by the tax is soil depletion, which is the effect of peanut cropping. Ideally, therefore, it is soil depletion that needs to be taxed, and not the production giving rise to the depletion. The problem is akin to taxing fuel with the aim of diminishing air pollution. Practical considerations concerning the collection of the tax, in these cases, outweigh the theoretical desirability of levying the tax on its target instead of a proxy. However, a price has to be paid for adopting the tax solution that appears the more practical. Peanut cropping is not the exclusive cause of soil depletion. Nor are all methods of peanut farming equally depleting. Furthermore, the direction of technological innovation prompted by an environment-related tax is distorted by the use of a proxy. Unfortunately, this last point is lost, given the short-run (comparative static) nature of the SAM employed.

The short-run (comparative static) orientation of the SAM approach and its focus at the micro level call for a final observation. This approach neglects a dynamic (historic) and macro perspective on soil conservation and the process of desertification. None the less, macrodynamic factors

Soil Conservation and Sustainable Development: the Sahel

such as population growth and low levels of economic development play a crucial role in the failing relationship between environment and economy in the Sahel. Rapid population growth has put great pressures on the fragile land. Poverty, in turn, is hindering technological relief from these pressures, while permitting patterns of exploitation of the land that are subject to increasingly short time-horizons. To make matters worse, the record shows a strong positive correlation between the rate of population growth and poverty, thus making for a vicious cycle. To break this cycle will require policies of a wider focus, in addition to those discussed in the micro context of this paper.

Notes

1. I should like to thank Irma Adelman of the University of California and Ko Doeleman of the University of Newcastle, New South Wales, Australia for their helpful comments and suggestions.
2. Three compounds from the adjoining section of Keur Ismaila are included with the compounds of Keur Magaye.
3. For the relatively poor farmers in the sample villages, the hectarage elasticity could be much lower because of the lack of alternatives to peanut cultivation. It is conceivable that the hectarage response could approach zero and that the only impact of the tax would be a reduction in rural income.
4. At the time of the study, 1,000 CFA francs were equal to over US$3.
5. Peanut production per hectare was averaged at 1,000 kg for this experiment.
6. A more idealistic scenario could be envisaged in which compound members identify strongly with one another and with the compound, so that the costs and benefits of land are essentially distributed equally throughout the group. In this case, even the most insecure managers would have an incentive to maintain the soil quality of the land and make improvements in the land where possible. Even if they did not directly reap all the benefits of such care to the land, they would know that someone whose well-being they valued would benefit. However, this scenario does not appear to be borne out by the facts on land improvement.

8 The Green Revolution, Biotechnology and Environmental Concerns: A Case Study of the Philippines

R. Sathiendrakumar and W.K. Norris

THE GREEN REVOLUTION IN THE PHILIPPINES

Rice is, in terms of tonnage, the dominant crop produced for domestic consumption in the Philippines. In fact, about one-quarter of cultivated land in the Philippines is devoted to rice production.

Rice is grown in all the regions in the Philippines, with the major regions of rice production (over 60,000 metric tonnes per region) being Ilocos, Cagayan Valley, Central Luzon, Southern Tagalog and Ricol in the Luzon area, Western Visayas in the Visayas area and the Central Mindanao area (see Figure 8.1). There are three distinct rice environments: irrigated lowlands, rainfed lowlands and rainfed uplands. Statistics, however, are classified only by irrigated and rainfed areas (both lowlands and highlands). In 1989, 59 per cent of the total rice harvested area was irrigated, and yielded about 69.7 per cent of the total pallay output (pallay is unprocessed rice, known as 'paddy' in some parts of Asia).

The Green Revolution in the Philippines commenced in 1966 with the release of the semi-dwarf high-yielding varieties (HYVs) of rice by the International Rice Research Institute (IRRI), which is located in the Philippines. The Green Revolution aimed at increasing agricultural production in order to overcome the problem of shortages in food supply in the LDCs, which was mainly caused by the pressure of increased population and the scarcity of land suitable for agriculture (Ranadhawa, 1974; Mellor, 1976; Ghatak and Ingersent, 1984; Hayami and Ruttan, 1985). The first Green Revolution in the Philippines was, at least in the short run, of the land-saving type and has helped to overcome the population pressure on the land. The HYVs were rapidly adopted in the Philippines in the late 1960s and early 1970s. This resulted in the average annual growth rate in

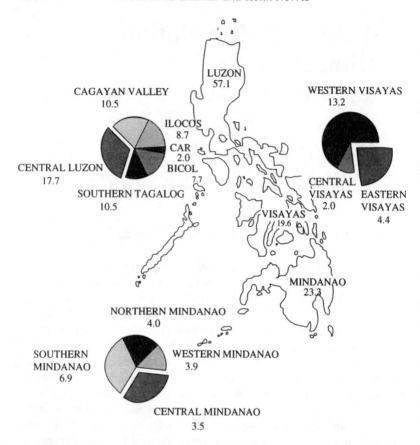

Figure 8.1 Percentage distribution of production by geographical region, calendar year 1980–89 (average). (*Source*: Philippines, Bureau of Agricultural Statistics, 1990).

rice production increasing from 2.2 per cent in the period 1955–65 to 4.5 per cent in the period 1965–80 (Pingali, Moya and Velasco, 1990). During this first period the growth in output (4.5 per cent) exceeded the growth in population (2.9 per cent). The pattern was similar in other south-east Asian countries and in parts of southern Asia (Rosegrant *et al.*, 1985).

The successful adoption of the first Green Revolution in the Philippines and other Asian countries is explained by the gain in yield. However, several authors (e.g. Bhalla and James, 1986) have argued that the benefits therefrom were unevenly spread because complementary inputs were

The Green Revolution in the Philippines

required, such as irrigation, fertilizers, pesticides and insecticides and farm machinery, that have a high capital intensity.

Rice production

Farm income from rice production (Y) is given by the product of acreage (A), yield per hectare (Q) and farmgate price for unprocessed rice or pallay. The price of rice is given, and the limit of cultivable land has been reached. Thus, in order to increase farm income (Y), the only variable under the farmer's control is the productivity of land (Q). If growth in farm income is to be sustained in the long run, the growth in the productivity of land has to be sustained accordingly. The question that has to be posed is whether the increase in rice productivity brought about by the first Green Revolution in the Philippines is sustainable in the long run. This chapter will consider the impact of the use of chemical fertilizers and the prospect of continued yield increase in rice production in the Philippines.

Despite the early increase in productivity in the Philippines in the 1970s, there has been a decline in the rate of growth of productivity since then. During the period 1980–89 the annual average growth rate in rice production was 2.4 per cent (Table 8.1). Of this, 0.1 per cent was due to an increase in the area harvested and 2.3 per cent to an increase in productivity. The rate of population growth during the same period was 2.5 per cent. Thus the growth rate of productivity lagged behind population growth during the 1980s. This was reflected in rice imports in 1988 (181,167 metric tonnes) and in 1989 (219,765 metric tonnes). It appears that the rate of growth of output may have peaked in the Philippines, as it has in other Asian countries (Oram, 1988).

Adoption of HYVs

In 1989, 94 per cent of the irrigated area devoted to rice and 80 per cent of the rainfed area was under HYVs. The annual average increase in the acreage in the irrigated area of HYVs was 3.5 per cent between 1980 and 1989. In the same period there was a decline of 1.1 per cent per annum in the area of HYVs harvested in the rainfed area (see Table 8.2). After accounting for the decline in the extent of rice cultivation in the rainfed area of 2.8 per cent during this period (see Table 8.1), we can see that the proportion of HYV in the rainfed area between 1980 and 1989 actually increased by 1.7 per cent.

Table 8.1 Area harvested, production and yield per hectare for irrigated and rainfed rice in the Philippines, 1980–89

Year	Area harvested (thousands of hectares)			Production (thousands of metric tonnes)			Yield per hectare (metric tonnes)		
	Irrigated	Rainfed	Total	Irrigated	Rainfed	Total	Irrigated	Rainfed	Total
1980	1,609	1,862	3,471	4,507	3,140	7,646	2.8	1.69	2.2
1981	1,656	1,763	3,419	4,788	3,123	7,911	2.89	1.77	2.31
1982	1,741	1,610	3,351	5,343	2,990	8,334	3.07	1.86	2.49
1983	1,668	1,387	3,054	4,888	2,406	7,295	2.93	1.74	2.39
1984	1,755	1,408	3,162	5,136	2,993	7,829	2.93	1.91	2.48
1985	1,838	1,469	3,306	5,821	2,985	8,806	3.17	2.03	2.66
1986	1,878	1,586	3,464	5,980	3,267	9,247	3.18	2.06	2.67
1987	1,852	1,404	3,256	5,809	2,731	8,540	3.14	1.94	2.62
1988	1,956	1,437	3,393	6,106	2,865	8,971	3.12	1.99	2.64
1989	2,064	1,434	3,497	6,592	2,867	9,459	3.19	2.00	2.70
Annual average increase between 1980 and 1989	2.81	−2.86	0.08	4.31	−1.01	2.39	1.46	1.89	2.30

Source: Philippines, Bureau of Agricultural Statistics (1990).

Table 8.2 Production, yield per hectare and area under HYV and traditional varieties of rice in the Philippines, 1980–89

	High-yielding varieties (HYV)							Traditional varieties (TV)						
	Irrigated			Rainfed				Irrigated			Rainfed			
Year	Area (thousands of hectares)	Production (thousands of metric tonnes)	Yield per hectare (metric tonnes)	Area (thousands of hectares)	Production (thousands of metric tonnes)	Yield per hectare (metric tonnes)		Area (thousands of hectares)	Production (thousands of metric tonnes)	Yield per hectare (metric tonnes)		Area (thousands of hectares)	Production (thousands of metric tonnes)	Yield per hectare (metric tonnes)
1980	1,418	4,107	2.90	1,278	2,414	1.89		191	400	2.10		584	726	1.24
1981	1,478	4,397	2.97	1,237	2,466	1.99		178	391	2.20		527	656	1.25
1982	1,611	5,048	3.13	1,215	2,491	2.05		130	296	2.27		395	500	1.27
1983	1,537	4,570	2.97	1,048	1,980	1.89		130	318	2.44		339	427	1.26
1984	1,634	4,859	2.97	1,119	2,330	2.08		121	277	2.29		288	363	1.26
1985	1,714	5,495	3.21	1,173	2,577	2.20		123	326	2.64		296	407	1.38
1986	1,754	5,665	3.23	1,268	2,797	2.21		125	315	2.53		318	470	1.48
1987	1,713	5,465	3.19	1,095	2,268	2.07		139	344	2.48		310	463	1.50
1988	1,819	5,755	3.16	1,135	2,398	2.11		137	351	2.56		301	467	1.55
1989	1,939	6,264	3.23	1,153	2,438	2.12		125	328	2.62		281	429	1.53
Annual average increase between 1980 and 1989	3.54	4.80	1.20	−1.14	0	1.28		−4.6	−2.18	2.49		−7.81	−5.8	2.36

Source: Philippines, Bureau of Agricultural Statistics (1990).

Adoption rates of HYVs are now very high. Given the scarcity of land, it follows that significant further increases in yields will be possible only if either the area under irrigation is increased, or new HYVs such as hybrid varieties with greater yield potential are developed, or better management or agronomic practices are adopted. These three possibilities are discussed below.

Increasing the area under irrigation will increase the rice yields per hectare (Table 8.1) and therefore will increase total rice production in the Philippines. On the other hand, developing a small-scale fully developed irrigation system in 1985 was expected to cost around US$12,000–15,000 per hectare (Hrabovszky, 1985). The aggregate lending and assistance for irrigation investment in south-east Asia by the four main financial donors for irrigation development – the World Bank, the Asian Development Bank (ADB), the United States Agency for International Development (USAID) and Japan's Overseas Economic Corporation Fund (OECF) – has been declining in real terms. The decline was from US$648 million (at 1980 prices) in 1977–79 to US$211 million (at 1980 prices) in 1986–87 (Rosegrant and Pingali, 1991).

The recent damage caused by natural calamities such as the earthquake in 1990 and the volcanic eruptions of 1991, coupled with the fall in real per capita income, has further exacerbated the increasing public and foreign debt carried by the Philippines. This rise in public and foreign debt, together with concerns about the environmental implications of irrigation projects and the fall in real world rice prices, has aggravated the decline in lending to the Philippines by the major donors (Rosegrant and Pingali, 1991).

A promise of improvement in yields is based on breakthroughs in genetic engineering research in rice breeding, which screens rice varieties for characteristics such as high yields, high resistance to pests and diseases, high responsiveness to high-input agricultural systems and high stability. Research aims to identify the genes responsible for these characteristics and to transfer them into one plant. Work in the Philippines on the development of hybrid rice varieties commenced about ten years ago and has been undertaken by the IRRI together with the Philippines Rice Institute. According to the breeders at the IRRI it will take another four to five years to get tangible results.

A further possibility may be to transfer the symbiotic nitrogen-fixing capacity of leguminous crops onto rice. For this to be a success, the new variety should have higher energy production capabilities than the existing varieties, since a part of the energy from the rice plant is utilized by the bacteria to fix nitrogen. In other words, the energy requirements for production and nitrogen fixation are greater than for production alone, so that after nitrogen fixation the yield of the new variety may be the same as or less than that of the old one. The success of this form of genetic engineering will depend on the reduced cost of production of the new variety in terms of lower fertilizer requirements. The IRRI is not working on this aspect at the moment.

Soil nitrogen continues to be the major source of nitrogen in many rice-producing areas. Flooding the ricefields favours the rice environment by increasing the availability of nutrients, particularly phosphorus, iron and manganese. It also maintains the soil nitrogen fertility as well as stimulating nitrogen fixation. Soil nitrogen is partly replenished because the rice plant has a self-nourishing effect. Thus a part of the soil nitrogen is recovered by biological nitrogen fixation (Watanabe, 1991). None the less, the increase in high yields brought about by the adoption of HYVs in the Philippines also increases the depletion of nitrogen from the soil if proper supplements of nutrients are not made. These supplements are available in the form of chemical fertilizers or biofertilizers. The biofertilizers may be either of the symbiotic type or of the non-symbiotic or free-living type. The symbiotic type consists of a specific plant species and a bacterium. These two together fix atmospheric nitrogen in a form assimilable and utilizable by the plant. The plant component (e.g. azolla-anabena and legume-rizobium) is later used in the ricefield as green manure or compost.

In the case of a non-symbiotic or free-living type of biofertilizer, the nitrogen-fixing organisms form a loose association with the crop plant. These organisms are found within the habitat or within the root zone of the rice plant. The organism fixes nitrogen from the atmosphere and excretes nitrogenous compounds which are utilizable by the rice plant. Examples of these organisms are the blue-green algae, rhizosphere bacteria such as azoroporillum, and clostridium.

Finally, better management or agronomic practices will not only reduce the yield gap between research stations and the farmers' fields but also stem the stagnation or decline in yield potential under intensive cultivation of rice, especially in the irrigated areas. A survey carried out in Laguna and Central Luzon showed that the current yield gap between the average farmer and the experimental stations is 1.2 metric tonnes per hectare (Moya and Pingali, 1989). Furthermore, the high yields obtained from the long-term fertility trials at the IRRI are exhibiting long-term decline (Rosegrant and Pingali, 1991). Therefore, out of the three approaches discussed above, that of better management or agronomic practices has the greater scope for increasing the yield potential in the short or medium run.

The labour market impact of the Green Revolution

The adoption of new technologies, such as HYVs, has implications for the labour market. Both the level and the distribution of employment and earnings are affected.

It is not possible, a priori, to predict the impact of a technological change on the level of employment in the sector directly affected, or in other sectors. Consider the impact of the adoption of HYVs within the rice-farming sector. Three effects of the new technology on employment may be considered: (a) whether the new technology is labour-saving-biased or not, i.e. whether substitution of capital and fertilizers for labour occurs; (b) whether the decline in total cost that results from an increase in labour productivity is a cause for concern, i.e. the extent to which the same output can be produced with less labour; and (c) whether consideration must be given to the increases in demand for rice that result from the fall in its real price consequent on the reductions in cost. The combined direct outcome of these effects is uncertain within rice farming. Moreover, like any other technological change, HYVs, by influencing the demand for other inputs, will affect employment elsewhere. These indirect spillover effects will largely be ignored here, although the evidence suggests that they may be significant in scale compared with the direct effect (Balisacan, 1991).

A substantial number of case studies on the employment effects of the Green Revolution have now been undertaken. In the case of the Philippines, the results of three studies can be cited. The first considered the employment of hired labour at 62 farms between 1966 and 1975. This was the main period during which HYVs were being adopted. On HYV farms, average hired labour per hectare averaged 85 workdays per year, whereas on farms cultivating traditional varieties employment averaged 51 workdays per year (Barker and Cordover, 1978).

The increase in the demand for labour inputs largely arose from the additional labour needed for seeding, weed control, the application of fertilizer and pesticides and additional harvesting. The amount of labour needed for land preparation declined, offsetting about one-half of the extra labour needed for crop care because of an increased use of tractors. It has been a matter of some debate whether increased mechanization was a result of the use of HYVs *per se* or whether it was an independent change. The conclusion drawn from this debate is that the use of tractors was not a direct result of the adoption of HYVs and thus the technology was not labour-saving-biased (Griffin, 1978).

The Green Revolution in the Philippines 203

A second study, also undertaken under the aegis of the IRRI, is based on an analysis of daily farm records between 1975 and 1980 at two locations, Iloilo (Western Visayas) and Pangasinan (Central Luzon) (Barlow, Jayasuriya and Price, 1983). The project was based on a linear programming model. The objective function to be maximized was the cash surplus of farmers, with the need to meet subsistence requirements appearing as a constraint. The new varieties permitted either a second rice crop each year or a second non-rice crop. Simultaneously, some increased use of mechanization occurred. The net effect on the demand for labour, both family and hired, was uniformly positive on all Iloilo sites, but negative in three of the five Pangasinan sites where tractors and threshers were introduced. As argued above, the increasing use of the latter is not directly attributable to the introduction of HYVs.

One finding of this case study was that the adoption of HYVs leads to some reduction in peak demands for labour. The precise way in which the demand for labour varies over the year depends upon which combination of crops is grown, but a smoothing of labour demands was generally found, although in some cases a second peak demand for labour is introduced. Work undertaken by Price and Barker (1978) has also concluded that more labour is employed in the cultivation of HYVs than, *ceteris paribus*, in traditional varieties, and that there is a levelling of seasonal fluctuations in the demand for labour. Thus the net employment effect of the adoption of HYVs within rice cultivation has been to lead to a higher level of employment.

The effect on real wages is less clear cut. The real wage in rice farming, i.e. the money wage paid to labourers deflated by the Consumer Price Index, has fallen over the whole period since the introduction of HYVs and, in 1990, was about 70 per cent of its level in 1966. While no econometric analysis has been undertaken, it seems certain that in the absence of the Green Revolution the real wage would have fallen further. The major cause of the decline in the real wage was the failure of the demand for labour to keep pace with the rate of growth of the rural labour supply. As has been shown, the introduction of HYV led, *ceteris paribus*, to an increase in the demand for labour and hence will have ameliorated the decline in the real wage. It should be noted that real wages in the same period fell not only in agriculture: real wages of both skilled and unskilled workers in the rest of the economy are also lower than they were in 1966.

Finally, it can be said that the distribution of the wage gains from the Green Revolution may have been regressive, at least in the beginning. There was a tendency for the adoption of HYVs to have occurred earlier on large farms. However, the rate of diffusion of HYVs was so rapid and the lags in adoption between large and small farms so small that any distributional changes will have been transitory (Barlow, Jayasuriya and Price, 1983; Balisacan, 1991). In general, the adoption of HYVs has been impeded more by geographical factors – mainly the availability of irrigated land – than by institutional factors. The Green Revolution has contributed, therefore, to increases in wage differentials between regions by raising the demand for labour to a greater extent on irrigated land. However, regional wage differentials did not persist. This issue has been investigated by Otsuka, Cordover and David (1990). They find that in the Philippines, unlike the situation in India for example, the adoption of HYVs gave rise to interregional migration flows which tended to equalize wage rates.

CHEMICAL FERTILIZER USAGE AND THE ASSOCIATED ENVIRONMENTAL PROBLEMS

Fertilizer usage in rice cultivation

Rice producers are now the major users of fertilizers in the Philippines. HYVs need more fertilizer and have thus brought about a substantial increase in consumption, as can be seen from Table 8.3. Consumption increased no less than sevenfold between the early 1950s and the late 1970s. This growth in fertilizer use was due entirely to increases in the use of chemical fertilizers. Nitrogen is the major nutrient component of the

Table 8.3 Consumption of chemical fertilizer nutrients (N, P_2O_5 and K_2O) in Philippine rice production for four different periods

Period	Consumption (thousands of metric tonnes)
1950/51–1954/55	36.7
1960/61–1964/65	89.7
1970/71–1974/75	234.5
1975/76–1979/80	275.7

Source: Based on Rosegrant and Roumasset (1988), Table 1.

fertilizer used in rice production. The responsiveness of rice to the application of nitrogen depends on the prevailing production environment. Notably, the responsiveness of HYVs to nitrogen is heavily dependent upon the availability of complementary inputs such as water. The timely availability of water increases the profitability from higher rates of nitrogen application. This explains the higher usage of urea in irrigated rice-growing areas as opposed to rainfed areas (Table 8.4).

Ammonium sulphate, which was the major source of nitrogenous fertilizer in the 1950s and 1960s, was replaced by urea from the 1970s onwards. The amount of nitrogen available from urea is double the amount available from ammonium sulphate.[1] From the 1970s onwards the price of nitrogen obtained from urea was lower than that of nitrogen obtained from ammonium sulphate. The relative cheapness of urea compared with ammonium sulphate in terms of its nitrogen content encouraged the government to import urea and the farmers to use it (Table 8.5). The entire supply of urea in the Philippines is met by imports. Since the mid-1970s, controls have been used to restrict the amount of urea imported, and the balance of the nitrogen requirement was met by the use of domestically produced ammonium sulphate. Import controls on urea were aimed at protecting local producers of ammonium sulphate who continued to operate inefficient and obsolete plants (Balisacan, 1989a). In the 1980s about 80–90 per cent of the ammonium sulphate used in the Philippines was imported, the remainder being produced by the Philippine Phosphate Fertilizer Corporation (Philphos).

During the 1960s and 1970s, local fertilizer manufacturers had gradually shifted from production to importing. These local firms, with the exception of Philphos, ceased production completely in 1984 shortly after the government was forced to withdraw cash subsidies granted to domestic producers to encourage import substitution (Balisacan, 1990). The merit of encouraging domestic production of fertilizer in the Philippines is not clear, as it is hard to envisage the country developing a comparative advantage in its production. The objective of fertilizer policy in the Philippines since the 1950s has been first to provide low-priced fertilizers to farmers and second to protect domestic fertilizer manufacturers (Balisacan, 1989a). These two objectives are clearly conflicting, and in the case of the Philippines the major beneficiaries have been the local fertilizer manufacturers.

Since the 1970s rice crops have been the single most important user of chemical fertilizer in the Philippines, where fertilizer adoption in rice is now synonymous with the adoption of the modern-variety–fertilizer–irrigation technology, because of the high complementarity between

Table 8.4 Inorganic fertilizer usage for rice (irrigated – rainfed), 1988 and 1989

Fertilizer used (quantity and value)	1988			1989		
	Irrigated	Rainfed	Total	Irrigated	Rainfed	Total
Total fertilizer (metric tonnes)	306,238	118,072	424,310	340,381	154,673	495,054
Average per hectare (kg)	191.5	142	174.5	183.5	168	178.5
Urea (metric tonnes)	178,121	58,588	236,709	187,295	76,591	263,886
Average per hectare (kg)	111.5	70.5	97.5	101	83.5	95
Ammonium sulphate (metric tonnes)	14,189	10,321	24,510	23,178	17,262	40,440
Average per hectare (kg)	9	12.5	10	12.5	19	14.5
Import price of urea (cost and freight) (US$)	–	–	157.64	–	–	139.08
Import price of ammonium sulphate (cost and freight) (US$ per metric tonne)	–	–	94.75	–	–	103.05
Value of urea used at imported price (US$ million)	–	–	37.31	–	–	27.19
Value of ammonium sulphate used at current price (US$ million)	–	–	2.32	–	–	4.17

Source: Philippines, Bureau of Agricultural Statistics (1990).

Table 8.5 Relative prices of nitrogen from urea and ammonia sulphate, 1980–89
(pesos per metric tonne)

	1980	1981	1982	1983	1984	1985	1986	1987	1988	1989
Retail price of urea	2,221	2,612	2,575	3,434	5,546	5,265	3,137	3,009	3,937	4,154
Price of nitrogen from urea	4,935	5,805	5,722	7,632	12,324	11,701	6,971	6,686	8,748	9,230
Retail price of ammonium sulphate	1,697	1,987	1,930	2,404	3,124	3,096	2,284	2,274	2,697	2,958
Price of nitrogen from ammonia sulphate	8,079	9,461	9,192	11,448	14,876	14,745	10,875	10,831	12,842	14,084

Source: Philippines, Bureau of Agricultural Statistics (1990).

fertilizer, modern rice varieties and irrigation (Hayami and Ruttan, 1985). The application of fertilizer to rice crops in the period between the mid-1960s and the early 1980s increased total rice production by approximately a third (Balisacan, 1989a). Over the same period, the use of fertilizer grew by a factor of 3.

Price and income effect of fertilizer policies

Although one avowed aim of the fertilizer policy was to lower prices for farmers, they in fact pay more than the world price for fertilizer. In the 1960s and early 1970s the average implicit tariff on fertilizers was 36 per cent. Between 1972 and 1982, with the introduction of cash subsidies to fertilizer importers, the average implicit tariff was reduced to 20 per cent. The annual average implicit tariff between 1960 and 1987 was estimated to be 26 per cent.

The responsiveness of fertilizer application depends on the price elasticity of demand for fertilizers, which has been estimated to be around -0.5 per cent for short-term period and small price changes (less than 30 per cent) and around -0.8 per cent for large price changes (Balisacan, 1989a). Using these elasticities and the annual average implicit tariff on fertilizer, Balisacan estimates the consequent average annual decline in real farm income in the Philippines between 1960 and 1987 at 14.4 pesos per hectare for irrigated rice, 9.3 pesos per hectare for rainfed rice and 3.3 pesos per hectare for upland rice.

Fertilizer policy in the past has given positive protection to fertilizer manufacturers and importers. Between 1960 and 1987, the average nominal protection rate (implicit tariff plus cash subsidy) of 42 per cent was much higher than the legislated implicit tariffs of 16 per cent in the 1960s and 35 per cent in the 1970s and mid-1980s. The implicit tariff together with the cash subsidy has enabled the fertilizer importers to earn quasi rents (i.e. the amount of revenue received by a firm over and above what is needed to cover its variable cost) (Balisacan, 1990).

For the period 1960–86, the losses to farmers from artificially high fertilizer prices is estimated to have averaged about 228 million pesos per year. These losses were partly offset by gains by local manufacturers. Thus in the Philippines up to 1987, in contrast with many developing countries, much of the fertilizer subsidy did not accrue to the farmers but benefited the fertilizer importers and manufacturers. 'The prices paid by farmers for their fertilizers were, by and large, higher than what these prices would have been in the absence of regulatory policies on international trade and domestic distribution' (Balisacan, 1989b, p. 35).

Furthermore, there are other factors, such as domestic distribution costs, the exchange rate and trade policies, that have an influence on the price of fertilizer to the farmer. Improving the effectiveness of total distribution is one example: the distribution costs represent about 40 per cent of the retail cost of fertilizer in the Philippines.

Environmental problems

It has been argued (e.g. in Markandya, 1991) that the increase in productivity achieved by the widespread use of fertilizers and pesticides has resulted in environmental damage. The intensive use of fertilizer has possible adverse long-term effects on soil structure, crop productivity and off-farm pollution. Off-farm pollution caused by nitrogenous fertilisers takes two forms: pollution of run-off water, and ammonia volatilization (Smiddle, 1972; Blackmer, 1988). Ammonia volatilization releases a substantial part of the fertilizer nitrogen into the atmosphere (30–40 per cent). The factors responsible for the amount of nitrogen lost from a field by run-off or leaching are the quantities of water that move through the surface soils and the amount of nitrates available when the movement of water takes place. In tropical soil in which the absorptive and nutrient-fixing capacity of the clay content is low, the soil leaching may equal that removed by crops and may contribute to the pollution of shallow aquifers which are also used for drinking purposes (Tivy, 1990).

The extensive use of fertilizer in irrigated rice production does not lead to an increase in the salinity of the rice-fields, because rice is grown in flooded conditions. The evidence from the experimental rice-fields of the IRRI which have been treated with large amounts of inorganic fertilizers over the past 15 to 20 years supports this. However, salinity may be a problem in the upland rainfed areas. Unfortunately, no systematic study has been undertaken in the Philippines on this subject.

The contamination of groundwater by agricultural chemicals, including inorganic fertilizers, is of potential concern to public health and welfare authorities. The leaching of nitrates from the fertilized field to groundwater is relatively slow, and much of the reservoir of nutrients now moving through the soil may not yet have reached the groundwater. The extent of nitrate leaching is thought to be directly proportional to the level of nitrogen input. Studies show that input is not sufficiently deterred by price and that farmers tend to respond to decreased fertilizer efficiency by applying more of it (Schepers and Hay, 1988), thereby adding to nitrate leaching.

In the United States it has been shown that nitrogenous chemical fertilizers create negative externalities in the form of nitrate leaching into the groundwater (D'Itri and Wolfson, 1988). For example, in Iowa nitrates and pesticides used by the farmers are found in many groundwater aquifers. In the shallow aquifers especially, nitrate concentrations are found to be above the maximum contamination limit of 10 mg per litre of nitrate for public drinking-water. In rural areas in the Philippines, groundwater for domestic consumption is drawn mostly from shallow wells. The highest nitrate concentration recorded by the Soils and Water Science Division of the IRRI is 2.8 parts per million (ppm). Water with 10 ppm of nitrate is considered unsafe for human consumption. While these findings appear reassuring, nitrate readings are expected to go up with the passing of time. Discussions with the officers in the Soil and Water Science Division of the IRRI confirm a concern for the safety of drinking-water.

In the rainfed areas, where rice is not grown under flooded conditions, the extensive use of chemical fertilizers may contribute to the pollution of rivers and lakes in the Philippines, thereby creating external costs for users. These external costs establish a case for the examination of the use of biofertilizers as an alternative to chemical fertilizers.

In addition to its adverse environmental impact, the use of inorganic fertilizer raises an economic concern. Fertilizer is mainly imported and the country is constrained by balance of payments problems. Moreover, efficiency in the application of nitrogen fertilizer is fairly low. At present 30–40 per cent of the nitrogen applied is lost to the upper atmosphere. In absolute terms this amounts to a loss of about 44,000 metric tonnes of nitrogen from urea and 3,400 metric tonnes of nitrogen from ammonium sulphate at the 1989 level of application. It has been estimated that half this loss could be eliminated by incorporating the fertilizer in the soil instead of the present practice of broadcasting it. We have calculated at about US$8.1 million the amount of foreign exchange that could thus be saved. Given the difficult economic situation facing the Philippines, it may become impossible to afford the use of inorganic nitrogenous fertilizers on the present scale.

Unemployment, underemployment and rural poverty caused by the use of chemical fertilizers

Almost 200 years ago, Malthus raised the potential conflict of rising population growth rates with limited resources. Malthus's proposition may be quite pertinent to the Philippines. The population growth rate in the recent past has been around 2.5 per cent per annum. The high growth in population coupled with the possible soil degradation brought about by the use of large amounts of chemical fertilizers are, at least in part, responsible for the increase in poverty in the rural and the urban sectors. The stagnation in the growth in rice yields in the irrigated areas and the reduction in rice yields in the rainfed areas during the past five years may be associated with soil exhaustion or degradation following the continuous use of chemical fertilizers.

Agricultural lands are a scarce resource in the Philippines and they should be used in a manner which is capable of sustainable high production rates. For this reason, we suggest that the usage of chemical fertilizer needs to be kept to a minimum and should be complemented with the use of biofertilizer. Failing sustainable production, reduced rice yields coupled with the limited agricultural land available for rice cultivation will contribute to increased poverty in the rural agriculture sector and encourage rural–urban migration in the future.

ALTERNATIVES TO CHEMICAL FERTILIZERS

Economic growth in the Philippines is constrained by the level of public and foreign debt. At the same time, economic development is required to reduce the problems of poverty and starvation among a population that continues to grow quite rapidly. This scenario requires the Filipinos to strive for sustainable growth in agricultural production with reduced dependence on imported inputs such as urea. Sustainable development 'involves maximizing the net benefits of economic development, subject to maintaining the services and quantity of natural resources over time' (Pearce and Turner, 1990, p. 24).

The impact of chemical fertilizers in the cultivation of HYVs of rice in the Philippines may no longer be sustainable in the long run. The rice field is an asset. The immediate objective is to maximize the output from this asset. The continuous use of chemical fertilizers, in addition to contributing to the problem of pollution, has long-term adverse effects on the soil structure. If adequate steps are not taken to correct this problem, the soil quality is at risk. Given the scarcity of land in the Philippines, it is imperative that the productivity of rice lands be increased to overcome the problem of poverty. Yet the desire to increase rice production in the short run by the use of inorganic fertilizers may lead to its long-run decline. Under the circumstances, biotechnology has the potential to reconcile the conflict between the short-run growth in rice production and the long-run depreciation of the rice lands, i.e. developments in biotechnology may improve the productivity of rice production on an ecologically sustainable basis and offer farmers alternatives (Dorfman, 1985).

Biotechnology and rice production

To offer rice farmers alternatives to the use of inorganic fertilizer, genetic engineering can be used in different ways. One way, as we have seen, is to incorporate nitrogen-fixing genes into the rice crop in order to render it less dependent on chemical fertilizers. Another way is to develop crop varieties that are suited for lands of low fertility or lands of high salinity. The aim is to cultivate those marginal lands which previously were not suitable for rice production. Furthermore, genetic engineering seeks to increase productivity by improvements in the resistance of rice to drought and pests. In contrast with some technologies such as mechanization, which are characterized by lumpiness in the use of capital, the adoption of the varieties developed by genetic engineering is scale-neutral (Hayami, 1984) and will not therefore be monopolized by the large commercial farms. The Philippine experience confirms this hypothesis of scale neutrality in the use of HYVs. Ruttan (1977), after reviewing the adoption of HYVs in other parts of Asia, also confirmed their scale neutrality.

One of the problems encountered when genetic engineering attempts to incorporate nitrogen-fixing genes into the rice plant is the energy requirement of the fixation of nitrogen by bacteria. The bacteria obtain the energy from the rice plant. Therefore, part of the energy that could have contributed to rice yield may have to be sacrificed for the sake of nitrogen fixation. The current outlook is that genetic engineering research aimed at protecting the environment and at the same time increasing rural income will need a long period for development and require considerable research effort and substantial expenditure. The Philippines lacks the scientific and financial resources to support genetic research on a significant national scale.

An application of biotechnology not involving genetic research concerns the use of biofertilizers in rice cultivation. HYVs of rice deplete the soil of its stock of nitrogen reserves. Biofertilizers, through a microbial process, help to convert atmospheric nitrogen into a plant-useable form. One technique is to supply biofertilizer to the rice plant by means of the application of available green manure or compost. Another method is to intercrop rice with nitrogen-fixing crops such as legumes, or to follow a cropping pattern for the rice field which includes the growing of legumes.

Biofertilizers are unlikely to replace completely the inorganic nitrogenous fertilizer in use at present because the nitrogen requirements of HYVs exceed those currently supplied by, for example, legumes (Watanabe, 1991). The development of rice production systems which enable biofertilizers to act as a complement to inorganic nitrogenous fertilizer are considered more promising.

Green manures as biofertilizers

The development of HYVs which are early maturing (100 days) has made double cropping of rice possible both on irrigated land and in the rainfed, partially irrigated lowland areas in the Philippines. Typically, one crop of rice is grown between June/July and September/October while the second crop is grown between October and January. In rainfed areas one crop of rice is grown between June/July and September/October. Cultivation patterns allow the rice lands to be used for the cultivation of a green manure crop whenever the land is not occupied by rice. These crops need to have a very low water requirement in order to survive the dry period when rice cannot be grown. In the case of rainfed, partially irrigated lowlands, these crops should also have a short growing season to fit in between the two rice crops.

Green manuring, in addition to providing nutrients to the rice soil, also helps in improving the soil structure. Research has revealed the benefits of green manuring in increasing rice yields and improving the sustainability of rice production (IRRI, 1988). The IRRI has concentrated on four green manures for use in the Philippines: azolla, sesbania, indigo and leguminous cash crops. In each case there are circumstances and factors that have thus far inhibited the adoption of green manures on a major scale.

Azolla

Azolla has been used as a source of nitrogen for the rice plant under the irrigated conditions in Vietnam and China (Rosegrant and Roumasset, 1988). It is an aquatic fern that lives in symbiosis with a nitrogen-fixing blue-green algae and can be grown separately and then incorporated into the rice field or, alternatively, grown as an intercrop with rice. Because azolla growth is dependent on the availability of water, it is useful as an intercrop only in irrigated rice fields.

Studies conducted at the IRRI have shown that azolla grown separately as a green manure is not cost efficient. The major reason for this is the opportunity cost of the land, accounting for about 62 per cent of the cost of azolla cultivation (Rosegrant and Roumasset, 1988). Also, azolla grown separately needs considerable extra labour to transport the green manure to the rice field.

The viability of azolla grown as an intercrop depends, in part, on the alternative source of nitrogen that is used. As has been shown, fertilizer policy in the Philippines has maintained the domestic price of nitrogen obtained from ammonium sulphate above that obtained from urea. Rosegrant *et al.* (1985) found that the cost of nitrogen from ammonium sulphate was in fact about 60 per cent higher. In most environments, azolla is competitive with ammonium sulphate. The cost of nitrogen from azolla is competitive with that from urea only in very well irrigated areas. However, the external environmental costs of urea or inorganic fertilizer are not considered in this comparison. While this remains the case, the view among agronomists appears to be that azolla does not offer serious possibilities in the Philippines. Apart from its irrigation requirements, azolla needs phosphorus-abundant soils that are rare in the Philippines. It has been concluded that azolla is only viable in mixed fish/rice cultivation, which is also rare in the Philippines.

Sesbania

A second plant that can act as a biofertilizer is sesbania. This plant overcomes a drawback in the use of green manure, namely that whenever fertilizer is applied to the green manure, the bacteria in the root nodules of the green manure stops fixing nitrogen from the atmosphere. The drawback is particularly acute in irrigated lowlands where two rice crops per year are grown. However stem-nodulating green manure species such as sesbania, which is a pre-rice green manure, can overcome this problem. Sesbania, because of its stem-nodulating character, is able to fix nitrogen from the atmosphere even if the soil contains high levels of nitrogenous fertilizer. The bacteria are not in contact with the soil and are able to fix atmospheric nitrogen even in the presence of high soil nitrogen. Sesbania is broadcast with the dry season rice crop and then it is ploughed under before the wet season crop. In the Philippines a locally designed, animal-drawn implement shows promise for incorporating the green manure (Garrity, 1990).

Sesbania is grown as a green manure in several Asian countries, notably India and China. To date, the area cultivated in the Philippines is negligible. Experience from other countries suggest that the use of sesbania as a green manure increases labour demand (compared with demand when inorganic fertilizers alone are applied). Where sesbania is grown in the same fields as rice, the additional demand for labour inputs arises from seeding and from the incorporation of the green manure into the soil. The authors are not aware of estimates of the labour involved in seeding. Nor is there a firm guide to the labour required to incorporate sesbania into the soil. Garrity and Flinn (1987, p. 118) concluded that 'the man-animal time required to incorporate (sesbania) varies ... Sesbania incorporation increases the time spent in land preparation for transplanted rice in India from 20 per cent to 50 per cent. The additional effort in northern Bangladesh was reported to be 100 per cent.'

Sesbania is not a green manure that is likely to be widely used in any country. It succeeds only in related environmental niches. This apart, the main obstacle to its cultivation is the availability of seed. Thus the existence of a domestic seed industry is a necessary condition to be met before the negligible use of sesbania in the Philippines can expand.

Indigo

The third of the green manures tried in the Philippines is indigo. It is estimated that about 10,000 hectares of land are devoted to rice/indigo cultivation. This is a trivial amount when measured against the approximately 3 million hectares under rice cultivation. On the other hand, the practice of rice/indigo cultivation has been independently adopted by farmers. Recently, research and agricultural institutes have developed an interest in this green manure. The use of indigo as a source of biofertilizer is restricted to rainfed lowlands. The small area currently cultivated is in north-west Luzon.

Indigo fits well with the rainfed or partially irrigated rice-based cropping system. It is planted at the end of the wet season (post-rice crop) as an intercrop with many other upland food and cash crops, including maize, mungbean and tobacco. As a rule, indigo is grown after the harvest of the dominant upland crop, and then is ploughed under by animal power during land preparation for the wet season rice crop in July. Owing to its slow initial growth, indigo is not suitable as a short duration pre-rice green manure. Therefore, it is not used for cultivation in irrigated rice fields.

According to Garrity *et al.* (1989), yield losses are not observed in maize or dry-season wheat crops when intercropped with indigo. It is claimed that the requirement for inorganic fertilizer is reduced by one-half to two-thirds when rice is preceded by indigo. Based on an application of 60 kg of nitrogen per hectare for rainfed rice, the savings in terms of nitrogen will be about 30 to 40 kg of nitrogen per hectare. If the lower figure of 30 kg is assumed, the savings would be equivalent to about 70 kg of urea per hectare. In terms of the 1989 average retail price of urea, this would amount to around 300 pesos per hectare.

Against this saving has to be set the cost of additional inputs of labour. These arise in the following way. The largest additional demand comes from harvesting for next season's indigo seed. Harvesting tends to be non-competitive with other activities, however. Extra labour is also required for the seeding of indigo into the main crop and for harrowing. It is estimated that intercropping rice with indigo adds eight to ten workdays to labour requirements per hectare per year. Garrity *et al.* (1989) have costed this at about 350 pesos, using the standard hired labour rate of 40 pesos a day.[2] In fact, some of the additional labour will be family labour that would otherwise not be utilized and thus has a low or zero opportunity cost. The partial replacement of nitrogen from fertilizer with nitrogen from indigo also leads to a small offsetting saving of labour used to transport and broadcast fertilizer. On balance, this suggests that indigo intercropping was approximately cost neutral in terms of the prices and wages prevailing in 1989. Given the excess labour supply in the rural sector, it seems likely that the ratio of fertilizer prices to the hired labour wage will increase in the future, rendering indigo cultivation more profitable in the areas to which it is suited.

The IRRI recently initiated a research programme on the use of indigo as an intercrop and it is thought that indigo cultivation may spread from its present concentration in north-west Luzon. There remain obstacles to its diffusion, however. Officers of the Department of Agriculture favour the now traditional use of fertilizers. The use of indigo as a biofertilizer, like other new technologies, probably requires a champion to further its cause and break the conventional wisdom. As it is a Philippine technology, the IRRI's work may help to advance this cause. None the less, five years is probably a minimum timetable for the adoption of indigo on a significant scale.

Leguminous crops

A final alternative method of biofertilization involves the cultivation, in rainfed areas, of leguminous cash crops such as mungbean, cowpea and black gram. These require reduced amounts of water and are grown during the period between November and May, when the land is not occupied by rice. In the case of rainfed, partially irrigated rice fields, the rice fallow period between February and May can be utilized for the cultivation of quick-growing varieties of leguminous crops. A system of multiple cropping with leguminous cash crops not only increases the profitability of agriculture to the poor farmers in the Philippines but also reduces the reliance on chemical fertilizer required for the rice crop to follow. Research has revealed that leguminous cash crops can provide from

25 to 60 kg of nitrogen per hectare (Kulkarni and Pandey, 1988). The system is not feasible in irrigated lowlands because of possible waterlogging which the leguminous cash crop is not able to withstand. The main constraint on the cultivation of leguminous crops in rainfed areas is the local market demand for the crops themselves. Unlike the biofertilizers discussed earlier, legumes have cash value and are grown for sale. The cultivation of legumes adds to labour requirements, but mainly at a time of the year when labour is underutilized. The viability of growing leguminous crops rests especially on the possibilities to realize the market value. The development of vegetable markets is a matter of some priority, therefore, because such markets not only assist in improving farming incomes and in extending the dietary choice of the population, but also have some potential for limiting the adverse environmental effects of the use of inorganic fertilizer.

While considerable research into green manures has been carried out in the Philippines and while some crops, particularly indigo and legumes, seem to have some potential for further adoption, the present scale of cultivation is very small. Obstacles include the attitudes of extension workers, the subsidization (until 1991) of fertilizers, and (in the case of sesbania) the lack of a seed industry. There are also agronomic difficulties in obtaining an even spread of green manures in rice fields and in the rate of decomposition of green manures in the fields.

SUMMARY AND CONCLUSIONS

Farmers in the Philippines have been quick to adopt HYVs and their cultivation is more or less universal in suitable areas. The cultivation of HYVs requires a substantial use of fertilizers, and fertilizer policies in the Philippines have attempted both to stimulate domestic production and to reduce the price to farmers. These policies have been subject to frequent changes, of which it is not easy to understand the rationale. Currently all urea, the most widely used fertilizer in rice production, is imported.

The Philippines has a large external debt and a deficit on the current account of the balance of payments. The need to limit the rate of growth of imports is a constraint on economic development. Thus, *ceteris paribus*, a reduction in the use of fertilizers in rice production would improve the prospects for economic growth and, in the long run, would also appear desirable, as sustained application is leading to soil degradation and to water pollution.

Labour, particularly hired labour, has benefited from the Green Revolution as labour requirements per hectare are significantly higher in the cultivation of HYVs than in that of traditional varieties. In the 20 years or so since the Green Revolution occurred, the rate of growth of the population, and of the labour force, has been relatively rapid. Thus labour supply in rural areas has grown faster than the demand for labour, and the real wage in the production of rice, and in agriculture generally, has declined. Despite high levels of urban migration, unemployment and underemployment rates are nearly 50 per cent in rural areas, and over one-half of rural households receive incomes below the poverty line.

Given the severity of the incidence of underemployment and of poverty, it would be of enormous social benefit if new agronomic practices could entail a reduction in the use of fertilizers and an increasing input of labour, provided that additional labour costs were justified by increased output or by reductions in inputs other than labour. Two current developments in biotechnology help to achieve this aim. The first is an attempt to develop varieties of rice that will fix nitrogen in the plant, thus reducing the need for nitrogen obtained from the soil. Although this is an exciting prospect, such new varieties are not expected to be ready for implementation in the Philippines in the foreseeable future. The second development is research into the nitrogen-fixing properties of certain plants that can either be used as green manures or be grown as cash crops with rice. Of the former, sesbania and indigo have some potential for adoption in the Philippines. Work under experimental conditions at the IRRI has shown that the cultivation of either plant as a green manure enables substantial savings to be made in fertilizer while maintaining rice yields per hectare. The use of green manures requires more labour than does the use of fertilizers, but calculations suggest that, at 1989 price/wage ratios, there is no net addition to cost. At present, however, sesbania is used on a negligible scale, while only about 10,000 hectares of rice paddy are intercropped with indigo.

A number of obstacles stand in the path of the widespread adoption of either crop. Green manures need to be as reliable as fertilizers in order to be adopted by farmers, and at present this is not the case. Growth may be uneven and there are no seed industries. Nor is it likely that green manures will be adopted on a very large scale in the Philippines. Rather, agronomists write in terms of these crops finding 'environmental niches'. None the less, there does seem to be some scope for a significant increase in the cultivation of green manures. It is to be regretted in this regard that present policies are not aimed at reducing the dependence on chemical fertilizers.

In order to increase the efficiency of long-term agricultural production, it is recommended that policy be designed to internalize the external cost of the use of inorganic fertilizer so that this cost will be borne by the producers (Tietenberg, 1984; Pearce and Turner, 1990). Such a price-based policy may provide incentives for environmentally friendly technologies to be developed in the future. In the Philippines, which is a net importer of fertilizers, the true cost of inorganic fertilizer is the price of production or import (whichever is the lowest) plus the internal distribution costs plus the negative externalities created by the use of chemical fertilizers to the land and to third parties via the pollution of rivers, lakes and wells. Unfortunately these negative externalities are not easily quantified.

In the past, especially up to 1987, the use of tariff or non-tariff barriers such as import quotas or licences was aimed, not at discouraging the use of chemical fertilizers, but at encouraging the domestic production of fertilizer. Nevertheless, this policy may have had an inadvertent environmentally beneficial effect. The fertilizer policy subsequently pursued, prior to April 1991, included a subsidy component to the farmer. This was to encourage the adoption and use of fertilizers in agriculture in the Philippines, as it was considered to be a yield-increasing input in the short run. As we have seen, the subsidization of fertilizer prices may also have had significant environmental impacts, although it seems unlikely that these were taken into consideration. Lutz and Young (1990, p. 3) write in this regard: 'environmental effects should not, a priori, be assumed to be negligible and excluded from agricultural policy analysis'.

If environmental effects are to be included in agricultural policy, this raises the prospect of intervention and, notably, the use of taxes and regulations. Consider first a regulatory approach, based on setting standards for the allowable levels of nitrate leaching into the drains from cropland. Standards are set to prevent increases in the levels of nitrates in the groundwaters and rivers which may cause health and other risks. The factors influencing the rate of leaching are the amount of fertilizer applied, the volume of water involved, the soil texture and the organic content of the soil. Thus one of the major problems of regulation would be to set fertilizer application rates and times for different ecological conditions that will minimize this leaching. Furthermore, in order for the standards to be effectively enforced, the government would have to monitor the nitrates in the drains of each farm periodically, preferably immediately after an application of fertilizer. In view of these technical difficulties, as well as in view of weak institutions, a lack of administrative skills and the very low literacy rate of farmers, the enforcement and monitoring cost of regulation would probably be high and its effectiveness limited.

A second approach is based on the use of taxes. Taxes on those inputs which are considered to be environmentally damaging could be used so that the policy instrument would encourage the farmers to adopt practices which are environmentally favourable and which contribute to the sustainability of rice output. Tax revenue, in turn, might be used to pay for (mitigate) at least part of the pollution damage. It may be possible to vary the tax when nitrogen use exceeds prescribed amounts per acre in the rainfed and partially irrigated areas. This is to ensure that the nitrogen required to supplement the biotechnology in these areas is made available at the pre-tax rate, but that the excess use of nitrogen, i.e. over and above the requirement to supplement biotechnology, is taxed. A differential tax scheme may further encourage the adoption of green manures, because the object of this tax scheme is to minimize the income losses from higher input prices in the short run while maintaining a high sustainable yield increase in the long run. It has been proposed to administer a differential tax through the levy of a general tax on fertilizer with a refund given for use of nitrogen input below a given level (Weinschenk, 1987). As seen above, the adoption of biotechnology induced by the higher tax on chemical fertilizer will have potential beneficial employment effects in the agricultural sector in the Philippines in addition to its environmental contribution.

A tax on fertilizers would also provide an incentive for farmers in the Philippines to adopt improved agronomic practices, such as fertilizer placement rather than the present method of broadcasting the nitrogenous fertilizers. The latter method is labour saving compared with placing the fertilizer, but broadcasting inevitably leads to nitrogen losses through volatilization. Hence fertilizer placement will lead to allocative efficiency in the short run and also will be labour augmenting. Moreover, tax-augmented fertilizer prices will make additional research efforts possible in respect of genetic engineering, with the prospect of new HYVs utilizing atmospheric nitrogen and of a further reduction of the dependence on imported, inorganic fertilizer.

Notes

1. One metric tonne of urea contains 0.45 metric tonnes of nitrogen. One metric tonne of ammonium sulphate contains 0.21 metric tonnes of nitrogen.
2. Labour needed during the rice-harvesting period is paid at a higher rate, normally in kind. Harvest labourers are entitled to a proportion, usually one-sixth to one-tenth, of the rice they harvest.

9 Technology–Environment–Employment Linkages and the Rural Poor of Bangladesh: Insights from Farm-Level Data

M. Alauddin, M.K. Mujeri and C.A. Tisdell

It is now widely accepted that sustainable development is a desirable goal. While the concept of sustainable development has a variety of interpretations, discussions of the concept bring attention to the fact that short-term benefits from economic change may well be at the expense of future attainable levels of production or income. But sustainable development is not solely concerned with the question of sustaining GDP or growth of GDP. The concept of sustainability allows several interpretations (Douglass, 1984; Tisdell, 1988, 1990, 1991, 1993). It inevitably also involves the question of sustaining livelihoods and employment and raises broad questions about the measurement of welfare and appropriate goals for humanity.

As a result of this new focus in development policies, greater emphasis is being placed on environmental and ecological factors as constraints on economic production and as important factors to be taken into account in assessing new projects and economic changes. In particular, economic, environmental and natural resource interdependencies as well as resource depletion problems are stressed as being important (cf. World Commission on Environment and Development (WCED), 1987). As a result, more account is being taken of environmental externalities and more accurate estimates of the real social benefits and costs of economic developments are being made. This is an important change because the nature of decisions depends not only upon how well data are analysed but most importantly on what data one chooses to collect for analysis.

This chapter provides farm-level evidence of environmental problems, which are to a large extent a result of technological change, and related economic issues in rural Bangladesh. While serious environmental problems do occur in urban areas, most of Bangladesh's population live in rural areas and depend on these for their livelihood.

The focus of this chapter is on major economic and associated environmental changes which have occurred in three rural villages of Bangladesh. These changes have all been associated with the adoption of new technologies and have had effects on production, employment, income distribution and the distribution of opportunities to earn a livelihood. These effects were documented to the extent possible in the limited time available for the collection of primary data from the villages. The specific results for these three case studies follows an overview of environmental and employment issues in rural Bangladesh.

STATE OF THE RURAL ENVIRONMENT AND TRENDS

Rural Bangladesh has always had to cope with a large number of natural environmental problems which not infrequently have resulted in disaster for Bangladeshis. Because of Bangladesh's geographical location, these problems include floods, droughts and cyclones. To a considerable extent economic growth has magnified the effect of such natural disasters and has added to environmental problems in rural Bangladesh. This can be attributed both to economic developments and population growth in Bangladesh and to externalities or spillovers from economic developments and change in nearby countries such as India and Nepal. These have resulted in an increase in land and water use which in turn has had serious environmental effects, as will be discussed below in some detail.

The following types of environmental deterioration have occurred in rural Bangladesh:

(a) There have been increased fluctuations in flows of rivers and streams, resulting both in more severe flooding and reduced availability of water during drought or dry periods. There are also adverse consequences for navigation, fish stocks, siltation, water quality and so on.
(b) There is reduced availability of water (at least, in critical periods), for instance because of greater demands on groundwater to irrigate high-yielding varieties (HYVs) of crops which have been widely adopted following the Green Revolution and have reduced water quality, e.g. salinization caused by reduced inflows of fresh water into rivers and streams, or the leaching of nitrates into groundwater through using artificial fertilizers in the growing of crops.

(c) There has been a decline in soil quality resulting from, for example, the increasing incidence of multiple cropping. The humus content, structure and nutrient content of the soil can suffer from such practices. The use of artificial fertilizers can also increase soil acidity and destroy useful flora and fauna in the soil. Shahidul Islam (1990) points out that in Bangladesh 'with most crops not only the edible portion (grain) is harvested but also the straw for fuel and fodder. The result is that soils are low in organic matter and depleted of nutrients ... Bangladesh soils are hungry, sick and almost lifeless, they are in urgent need of organic matter to put life into them ...'

(d) Increased pesticide use, especially of insecticides, has had adverse environmental consequences. These include the destruction of natural enemies of pests, the induced resistance of pests to pesticides, an adverse impact on fish and, because chemicals such as DDT continue to be used, adverse consequences for wildlife. While the use of pesticides is low in Bangladesh as compared with more developed countries such as Japan (cf. Ameerul Islam, 1990), the proportion of environmentally hazardous pesticides used, such as insecticides including organochlorides, is higher in Bangladesh. Controls on their use are also less stringent than in more developed countries.

(e) Deforestation has increased as a result of logging, the extension of agriculture in some areas and the intensification of slash-and-burn agriculture which has reduced the length of the rotation cycle in *Jhum* (shifting) cultivation. Deforestation has a number of serious environmental consequences, such as greater fluctuations in river flows, more rapid erosion of soil and loss of wildlife, which in turn results in reduced genetic diversity.

(f) With the introduction of Green Revolution technologies and the increased adoption of HYVs, there has been a loss of traditional varieties of crops. These were usually well adapted to *natural* local conditions. As a result, there is greater dependence on fewer varieties, the productivity of which may not be sustained or which can be sustained only under artificial environmental conditions such as are achieved by irrigation and the use of artificial fertilizers. Thus the genetic base of agricultural production is being reduced and this may make it difficult in the long run to sustain agricultural productivity (see, for example, Alauddin and Tisdell, 1991b).

(g) Substantial reductions of stocks of inland fish have occurred in Bangladesh as a result of environmental changes (such as reduced water availability and quality in streams and rivers, the draining and filling of water bodies) and to some extent the greater use of chemicals in agriculture associated with the adoption of Green Revolution technologies. This has serious nutritional consequences because freshwater fish are the most important source of animal protein for Bangladeshis.

(h) Indigenous wildlife continues to disappear, mainly as a result of over-harvesting and habitat alteration. Habitat alteration is brought about by the expansion and intensification of economic activity and by rising population levels in Bangladesh.

The incidence of environmentally related diseases is high in Bangladesh and is to a large extent linked to faecal pollution of surface water. In Bangladesh, 80 per cent of illness has in fact been attributed to waterborne diseases (Haque and Hoque, 1990). This incidence remains high in rural areas and, apart from the suffering it causes, it is a serious economic impediment. In some parts of Bangladesh the extension of the area under irrigation has added to the incidence of malaria by providing a more suitable environment for mosquitoes to breed.

Several of these issues will be confirmed by farm-level studies in three villages in different parts of Bangladesh.

THE SURVEY AREA

The three villages studied are Durgapur, located inland in the west of Bangladesh in greater Kushtia, and two villages located in eastern Bangladesh in the Chittagong Hill Districts (see Figure 9.1). The natural ecological and geographical conditions prevailing in the first village differ substantially from those in the other two villages. Durgapur is located on shallow floodplains with a distinct wet and dry season owing to the monsoon. The other two villages are located in hilly areas and the annual rainfall is heavier and extends over a longer period of the year, even though a dry period does occur (cf. Haroun Er Rashid, 1991).

If there were any substantial forests in the neighbourhood of Durgapur, they have long since disappeared and there is a severe shortage of fuelwood in the area. In the villages in the Chittagong Hill Districts, some forest remains but is disappearing as a result of economic and technological change and development, the sustainability of which is being questioned in this chapter.

There is one important difference between the two villages studied in the Chittagong Hill Districts. In one case shifting agriculture is practised

Bangladesh: Technology–Environment–Employment Linkages 225

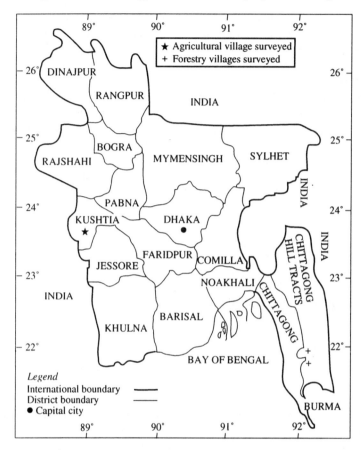

Figure 9.1 Map of Bangladesh showing survey areas.

by tribal people, whereas in the other settled agriculture is practised by Bengalis. Thus the effects of two different farming systems can be considered. However, both groups farm on forestry land in return for providing 80 days of free labour per year to the Forestry Department.

It should be noted, however, that the three villages do not characterize the environmental and changing environmental conditions of all rural villages in Bangladesh. A larger group of villages would need to be studied to obtain information on such matters. For example, a floodplain village in which irrigation permits a continuous or almost continuous supply of rice or similar crops throughout the year has not been included. The village of

Durgapur, although it is located on shallow floodplains, does not use irrigation to any significant extent and its cropping pattern is much more diversified than that of most villages on the floodplains in Bangladesh. None the less, it has been influenced by Green Revolution agricultural technologies.

In considering the experiences of Durgapur and the two villages from the Chittagong Hill Districts, we shall pay particular attention to variations in exchange and non-exchange income obtained from the use of natural resources. As a result of technological change and economic growth, it often happens that, while exchange income increases, non-exchange income falls (Alauddin and Tisdell, 1989). This trend can be especially disadvantageous to the poor and the underprivileged, including the landless, women and children.

AGRICULTURE IN AN UNFORESTED AREA ON THE FLOODPLAINS

The survey area and methodology

The data used in this study were derived from a direct sample survey. The collected data relate mainly to the crop year 1990/91. Through a direct questionnaire method, farm-level data were collected with the aid of research assistants during September 1991. The survey village of Durgapur was selected deliberately, on account of: (a) its relatively long tradition with HYV technology; (b) its easy access by road or train from the District headquarters and capital city of the country; and (c) the fact that one of the authors of this report had prior knowledge of the village. A further reason was that the village is known to have had a very diversified cropping pattern, and a large number of rice varieties were cultivated prior to the Green Revolution. The extent to which crop diversification has been sustained following the Green Revolution is one of the foci of interest.

Geographically, Durgapur belongs to the floodplains classified as shallow flooded areas (Hossain, 1991). The area generally occupies the higher sites in the landscape, which are inundated to a depth ranging from a few centimetres to less than a metre during the monsoon.

The year 1990/91 was a fairly normal one for the village. Technologically the village is advanced when compared with many other villages in the area. In all, 84 landowning and 38 landless households were directly interviewed. These numbers constituted 80 per cent of households in the respective categories. The number of farmers interviewed corresponded to

Landowning households

Socioeconomic characteristics

Table 9.1 provides information on selected socioeconomic characteristics of the 84 landowning households interviewed in Durgapur. Several observations may be made from Table 9.1. Average household size is positively associated with the size of the farm, ranging from 4.95 for the marginal farms to 7.50 for the large landowners. Agriculture is the major occupation of the overwhelming majority (85 per cent) of the households. Occupational distribution does not seem to differ significantly across farm size. The literacy level of the head of household increases significantly with farm size. The average area owned per household in the village is more than four times the average farm size for the marginal farmers (0.54 acre). The incidence of sharecropping is insignificant, as indicated by the average sizes of owned and operated holdings. As expected, the average number of plots as well as their size are positively associated with the size of the farm.

From Table 9.2, which provides information on cropping pattern, average production per acre and labour intensity per acre for various crops, it can be seen that:

(a) Jute is the major crop cultivated in the village (more than 33 per cent of total land cultivated) followed by local *aus* (23 per cent) and *aman* HYV rice (16 per cent). The total area under cereals (rice and wheat) is nearly 51 per cent, of which HYVs cover about 20 per cent of the total area cultivated. Pulses (7 per cent) and sugar cane (6 per cent) are the other major crops. This cropping pattern is very different from that for Bangladesh as a whole.
(b) Of the HYVs, *aman* HYV is the major crop cultivated by most of the farmers; its average yield is more than twice that of the corresponding local variety.
(c) In terms of labour intensity per acre, sugar cane ranks first with 120 workdays followed closely by *boro* HYV rice (118 workdays) and jute and *aman* HYV with 114 and 113 workdays respectively. Local varieties of *aus* and *aman* have more or less similar labour requirements per acre. Pulses and oilseeds do not feature prominently in terms of employment generation even though they are nutritionally important. These two crops have labour requirements per acre of 15 and 18 workdays respectively.

Table 9.1 Socioeconomic characteristics of landowning classes: Durgapur, 1990–91

Farm size (acres)	No. of samples	Average family size (no.)	Major occupation (%)		Literacy level of head of household (%)		Average area owned per household (acres)	Average area operated per household (acres)	Average number of plots per household	Average size of plot (acres)
			Agriculture	Others	I1	L2				
Less than 1.00	21	4.95	85.71	14.29	85.71	14.29	0.54	0.57	2.38	0.23
1.00–2.50	29	5.62	75.86	24.14	86.21	13.79	1.55	1.53	3.41	0.45
2.50–5.00	26	6.65	92.31	7.69	61.54	38.46	3.38	3.36	4.15	0.81
5.00 and above	8	7.50	87.50	12.50	37.50	62.50	7.33	5.31	6.38	1.15
Total or average	84	5.95	84.52	15.48	73.81	26.19	2.41	2.21	3.67	0.66

1. I = illiterate.
2. L = literate.

Table 9.2 Cropping pattern, production and labour requirements in crop production: Durgapur, 1990–91

Crop	Area under crop		Average production per acre (maunds)[1]	Average labour requirements (workdays)
	Total (acres)	%		
Jute	73.05	33.42	29.56	114
Aus local	50.58	23.14	21.09	99
Aman HYV	35.26	16.14	46.09	113
Aman local	16.03	7.34	21.94	103
Pulses	15.71	7.19	6.57	15
Sugar cane	13.32	6.10	260.05	120
Boro HYV	5.89	2.70	26.85	118
Wheat HYV	3.48	1.59	25.66	50
Oilseeds	3.22	1.47	5.18	18
Others	1.99	0.91	–	50
Total	218.53	100.00	–	–

1. Maund equals approximately 37 kg.

Table 9.3 presents survey results on the distribution of land, cropping intensity and the percentage of HYV area in rice cultivation by size of farm. It highlights the following:

(a) the distribution of operated land over farm size is highly unequal: the lowest 25 per cent operate only 6.4 per cent of the land while the top 10 per cent operate nearly 23 per cent of the land – the distribution of cultivated area follows a broadly similar pattern;
(b) the average intensity of cropping is 117 per cent, which is appreciably lower than the national average (more than 150 per cent), while the variations in intensity of cultivation among farm size shows the highest for the top landowning class (128 per cent);
(c) on average, more than 49 per cent of total cultivated land is under rice, and this percentage increases with farm size: in terms of percentages of land, marginal farmers (less than 1 acre) allocate 37 per cent to cultivation of rice, a proportion which increases to 57 per cent for the large farmers; and
(d) the smallest and the largest farms have the highest percentages of rice area under HYVs: 47 per cent and 46 per cent respectively.

Table 9.3 Land distribution, cropping intensity and intensity of HYV rice adoption by farm size: Durgapur, 1990–91

Farm size	Operated area		Cultivated area		Intensity of cropping (%)	Rice area as % of total cultivated area	HYV rice area as % of total area under rice
	Total (acres)	%	Total (acres)	%			
Less than 1.00	11.98	6.44	13.90	6.36	116.03	37.38	46.94
1.00–2.50	44.29	23.81	52.14	23.86	117.72	42.75	38.50
2.50–5.00	87.31	46.93	98.13	44.91	112.39	50.01	32.36
5.00 and above	42.46	22.82	54.35	24.87	128.00	57.42	45.66
Total or average	186.04	100.00	218.52	100.00	117.46	49.32	38.18

Table 9.4 Area under major crops by landowning classes: Durgapur, 1990–91 (acres)

Crop	Group 1 (marginal)		Group 2 (small)		Group 3 (medium)		Group 4 (large)	
	Area	Share in total area cultivated (%)	Area	Share in total area cultivated (%)	Area	Share in total area cultivated (%)	Area	Share in total area cultivated (%)
Jute	6.72	48.35	20.18	38.70	32.54	33.16	13.61	25.04
Aus local	2.51	18.06	11.88	22.78	22.87	23.31	13.32	24.51
Aman HYV	2.27	16.33	7.84	15.03	11.90	12.14	13.25	24.38
Aman local	2.27	16.33	7.84	15.03	11.90	12.14	13.25	24.38
Pulses	1.73	12.45	6.21	11.91	5.78	5.89	1.98	3.64
Sugar cane	–	–	2.31	4.43	6.44	6.56	4.57	8.41
Boro HYV	0.17	1.22	0.74	1.42	3.98	4.05	1.00	1.84
Wheat HYV	–	–	0.50	0.95	2.32	2.36	0.66	1.21
Oilseeds	0.25	1.79	0.66	1.27	1.65	1.68	0.66	1.21
Others	–	–	–	–	0.33	0.34	1.66	3.06
Total	13.90	100.00	52.15	100.00	98.13	100.00	54.35	100.00

Note: Group 1 refers to farm size class owning land less than 1 acre, Group 2 is class with land 1–2.5 acres, Group 3 with land 2.5–5 acres and Group 4 is the class with land of 5 acres and more.

In Table 9.4 we set out data on land area allocated to various crops by different classes of farmer. The following aspects are noted.

(a) Jute is the major crop in terms of area cultivated for all classes of farmers. The intensity of jute cultivation varies inversely with farm size. The marginal farmers allocate nearly half the land to jute cultivation, as compared with only a quarter by the large farmers. On the other hand, the percentage of the area allocated to cereals (rice and wheat) varies *directly* with farm size. While the lowest farmer group devotes 37 per cent of the cultivated area to cereal production, the highest farmer group allocates 57 per cent of their land area.

(b) The intensity of HYV adoption seems not to vary with size, even though it is appreciably higher for the largest group (17 to 18 per cent of cultivated land area for all farmer groups except the largest, for which the proportion is 27 per cent).

(c) Crops such as pulses and oilseeds seem to be more important for smaller farm size classes in relative terms although in absolute terms larger farm size classes cultivate more land under these crops. The annual crop, sugar cane, is cultivated more extensively by large farm size classes with the proportion in total area cultivated, which increases with farm size.

Table 9.5 shows the seasonal distribution of land cultivated by different classes of farmer. Small farmers tend to concentrate on cultivating their land more during the *aus* season. The percentage of total land allocated to crop production during the *aus* season appears to decrease with the size of the farm. The opposite seems to be the case during the *aman* season. Larger farmers allocate a higher percentage of land to *aman* rice

Table 9.5 Seasonal distribution of land cultivated by different farmer classes: Durgapur, 1990–91

Farm size (acres)	Annual	Season (% of cropped area)			Total
		Aus season	Aman season	Boro season	
Less than 1.0	–	66.41	18.15	15.44	100.00
1.0–2.5	4.43	61.49	18.53	15.55	100.00
2.5–5.0	6.56	56.45	22.64	14.34	100.00
5.0 and above	8.41	49.55	31.07	10.97	100.00
Average	6.09	56.58	23.47	13.86	100.00

Table 9.6 Labour requirements for major crops: Durgapur, 1990–91 (workdays per acre)

Farm size (acres)	Jute	Aus local	Aman local	Aman HYV	Boro HYV	Sugar cane	Pulses	Oilseeds
Less than 1.00	118	99	96	115	120	–	18	21
1.00–2.50	114	103	90	111	120	127	17	19
2.50–5.00	113	97	105	114	120	112	14	16
5.00 and above	112	96	108	114	111	123	14	17
Average	114	99	103	113	118	120	15	18

cultivation. During the *boro* season no discernible trend seems to be present, even though the percentage of land area allocated to *boro* crops by the large farms is marginally lower than that for other groups.

Table 9.6 gives labour requirements for major crops by different farmer groups. There does not seem to be any significant difference in intensities of labour use by the various groups, although in general smaller farm groups tend to apply more labour per acre of land. Possibly this is because the smaller groups are better endowed with labour resources than the larger ones.

Labour requirements of some of the major crops (e.g. jute, *aus* local, *aman* local, *aman* HYV, *boro* HYV and sugar cane) are significantly higher than those of other crops such as pulses and oilseeds. The transfer of land from the latter crops, even though undesirable from a nutritional point of view, increases employment.

Table 9.7 gives yields of major crops by farmer groups. It can be seen that the differences in yields per acre of different crops in relation to farm size do not reflect any discernible pattern. In most cases yields per acre for small farms seem to be higher when compared with those of large farms. The difference in yields between local and HYV crops (e.g. *aman* local and *aman* HYV) is quite substantial, indicating the high output potential of HYVs.

Landless households

At the time of the survey, 48 households out of a total of 150 were estimated to be landless. The incidence of landlessness (about a third) in the survey village is similar to the figure for Bangladesh as a whole (Bangladesh Bureau of Statistics (BBS), 1986).

Incidence and characteristics of landlessness

Table 9.8 sets out relevant information about landless households in the survey area. The bulk of the landless households did not inherit any land. Many landless householders' fathers were landless (about 42 per cent of the sample). Those households who inherited some land (18 per cent) became landless during their lifetime. A little over 34per cent of their land was sold off to maintain family, while the remainder was sold off to meet expenses for medical treatment, daughters' marriages or the purchase of draught animals, or as a result of defrauding.

Over 90 per cent of those from landless households work as agricultural labourers, very few having a subsidiary occupation. About 45 per cent of the sample households cultivated land under sharecropping

Table 9.7 Yields of major crops: Durgapur, 1990–91 (maunds per acre)

Farm size	Jute	Aus local	Aman local	Aman HYV	Boro HYV	Sugar cane	Pulses	Oilseeds	Wheat HYV
Less than 1.00	29.76	22.60	22.10	50.57	24.24	–	8.03	4.03	–
1.00–2.50	30.92	20.53	23.55	52.25	27.66	277	6.53	6.06	24.24
2.50–5.00	25.11	22.17	20.83	45.29	28.64	230	7.17	6.06	16.38
5.00 and above	32.44	19.07	21.43	36.23	19.56	273	4.55	4.55	36.36
Average	29.56	21.09	21.94	46.09	26.85	260	6.57	5.18	25.66

Table 9.8 Characteristics of landless households in a Bangladeshi village: Durgapur, 1990–91

	Characteristics	Number[1]	Percentage
1.	Average family size: 4.13	–	–
2.	Education level of head of household: literate	1	2.63
3.	Households not inheriting any cultivable land	31	81.58
	(a) Father did not inherit any land	13	41.94
	(b) Father became landless during his lifetime or did not give any land to his son	18	58.06
4.	Households who inherited/had land but became landless	7	18.42
5.	Total land area sold off in the process of being landless: 5.41 acres		
	(a) To maintain family: 1.84 acres		34.01
	(b) Result of daughter's marriage, medical treatment, defrauding and purchase of draught animals: 3.57 acres		65.99
6.	Major occupation		
	(a) Agricultural labour (including day labour)	35	92.11
	(b) Others (including rickshaw pulling, petty trading, service)	3[2]	7.89
7.	Subsidiary occupation		
	(a) None	32	84.21
	(b) Agricultural labour (including day labour)	0	–
	(c) Others (including rickshaw pulling, petty trading, service)	6	15.79
8.	Incidence of sharecropping		
	(a) Households operating land under sharecropping	17	44.74

Table 9.8 (continued)

Characteristic	Number[1]	Percentage
(b) Amount of sharecropped land per sharecropping landless household: 0.25 acre		
9. Households reporting:	29	76.32
(a) Bovine animals	2	5.26
(b) Goat, sheep	16	42.10
(c) Poultry birds	25	65.79
10. Homestead land		
(a) Number of households owning homestead land	18	47.37
(b) Average size of homestead land: 0.042 acre		
11. Number of households reporting indebtedness	32	84.21
Average loan: 1,194 taka		
Interest rates: range between 5 and 20 per cent per month		

1. Number of households interviewed: 38.
2. Including one disabled.

arrangements. The average amount of sharecropped land per household is very small (only 0.25 acre).

More than half the landless households in our survey do not have any homestead land. For those who have some, the amount is very small (0.042 acre on average) and not sufficient to have a vegetable garden. Over 75 per cent of those interviewed reported having some domestic animals and poultry. However, their number has, in most cases, remained much the same over the years, being limited by the shortage of grazing land and the inadequate supply of feed. Most households (over 84 per cent) reported indebtedness, the average size of loan being nearly 1,200 taka. The loan is mainly for consumption purposes, even though households also borrowed money for business reasons. It is noted that borrowers pay very high rates of interest, ranging up to 20 per cent per month.

We do not have any hard evidence to indicate whether landlessness is on the increase in the survey village. But the process through which marginal and small farmers in Durgapur and elsewhere became landless, e.g. distress selling (Rahman, 1982; Alauddin and Tisdell, 1989), suggests that, as demographic pressure and institutional arrangements (e.g. inheritance law) reduce the size of landholdings, adverse economic circumstances are more likely to force smaller owners to dispose of their land (Alamgir, 1980). As Cain (1983) argues: '... the lower end of the distribution is a sensitive indicator of change. There seems to be a process of polarization whereby the near landless are dispossessed and join the ranks of the landless while small farmers in turn become near landless.'

Seasonality of employment and wage rate

Macro-level studies on employment from crop production suggest that seasonal differences in labour demand since the Green Revolution have declined over the years (Chaudhury, 1981; Khan, 1985; Alauddin and Tisdell, 1991a). This is because cultivated land that was once left fallow or allocated to relatively less labour-intensive crops during the dry season is now used for crops such as *boro* rice or wheat, which are relatively more labour intensive. This is an important way in which the Green Revolution technology had added significantly to agricultural employment and may have reduced the seasonal incidence of poverty.

In the case of Durgapur, however, little *boro* rice is grown, because of the lack of irrigation. It is therefore not typical of the majority of floodplain villages, and one would not expect a reduction in the fluctuation in seasonal employment as a result of adoption of the Green Revolution technology in such villages, which still depend more on rainfed crops. None the less, its production of rainfed crops has been based on the Green Revolution technology. For example, an increasing proportion of *aman* HYV rice is being grown in the wet season and a limited amount of other HYVs are also being planted. Fertilizer and pesticide use has also increased. In this village, in fact, because of the dependence of crops on rainfall, Green Revolution technologies appear to have increased the demand for labour in the six months from April to October, with little impact on the demand during the remainder of the year. Thus the seasonal fluctuation in the demand for labour appears to have increased in this village, and this may be common for all rain-dependent villages (Alauddin and Tisdell, 1991a).

Table 9.9 sets out relevant information on overall labour demand from cereal and non-cereal crop production on a monthly basis. Also set out are

Table 9.9 Monthly distribution of employment from crop production and daily wage rate: Durgapur, 1990–91[1]

Month	Employment (workdays)			Percentage distribution of			Daily wage (taka)		
	Total	Cereal	Non-cereal	Total	Cereal	Non-cereal	Low	High	Normal
Baishakh	2,088	809	1,279	9.56	7.48	11.61	25	30	20
Jaishtha	3,383	1,274	2,109	15.50	11.78	19.15	20	25	15
Asharh	4,279	1,608	2,671	19.60	14.87	24.25	20	25	20
Sravan	1,291	1,291	0	5.91	11.94	0.00	15	20	15
Bhadra	5,621	3,596	2,025	25.75	33.25	18.38	30	35	20
Ashwin	1,569	1,020	549	7.19	9.43	4.98	15	20	15
Kartik	690	42	648	3.16	0.39	5.88	20	25	20
Agrahayan	1,240	637	603	5.68	5.89	5.47	20	25	20
Poush	968	432	536	4.43	3.99	4.87	20	25	20
Magh	326	3	323	1.49	0.03	2.93	25	30	20
Falgun	163	0	163	0.75	0.00	1.48	20	25	20
Chaitra	213	104	109	0.98	0.96	0.99	15	20	15
Total	21,831	10,816	11,015	100.00	100.00	100.00	–	–	–
Workdays per acre	100	97	103	–	–	–	–	–	–

Note: Baishakh, mid-April to mid-May; Jaishtha, mid-May to mid-June; etc. See also Figure 9.2.
1. For sample households only.

the relative shares in the respective totals. Table 9.9 further presents information on the monthly variation of average daily wage rates reported by the landless. The table suggests the following.

(a) Overall labour intensity is 100 workdays per acre, with little difference between labour intensity for the production of cereal and non-cereal crops. The overall figure is consistent with that for Bangladesh as a whole. The cereal crops are overwhelmingly dominated by rice, primarily *aus* rice. Non-cereal crops, on the other hand, are dwarfed mainly by jute and to a small extent by sugar cane, both being relatively labour-intensive crops.

(b) The seasonal distribution of employment in crop production is highly skewed. More than half the labour demand for crop production is concentrated during the months of Jaishtha (mid-May to mid-June), Asharh (mid-June to mid-July) and Bhadra (mid-August to mid-September). This is generally true for cereals as well as for non-cereals. The overall leanest months are Agrahayan (mid-November to mid-December), Magh (mid-January to mid-February), Falgun (mid-February to mid-March), Chaitra (mid-March to mid-April) and Sravan (mid-July to mid-August). The pattern is clear from Figure 9.2.

(c) There is significant monthly variation in daily wages. Wage rates are not consistently related to the distribution of employment from crop production activities. This could be because, when crop production activities are at a low ebb, opportunities for some other types of employment are available, e.g. rural housebuilding, petty trading, transportation services.

Environmental trends

Increased population, the extension of markets and the spread of new technologies have had considerable influence on the state of the rural environment in many parts of the world. Loss of soil quality and fertility, depletion of natural resources, adverse impact on surface water and groundwater, loss of trees and vegetation, shortage of fuel and energy, depletion of fish stocks and genetic diversity have occurred following the Green Revolution (Biggs and Clay, 1981; Repetto and Holmes, 1983; Conway, 1985; Jodha, 1985, 1986). Empirical studies have reported adverse environmental consequences of the new agricultural technology in Bangladesh. These include, among other things, the lowering of the

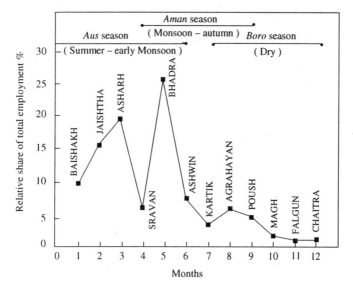

Figure 9.2 Percentage distribution of total employment by month: Durgapur, 1990–91.

groundwater level beyond the suction limit, loss of fish output in paddy fields and other water bodies, loss of natural soil fertility and increasing dependence on a narrow range of genetic materials (M. Ahmed, 1986; M.F. Ahmed, 1986; Hamid *et al.*, 1978; Jones, 1984; Tisdell and Alauddin, 1989; Alauddin and Tisdell, 1991b).

The objective of this subsection is to investigate whether there is farm-level evidence from Durgapur of such trends.

Environmental changes

Over the years population growth has led to the extension of human habitation into cultivable or cultivated areas. Marginal land area has been brought under cultivation by clearing jungles and bushes, destroying or at least disturbing wildlife habitats in the area.

During the past two or three decades, especially in the past 20 years, a number of fruit trees such as mango and jackfruit, *Khejur* palm (*Phoenix sylvestris*), *jam* (*Syzygium jambolana*) and *Tal* palm (*Boro ssus flabellifer*) have been cut down for fuelwood for use in the urban centres as well as for commercial brick kilns. Today one notices few fruit-yielding trees, whereas there was a multitude some 20 years ago. In the past few years, however, the relatively wealthy farmers in the village have begun to plant *Shisham* (*Dalbergia sissoo*). This has a high value for timber and is quite profitable to grow.

Livestock populations, including sheep, goats and poultry, have remained stationary or have even declined slightly. This is mainly due to: (a) the fact that there has been a significant decline in the area under fodder crops; (b) the virtual disappearance of pasture and grazing land; and (c) a shift from long-stem to short-stem varieties of rice cultivation (e.g. from *aman* local to *aman* HYV). Furthermore, the rapid depletion of trees such as banyan (*Ficus bengalansis*) and *Oshot* (*Ficus religiosa*) and acacia has led to a decline in the supply of the green leaves and fruit that are used as fodder for sheep, goats and cattle. If this trend continues, it is unlikely that the livestock and poultry population will increase in the future. Indeed, there are signs suggesting that it may decline further.

The indiscriminate use of pesticides and insecticides has affected fish output from the paddy. It was found during the course of fieldwork that fish production in the paddy fields has declined by about two-thirds over the past ten years. In our survey village, some farmers engage in the commercial culture of exotic fishes. Even though the yield is high, they are susceptible to diseases and do not taste good. The supply of indigenous species of fish has declined. The use of agrochemicals interrupts natural breeding and causes discontinuity in the lifecycle of fish. Furthermore, the reclamation of *beels* and low-lying areas for rice production has contributed to a reduction of habitat for local fish.

The new agricultural technology has had several other environmental spillovers. These include the following.

(a) Natural soil fertility has been affected adversely following the inappropriate application of chemical fertilizers, of which greater quantities are now needed to maintain crop yields. Because of a

stagnation (if not a reduction) in livestock populations, the supply of cow-dung for use as an organic manure has not increased. Furthermore, as a result of a decline in the availability of fuelwood, cow-dung is used more and more as a cooking fuel. Again, tree loss has resulted in a drastic reduction of leaves that could be used as green manure. The nutritional imbalance suffered by the soil has increased dependence on chemical fertilizers at the expense of renewable sources.
(b) The cropping pattern shown in Table 9.2 indicates that a wide variety of crops is grown in the survey village. This, however, masks the impact of Green Revolution technology in terms of the loss or narrowing of genetic diversity. New varieties of rice have driven out several indigenous varieties. Fewer strains of rice are now cultivated than before the Green Revolution. The cultivation of *aman* HYV rice relies almost exclusively on a single variety, i.e. BR-11. Furthermore, a variety of pulses and *Arhar* (pigeon pea) are no longer grown.
(c) The Green Revolution has tended to lead to a proliferation of weeds. The continued application of pesticides has led to the development of resistance to them. Moreover, toxic herbicides and insecticides have affected non-target populations, e.g. beneficial predators and parasites. For instance, the increased use of chemical agents has much depleted the frog and toad population. As a result, the need for insecticides has increased.

Under normal circumstances increased crop production tends to generate greater employment. Agricultural employment has increased slightly, even though there is very little agricultural activity during the dry season. Non-agricultural employment has increased somewhat as a result of infrastructural development. The improvement of the main road linking the village to the nearest urban centre has given some villagers the opportunity to engage in small businesses and provide transportation services. The precise quantification of non-agricultural employment is not possible. However, during the course of the field survey it was gathered that on average a labourer is currently employed for around 250 workdays a year. In Table 9.9 we presented an estimate of total agricultural employment of 21,831 workdays per year. Given that there are about 100–110 wage labourers in our sample, the per capita agricultural employment ranges between 200 and 220 workdays per year. Thus a total of 30–50 workdays of non-agricultural employment per head seems to have resulted from the changes that followed the Green Revolution and infrastructural development. The respondents indicated that annual employment opportunities per labourer, on average, increased from 150 to 250 workdays during the period. About 30–50 per cent of this increase resulted from expanded non-agricultural employment opportunities. However, this gain seems to have been achieved at the cost of a decline in environmental quality. Stagnant water owing to poor and inadequate drainage from roads, for example, has created breeding grounds for mosquitoes, with adverse effects on health.

Implications for rural poverty and sustainability

In the light of the preceding discussion about changes in the environment of our survey village, several questions emerge: What implications do these changes have for the rural poor? Are the increased employment opportunities adequate to compensate for the loss in environmental quality? Has the physical quality of life for the poorer sections of the rural community improved since the advent of the new technology?

It has been argued elsewhere (Conway, 1985; Lipton, 1985; Jodha, 1985, 1986; Alauddin and Tisdell, 1989) that the income of the rural poor (i.e. the landless and the near-landless) has two components: (a) exchange income: mainly wage income determined predominantly through the market mechanism; and (b) non-exchange income: determined predominantly by institutional and/or sociological systems in the rural community and usually obtained direct from nature without exchange. It includes, among other things, wild fruits, firewood, building and thatching materials, water from tanks and ponds for growing vegetables, free-range poultry and some grazing land for sheep, goats and cattle.

The sources of the exchange component depend on the general level of economic activity (both agricultural and non-agricultural). On the other hand, the extent of non-exchange income depends mainly on common access or low-cost access to natural resources. It is claimed that the non-exchange component of income could be important in the off-peak periods and is even more significant in abnormal years, when both the period of employment is shortened and falls in real income result in a reduced exchange entitlement (Sen, 1981). In other words, the exchange and non-exchange components complement each other, and in adverse circumstances the latter component acts as a buffer.

Our empirical evidence from the survey indicates that following the Green Revolution the scope and opportunity for exchange income has increased quite significantly. On the other hand, it seems that the scope for non-exchange income has declined. Resources for the supply of non-exchange income appear to be becoming scarcer in Durgapur. Population pressure and technological forces have depleted them, and the greater penetration of technological and market forces may have led to some resources being priced which were previously free or freely accessible. Although our database is limited and quantification must await further research, the evidence in Durgapur indicates a reduced availability of non-exchange income.

Fish stocks in the paddy fields have fallen or disappeared, and this implies that for many wage labourers or low-income earners an important source of supplementary employment (and therefore income), especially in the leaner months, has disappeared, and with it, in some degree, the opportunity to supplement the diet with animal protein.

The role of women, and to some extent children, in supplementing income from formal and informal sources has been undermined in a number of ways, for example: (a) restrictions on animal husbandry and free-range poultry; (b) loss of opportunities to collect wild fruits and firewood because of the dwindling supply of these resources: it now takes two to three times as long to collect fuel as in the past; and (c) loss of opportunities for rice processing (from paddy to rice) by traditional methods (*Dheki*). These have been replaced by labour-saving modern methods. The traditional labour-intensive method of oil milling has also died out.

Cropping patterns discussed earlier suggest that there is greater dependence on jute cultivation by the smaller farmers. This is partly because of the requirement for cooking fuel and thatching materials, and partly because they are better endowed with family labour. Moreover, jute cultivation does not require as much purchased input as does rice. Farmers, however, are finding it ever more difficult to access surface water for raw jute processing (retting, i.e. the process of separating jute pulp from its fibre). Although that is a general problem, it is more acute for the smaller farmers in the village. Currently processing is done in the standing water of the bigger ditches in the village, causing considerable environmental pollution.

Health risks resulting from environmental pollution can hardly be overemphasized. We saw earlier in this chapter that a very important cause of becoming landless is the need for medical treatment. If environmental change leads to the increased incidence of diseases such as malaria, it may increase the ranks of the landless.

Loss of genetic diversity raises doubts about sustainability of production. Dependence on a narrow range of genetic resources could be disastrous if there were a sudden widespread outbreak of any plant disease or pest or insect attack. Furthermore, any reduction in the availability of organic or green manure is likely to lead to a decline in soil quality. While wealthier farmers may be able to replenish soil nutrients by buying suitable inputs, the smaller farmers may find it difficult to do so because of their limited resources. This may result in the fertility of the soil of smaller farmers declining more rapidly. Constant effort is therefore required to sustain production and employment opportunities in modern agriculture (cf. Johnston and Cownie, 1969; Ishikawa, 1978) and there is always the possibility of the whole system collapsing.

AGRICULTURE IN A FORESTED AREA: CASE STUDIES FROM CHITTAGONG HILL DISTRICTS

In parts of Bangladesh, especially in the Chittagong Hill Districts, forestry and agriculture are inextricably linked. Technological, environmental and other changes have a profound influence on the way of life of the people in those areas. In this section we provide empirical evidence of changes since the introduction of the new agricultural technology in two forested areas of the Chittagong Hill Districts where agriculture and forestry interact. In doing so, we seek to provide answers to questions such as: How sustainable is economic development in the area? Is it becoming more difficult to maintain rice yields? What environmental problems are emerging, e.g. erosion, variability of water supply? Is there any evidence of growing difficulties in sustaining employment and livelihood, e.g. amongst particular groups? Is there any evidence of loss of genetic diversity in agricultural crop production? Is wildlife and its habitat being threatened by environmental and other problems?

Survey areas: description and method

Two relatively diverse areas in the Chittagong Hill Districts were selected for field work. One of the areas is dominated by shifting agricultural patterns while the other is characterized by settled agriculture. People in both areas engage to varying degrees in gathering, hunting and using free or common access resources in the forest.

Data were collected with the help of research assistants during August and September in 1991. The collected information relates to the 1990–91 crop year. Farmers and forestry officials were interviewed in both areas on the basis of questionnaires. In the settled agricultural area, 35 farmers were interviewed as against 31 in the area dominated by shifting agriculture. The former area is inhabited mainly by the Bengali people while the latter is inhabited by tribal people. None of the farmers in either of the two

Table 9.10   Some socioeconomic characteristics of households in a forested area with settled agriculture: Chittagong Hill Districts, 1990–91

Farm size (acres)	No. of sample	Average family size (no.)	Major occupation(%) Agriculture	Major occupation(%) Others[1]	Average cultivated area (acres)
Less than 1.00	4	7.75	100	0	0.63
1.0–2.5	13	7.00	100	0	1.38
2.5–3.5	15	9.20	93	7	2.93
3.5 and above	3	9.33	100	0	4.00
Total or average	35	8.23	97	3	2.19

1. Primary business.

areas owned any land of their own. Farmers cultivate Forestry Department land and keep the entire produce, on condition that they provide around 90 days of free labour for the plantation and maintenance of state forests.

Empirical evidence from the survey areas

Socioeconomic characteristics

This section outlines some socioeconomic characteristics of the villages surveyed in the Chittagong Hill Districts. Table 9.10 sets out relevant information on these aspects for the farmers in the area with settled agriculture. First, 80 per cent of the households operate farms of between 1.0 and 3.5 acres. Second, average family size for all farm size classes is higher than the national average and tends to rise directly with the size of the holding. Third, agriculture is virtually the sole occupation of the households.

Table 9.11 presents information on the basic socioeconomic characteristics of the respondents in the area with shifting agriculture. First, over 80 per cent of the sampled households have farms of less than 5.0 acres. A significant percentage of the households do not officially have any land. They are primarily agricultural labourers engaged in *Juhm* cultivation. Second, average family size tends to increase with the size of the land area cultivated. One observes that the average family size in settled agriculture is considerably higher than that for shifting agriculture. The degree of inequality in the distribution of land area cultivated is significantly higher

Table 9.11 Some socioeconomic characteristics of households in a forested village dominated by shifting agriculture: Chittagong Hill Districts, 1990–91

Farm size (acres)	No. of samples	Average family size (no.)	Major occupation (%) Agriculture	Major occupation (%) Others[1]	Average cultivated area (acres)
Landless	14	6.29	0	100	0
Less than 2.5	10	7.20	100	0	1.70
2.5–5.0	4	8.50	100	0	3.13
5.0 and above	3	10.33	100	0	5.00
Total or average	31	7.26	55	45	1.44

1. Primary agricultural labour engaged in shifting (*jhum*) cultivation.

in the village with shifting cultivation as compared with that with settled agriculture.

Income and employment changes since the Green Revolution

In Tables 9.12 and 9.13 we give the changes in income and employment opportunities since the introduction of the new agricultural technology, and make the following observations.

(a) Production in both areas is dominated by *aus* and *boro* rice. Virtually no other crops are grown. Average output per acre does not show any significant variations across farm size classes.

(b) Annual employment in terms of workdays increased for all farm size classes after the introduction of HYVs in both villages. For the settled agriculture area, it increased by nearly 6 per cent (from 268 to 283 workdays per acre). While employment on small farms (of less than 1.0 acre) increased by more than on all other farms, in terms of actual workdays they have the lowest level of employment per acre (260 workdays per year as against a range of 277 to 296 days for all other classes). The information from the village with shifting agriculture shows a similar pattern, even though the overall percentage increase in employment is more than double that for the settled agriculture area.

(c) In both areas the average daily income in nominal terms has shown similar levels of increase since the introduction of the new

Table 9.12 Income and employment changes due to HYV technology in a forested area with settled agriculture: Chittagong Hill Districts, 1990–91

Farm size (acres)	Average output per acre (maunds)			Average no. of days employed in a year			Average daily income (taka)			Average yearly income per household (in taka)		
	Aus	Boro		Before HYV	After HYV	% change	Before HYV	After HYV		Before HYV	After HYV	% change
Less than 1.0	19.30	22.83		238	260	9.2	13.89	29.86		3,750	8,300	121.3
1.0–2.5	17.68	24.00		263	277	5.3	15.78	39.57		4,158	10,962	163.6
2.5–3.5	20.20	24.04		278	296	6.5	28.01	72.88		7,789	21,573	176.9
3.5 and above	19.04	22.67		273	280	2.6	23.29	96.67		6,367	27,067	325.1
Average	19.38	23.77		268	283	5.6	21.70	58.31		6,114	16,007	161.8

Table 9.13 Income and employment changes due to HYV technology in a forested village dominated by shifting agriculture: Chittagong Hill Districts, 1990–91

Farm size (acres)	Average output per acre (maunds)			Average no. of days employed in a year			Average daily income (taka)			Average yearly income per household (in taka)		
	Aus	Boro		Before HYV	After HYV	% change	Before HYV	After HYV	% change	Before HYV	After HYV	% change
Landless	–	–		224	269	20.1	12.71	37.36		3,193	10,061	215.1
Less than 2.5	17.22	20.22		258	273	5.8	17.98	51.72		4,640	14,120	204.3
2.5–5.0	17.44	19.78		248	285	14.9	21.21	76.23		5,260	21,725	313.0
5.0 and above	18.00	23.75		277	297	7.2	25.00	100.00		6,930	29,700	328.6
Average	17.54	21.15		243	275	13.2	16.98	53.69		4,288	14,776	244.6

technology. However, while the incomes of households on the largest farms have quadrupled, the daily incomes of the lowest income classes have only doubled.

(d) In the village with settled agriculture, average annual household income has increased by 162 per cent compared with nearly 250 per cent in the shifting agricultural area. The increase for the marginal households (less than 1.0 acre) is 121 per cent in the former area, compared with over 200 per cent in the latter area. By contrast, the increase in annual household income for households with the largest farms in both areas is considerably higher (over 325 per cent) than those with small farms. However, if the movement in consumer price indices since the introduction of the new technology in the early 1970s is taken into account (Bangladesh Bureau of Statistics (BBS), 1990), one finds that the real household income has fallen in general and much more so for the lower end of the scale, so reducing their exchange entitlements.

Environmental change: trends and the current state

Over the years the environment in our survey villages has altered. In the shifting agriculture area, soil erosion has depleted soil fertility as a result of the loss of humus from topsoil. Farmers now have to apply more fertilizer to maintain yield. Weeds grow more rapidly. However, in the settled agriculture area the accumulation of humus from forests has increased soil fertility.

In both villages, the application of modern inputs has reduced fish production in paddy fields by 60–75 per cent following the Green Revolution. Irrigation is provided using surface water and groundwater. No problems seem to have been observed with groundwater. Access to surface water has become more difficult, however.

Ecological and environmental changes include changes in micro-level temperature, colder but shorter winters and hotter summers, an increase in air pollution and an increase in pests and insects. The substantial decline in pasture and grazing land resulting from population growth, human habitation and the extension of cultivation to marginal areas has led to a decrease in the livestock population. The number of rice varieties cultivated following the Green Revolution has gone down. At present most farmers seem to depend on two varieties, namely BR-11 and BR-6. Some varieties used in the past seem to be becoming extinct. Thus the new agricultural technology does not seem to have widened genetic diversity – if anything, it may have been narrowed somewhat.

For a significant percentage of the population in the villages surveyed, forests have always been a source of fuelwood, wild fruits, honey and herbal and shrub medicines. However, in recent years, the population in these villages find it more and more difficult to gather these commodities because of falling supply. It takes two or three times longer now to gather the same quantity.

In both areas the number of birds and ground-dwelling animals, e.g. tigers, deer and foxes, has declined significantly because of a loss of habitat and hunting. Illegal logging, human habitation and hunting have all disturbed wildlife habitat. There are insufficient forest guards. Illegal logging poses a serious problem for the conservation of forestry resources. It is difficult to round up illegal loggers who may have powerful political backers. Unless these activities can be minimized, prospects for the conservation of forests seem bleak.

There is little prospect of expanding forestry in the survey areas. Replanting is done only on a limited scale. Limited funds and the limited area available for forestry seem to be two major problems to increasing tree cover in the area. In the area of shifting cultivation the relative profitability of bamboo production (70 per cent of the area) means that other trees cannot be grown.

Shifting cultivation, especially on a short cycle (normally three years), causes soil erosion and the loss of topsoil. This affects the growth of forests and bamboos and results in a decline in productivity. Also, certain types of trees affect the ecological balance and micro environment. For instance, teak and mahogany plantations allow little or no undergrowth to develop, and this is vital for soil conservation. Furthermore, some trees with broad leaves, e.g. teak, may contribute to soil erosion in heavy rain. Finally, it was gathered during the course of the survey that donor agencies have sometimes insisted on planting varieties of trees that are unsuited to the local environment. In this case, it concerned the planting of eucalyptus, the seeds of which the donor had an interest in exporting.

Environmental change: implications for the sustainability of employment and development

Forests have been a source of supplementary income and sustenance for the population in our surveyed village in the Chittagong Hills for a long time. It seems clear that the growth of population, the depletion of natural resources and the penetration of technological and market forces have reduced low-cost or free access to non-exchange sources of income for the poorer sections of these communities. Earlier we provided some evidence that a marginal increase in employment has followed the Green Revolution. Even though nominal income and exchange sources have increased quite significantly, real income may have declined. Furthermore, the decline in the opportunities for earning non-exchange income in order to supplement exchange income may have led to a decline in the *overall* quality of life, while the traditional role of women and children in helping to supplement family income is thereby undermined.

Forestry resources are being used at a faster rate than they are being replenished. This has disturbed the agro-ecological balance and wildlife habitats. Thus the sustainability of some communities (e.g. tribal) may be threatened.

Employment growth seems unlikely to be sustained, owing to high population growth, the lack of markets for many goods and the lack of growth of non-agricultural employment. In order to increase employment and income and to achieve sustainable development, population growth must be reduced and the scope of forestry-related activities must be broadened. Major obstacles in achieving this would appear to stem from problems that include ethnic rivalry, cultivation practices (e.g. *Jhum* cultivation), lack of cooperation between administrative departments, and other factors. It was also found that illegal loggers enjoyed strong political and administrative support at almost every level.

CONCLUDING COMMENTS

In the three villages studied, the adoption of HYV crops and associated technology has increased the amount of labour used in agriculture at least marginally, and has increased yields substantially. Therefore the amount of income available through market exchange of commodities seems to have risen. Agricultural production has increased in all three villages, along with some extension of agricultural production in the Chittagong Hill tracts. One consequence has been a fall in the income available from non-exchange sources, e.g. food obtained by gathering, fishing and hunting, the collecting of thatch and fuel and the availability of free fodder and sustenance for livestock. This has particularly affected the ability of the poor to earn extra income. Notably, women and children from poor households have been less able to supplement family income from such non-exchange sources. However, in the time available for gathering of primary data we were not able to determine whether or not the income of the poor from exchange plus non-exchange sources had declined, nor what were the overall changes in their economic welfare following the adoption of new agricultural technologies.

While HYV crops have substantially increased yields, all villages report that, in order to maintain these yields, they are being forced to apply more artificial fertilizers. This indicates problems in sustaining yields as a result of falling soil fertility and possibly also growing problems owing to agricultural pests. Furthermore, local varieties of crops are disappearing as farmers come to concentrate on growing just a few HYVs. While this has no immediate negative consequences for agricultural yield and employment, in the long term widely used HYVs may fail ecologically as a result of susceptibility to a particular pest, exposing a risk of eventual serious consequences for production and employment. The preservation of genetic diversity, especially of hardy local varieties of such crops, could help to guard against such a disaster. It is also clear that, as a result of the introduction of modern agricultural technologies, Bangladesh is becoming dependent on the use of non-renewable resources such as oil, natural gas and fertilizer to maintain its agricultural production.

Although there has been a substantial increase in agricultural yields in Bangladesh, these have been at the expense of production in other sectors of the economy. In the Chittagong Hill Districts, for example, large reductions in the number of fish were reported, and there were substantial reductions in Durgapur too. Serious soil erosion and water problems were reported from the villages in the Chittagong Hill Districts, as well as in Durgapur. As a result of farming activity in the Chittagong Hill Districts, forest resources were reported to be dwindling. All such spillovers and opportunity costs need to be taken into account in quantifying the impact of new agricultural technologies on income and employment.

Crop yields in Bangladesh are still well below those of similar crops in more developed countries, and Bangladesh's rate of application of agrochemicals, for instance, has not nearly reached the levels prevailing in Japan. Possibly Bangladesh will increase its application of agrochemicals and intensify its agriculture further along lines common in more developed countries. But any further intensification of its agriculture will undoubtedly add to its environmental problems. It may be that methods of modern agriculture developed for temperate areas are not well suited environmentally to countries located in the tropics, such as Bangladesh. It seems that greater efforts should be made to adapt overseas technologies and develop new technologies which are more appropriate for the environmental and social conditions prevailing in developing tropical countries such as Bangladesh. Much more research needs to be done to resolve these issues and to take account of environmental spillovers and their employment effects.

Environmental/employment linkages arising from technological change and production variations in agriculture in Bangladesh have been identified for the particular villages studied but, to the extent that quantification has been possible, this is mostly for on-farm or on-site effects. There is a need, apart from strengthening on-farm or on-site estimates of environmental/employment linkages and associated sustainability characteristics, to go beyond this and to identify and quantify off-site effects. Such effects include production spillovers or externalities, backward and forward economic linkages and their environmental/employment effects, sustainability of production and employment, taking into account all such linkages and capacities for labour absorption in both rural and urban areas. Both the overall impact of production-induced environmental change on employment and distribution require further study (Markandya, 1991), including the marginalization of women and children and migration from rural to urban areas.

Finally, the chapter indicates the need for further research in developing new agricultural technologies appropriate for the ecological conditions and resource availabilities in Bangladesh, as well as the need to further study off-site employment and macro-dynamic effects of agricultural development.

10 Sustainable Development and Employment: Forestry in Malaysia

James E. Jonish

DEFORESTATION

In recent years, world attention has been focused on the deforestation of tropical areas, particularly the Amazon basin and south-east Asia. The envionmental consequences of deforestation include soil erosion, water pollution and increased quantities of carbon dioxide in the atmosphere, leading to the greenhouse effect and global warming. The main causes of this deforestation include: (a) poor forestry or logging practices; (b) harvesting above sustainable yields; and (c) forest cleared and replaced by agriculture or animal husbandry practices.

Many environmental non-governmental organizations and some developing countries have advocated an anti-tropical-hardwood campaign to limit its use and to boycott hardwood imports from countries which do not observe sustainable practices. Malaysia, particularly, has resisted these charges, arguing that it is the developed countries that are responsible for the greenhouse effect and global warming. It is perhaps significant in this regard that Malaysia is the world's third largest timber exporter and the largest tropical timber and wood products exporter, to the amount of approximately US$2.9 billion in 1990.

Malaysia's forestry sector and policies

Exports represent 13–15 per cent of Malaysia's total foreign exchange earnings. In the states of Sabah and Sarawak, revenues from logs account for some two-thirds of foreign exchange and 50 per cent of government revenues. Yet forestries and the wood-based industries employ a relatively small number of people in Malaysia, i.e. approximately 150,000. Experts estimate that Malaysia's harvest amounts to twice the sustainable yield, with Peninsular Malaysia being particularly at risk. This result is understated if environmental and ecological concerns are considered together with timber harvesting in any definition of sustainability.

This chapter examines Malaysia's policies and practices relating to land use for forest-based products and services. Sarawak receives especial attention for several reasons: (a) Sarawak has the largest remaining area of forest in Malaysia; (b) it has attracted international attention over its harvest practices and the protests of indigenous people; and (c) it has recently been visited by several delegations examining its forest management practices.

Technology plays several roles in the controversy. Some analysts view traditional clear cutting or 'swidden' (slash and burn) agricultural practices by indigenous people as environmentally benign and thus sustainable. Government officials claim they are primarily responsible for deforestation. Modern technology has its supporters and detractors. Inappropriate technologies ('cable logging', imported plantation practices, poorly designed transportation systems) are viewed as a major factor in deforestation by some analysts. Proponents of new technologies argue that technology interventions, such as introducing new species, developing new products and promoting techniques that reduce wastage of wood, would increase the sustainable yield of Malaysia's forest products.

The indications are that forestry practices such as annual harvesting ignore stated policy concerning sustainable annual forest yields, that monitoring resources are meagre and that the multiple objective use of Malaysia's forest resources is not considered in practice. Thus, sustainability has not been achieved in the narrow, single-objective context of timber harvesting. Sustainability in the broader economic-ecological sense of the social and amenity values, biodiversity and concerns over watersheds as well as timber harvesting, has been all but ignored. Harvesting in Sarawak amounts to three times the sustainable yield in narrow timber-harvesting terms and perhaps four to five times the sustainable yield in economic-ecological terms. Adequate supplies of timber may not be available in the mid-1990s to support the wood-based products industries in Peninsular Malaysia.

FOREST MANAGEMENT: CONCEPTS AND ISSUES

Multiple use of forests: optimal rotation

Forests, and particularly tropical forests, are valuable for other reasons besides timber production. Forest land services also include watershed catchment and erosion control, habitat protection, hunting and gathering, recreation and indigenous shifting cultivation. In tropical forests, the biodiversity of species is an additional source of value. In Malaysia's forests, which are 130 million years old, it is estimated that there are more than 2,500 species of trees and 8,000 species of plants (Nor, 1983).

If one leaves timber production aside, the value of the forest land services resides in the standing unharvested forest. These non-timber services are monotonically increased with the age of the forest, at either a constant, increasing or (more probably) decreasing rate. In any event, the generation of timber revenues through harvesting is at the expense of the amenity services of non-timber production.

Multiple use management is practised in most forested regions of the world. In many cases, the management rules adopted are arbitrary or *ad hoc*. One solution is to adopt multiple use practices on all parcels of land (Clawson, 1978). But this is inefficient, and instead economic principles of specialization may be applied to multiple use management. For instance, preserving hillside and riverside forests will aid water catchment and erosion objectives. Preserving some areas in perpetuity for non-timber uses, and allocating other areas for sustainable production, will aid both non-timber and timber objectives. An important social objective is the sustainability of economic activities which the timber and non-timber services provide. This includes the size and the time and spatial distribution of income and employment that are dependent on these timber and non-timber services.

While the introduction of amenity benefits yields results corresponding to intuition (increased rotation periods), multiple use harvesting policies are extremely complex. Operational models rarely yield 'even flow', unchanging harvesting policies as long-term goals and they are critically dependent on initial conditions and assumptions (Bowes and Krutilla, 1985). In addition, the valuation of the non-timber services is at a very elementary stage of development. In such instances, amenity values can be made as large or small as politically desirable.

Policy issues in tropical forest management

Other conceptual policy issues in forest management include the monitoring of the yield plan, the use of appropriate technology to minimize environmental side-effects while maximizing yield, and the ownership of forest lands.

Table 10.1 summarizes the steps in the production process for timber harvesting and processing. At each step the major environmental impact is identified, as well as some of the technology or management options to mitigate the damage caused. It is assumed in Table 10.1 that the annual

Table 10.1 Environmental and social concerns of timber processing at different stages of the production process

Issue	Environmental/social concerns	Technological and management considerations
Sustainable yield	Deforestation Loss of timber, amenity values Unsustainable employment	Multiple use objectives Determine/monitor annual quota
Production process		
Harvesting	Erosion, siltation Timber stand damage Safety/health	Monitor harvest practices
Transport		
Logging roads	Erosion Access to land	Monitor construction Monitor access
Water channels	Pollution	Debark at site Closed ponds
Sawmills		
Waste wood	Disposal Toxic preservatives	Improve recovery rate Use for pulp/paper Restrict dumping
Sawdust	Disposal	Use for pulp/paper Closed burning
Land stewardship		
Private	Amenity values ignored	Quotas with market-based incentives
Public	Non-optimal practices Short-run time perspective	Quotas with market-based incentives
Indigenous	Loss of cultural base, amenity values Economic activities lost	Indigenous lands set aside

permissible yield has been determined and concessions granted for harvesting. The determination and implementation of the sustainable annual yield is critical to the sustainability of the direct and indirect employment levels dependent on the timber and non-timber services provided by the forest.

Development and Employment: Forestry in Malaysia 261

In areas designated for commercial timber production, a selective cutting system is used in tropical forests: only trees of a certain species above a certain diameter and of a certain height are harvested. The remaining trees and species are left to regenerate. This contrasts with the temperate zone countries where trees in a plot ready for harvesting are likely to be of the same species and age.

Unfortunately, a selective harvesting system can cause considerable damage to the remaining stand of trees if poor harvest practices are used. In a study by the Food and Agriculture Organization (FAO, 1981b), moderate to severe damage was found to affect 40–70 per cent of the remaining species in tropical rainforests after harvesting.

Inappropriate harvesting of trees on steep slopes accelerates soil erosion and the flooding of lowlands. Failure to leave a band of trees along rivers and streams aggravates the water siltation problem, leading to concern over water quality and fish habitat. In these cases, appropriate harvest technology is known. The problem is one of monitoring the concessionaires working in the forests. Remote sensing or aerial photography provides one technology option for monitoring yield allotments.

The construction of logging roads raises two concerns, as Table 10.1 illustrates. One is that roads are often poorly constructed and maintained, accelerating water run-off and soil erosion. The other is more intractable: the logging roads provide access to previously closed areas into which landless settlers follow to clear forests for non-timber uses.

The forestry cycle, i.e. the number of years before re-entry on a sustainable basis, is subject to management practices and technology choices. Fast-growing species such as *Acacia mangium* or eucalyptus would shorten the cycle. 'Liberation thinning' (the practice of removing small non-economic species to allow more rapid growth of the remaining species) would increase annual incremental growth, resulting in a greater yield per hectare or a reduced cycle to harvesting, or both. It would also increase forestry employment, most probably of the indigenous population.

Cut logs are generally transported by road or water, although high-value tropical timber is sometimes removed by helicopter. Logs awaiting processing or loading for export are also stocked in rivers or streams, often for months at a time. With time, leaching of the logs occurs and water pollution results, comparable in strength to weak sewage. Yet simple technologies are available, including debarking the logs on site or storing them in closed ponds.

Sawmills can cause further environmental problems, notably regarding the volume and disposal of sawdust and waste wood. Disposal methods include incineration (often open burning), using sawdust as a form of landfill, with subsequent water pollution, or dumping it directly into rivers and streams. In addition, toxic preservatives used in sawmills are often burned or tipped into rivers. Laws usually prohibit all these practices, but they usually carry low penalties or are not enforced.

The control of waste wood is subject to technology and management intervention as well. The recovery rate of wood at sawmills depends on the quality of the logs, sawmill techniques and the supervision of workers using band-saws. Waste wood can be recovered for use in the pulp and paper industries. Alternatively, it can be used to provide energy on site.

In the timber harvesting and processing stages in Table 10.1, safety and health issues abound. Throughout the world forestry is a high-risk industry. In tropical forests, including those of Malaysia, the dangers are increased. In Sarawak it is estimated that there are seven fatalities per 1 million cubic metres of harvest (ITTO, 1990). Failure to use safety equipment is officially given as the reason. However, although workers receive good wages, they are paid on a piece-work basis. As a consequence, workers and managers both aim at achieving a high output, and harvesting often continues at night.

The final item in Table 10.1 concerns land management or stewardship. Is private management preferable to public management? If timber production alone is the concern, private managers have incentives to adopt efficient rotation. This assumes efficient operations large enough to take advantage of any economies of scale. It also assumes that the amenity and social values of the standing forest are small or that they can somehow be captured by the owners. This seems unlikely in the case of tropical forests. Thus, efficient forest management with significant amenity and social values is unlikely to occur under private sector management. On the other hand, public management systems also suffer from deficiencies. Elected government officials may have an even shorter time horizon than private owners, with the consequence that long-term amenity values will also be downgraded.

Finally, landownership and the use of forested areas is complicated by the rights of the indigenous population. While most countries have laws recognizing and protecting such customary lands, natural resource development often goes ahead while conflicting claims to landownership languish in the courts. Media attention has made agencies, administrations and governments more sensitive (perhaps also more antagonistic) to indigenous peoples' claims.

Development and Employment: Forestry in Malaysia

MALAYSIA'S FORESTRY SECTOR

The tropical rainforest of Malaysia presents a complex and varied ecosystem. The forest types are the lowland and hill *Dipterocarpus* forests, the mangrove forests along coastal areas and peat swamp forests in inland swamp regions. The tropical rainforest is generally taken to be synonymous with the species-diverse lowland and hill *Dipterocarpus* forests.

Land use and forest resources

As recently as 1960, 72 per cent of Peninsular Malaysia was covered in natural forests. In 1987 the proportion was 47.7 per cent. Most of these losses are due to the conversion of lowland forests to agricultural use, largely because under the Land Capability Classification (LCC) programme in the early 1970s all good soils and soils on low and flat terrain were classed as agricultural land. Forests were relegated to residual land use status, on poor soils or steeper slopes (Nor, 1983). States used the LCC as a justification for converting forests to agricultural or plantation uses. The patterns of land use in Malaysia in 1988 are shown in Table 10.2. Forests still cover a significant portion (62.4 per cent) of the country's land area owing to the remaining abundance of forest lands on Sabah and Sarawak.

Officials point out that the conversion of forest to agriculture in most instances involves the planting of another tree crop: rubber, oil palm, coconut or cocoa. Thus, plantation forests replace natural forests and

Table 10.2 Forest and land use, 1988 (thousands of hectares)

Region	Land area	Forests	Per cent	Rubber	Oil	% 'trees'
Peninsular Malaysia	13,160	6,280	47.7	1,559	1,569	71.4
Sabah	7,370	4,770	64.7	87	276	69.6
Sarawak	12,330	9,470	76.8	207	41	78.8
All Malaysia	32,860	20,520	62.4	1,856	1,887	73.8[1]

1. = 75.6 per cent if coconut and cocoa plantations (600,000 hectares) are added.
Sources: Malaysia, Ministry of Primary Industries (n.d.); Malaysia, Department of Statistics (1990).

provide some of the same protective values (i.e. water catchment) as the original forest. Under this interpretation, adding these monoculture tree crops to the natural forests gives a tree cover of almost 74 per cent. Obviously, however, biodiversity is lost, as are amenity values such as recreation etc.

With states using the LCC programme as a justification to clear forests for alternative uses, efforts made to protect forest reserves led to the political acceptance of the concept of a Permanent Forest Estate (PFE) in the National Forest Policy of 1978. Each state was to establish a PFE and manage it on a long-term basis, incorporating timber harvesting, environmental protection and amenity values. Under the National Forest Policy, a PFE was to establish: (a) protective forests (land to safeguard water supplies and soil fertility and to minimize damage by floods and erosion); (b) productive forests (land to supply all forest produce in perpetuity as required for agricultural, domestic and industrial uses); and (c) amenity forests (lands with unique flora and fauna for recreation, research and conservation). In practice, PFEs were established on land unsuited for agriculture and often after very time-consuming negotiations.

The distribution of PFE and state forest lands are shown in Table 10.3. Briefly, while 20.5 million hectares, i.e. 62 per cent of the country, remain forested, the establishment of the PFEs provides that 14.07 million hectares (43 per cent of Malaysia's land area) will remain as forest, if the policy is enforced. Nevertheless, a significant percentage of land will be converted to monoculture tree crops.

Table 10.3 Forest resources in Malaysia (thousands of hectares)

Region	Peninsular Malaysia	Sabah	Sarawak	All Malaysia
Permanent Forest Estate (PFE)	4,750	3,350	4,640	12,740
Productive	2,850	3,000	3,240	9,090
Protective	1,900	350	1,400	3,650
National parks and wildlife areas	590	490	250	1,330
State forest lands	940	930	4,580	6,450
Total	6,280	4,770	9,470	20,520

Source: See Table 10.2

Table 10.4 Annual production of logs and sawn timber (millions of cubic metres)

Area	Logs	Sawn timber
All Malaysia		
1980	27.9	6.2
1986	29.8	5.2
1990	39.8	8.4
Peninsular Malaysia		
1980	10.4	5.3
1986	8.5	4.0
1990	12.4	6.2
Sabah		
1980	9.0	0.54
1986	9.8	0.80
1990	9.0	1.73
Sarawak		
1980	8.4	0.35
1986	11.4	0.37
1990	18.4	0.51

Sources: Malaysia, Department of Statistics (1990); Malaysia, Ministry of Finance (1990)

Production, employment and revenues

The economic impact of the timber and wood-based products industry on Malaysia is considerable. The annual production of logs and sawn timber is shown in Table 10.4. As appears clearly, the bulk of timber harvests, in volume terms, is for the log export market. It is, however, now part of Malaysia's policy to encourage the development of the wood-based products industry, increasing value added and employment with primary and secondary processing and reducing the export of logs.

Estimates of employment in the timber and wood-based products industry are provided in Table 10.5. The employment impact of downstream processing is immediately apparent. With two-thirds the annual harvest of Sarawak, Peninsular Malaysia has twice the estimated number of employees as a result of the further processing carried out there, even though it has not developed its processing capabilities completely. Furniture

Table 10.5 Employment in timber and wood-based products industries: 1989 estimates[1]

Sector and no. of firms	Peninsular Malaysia	Sarawak	Sabah
Logging (n.a.)	21,000	3,520	16,462
Sawmills (1,035)	28,000	5,428	8,816
Plywood/veneer (60)	15,930	2,222	4,070
Moulding (122)	5,000	769	935
Furniture (2,000)	10,000	n.a.	n.a.
Total	79,930	39,939	32,283
All Malaysia	150,152		

1. Data available for Peninsular Malaysia. Estimates for Sarawak and Sabah from output/employment ratio for logging, employment/firm ratios for sawmills, plywood, moulding.

Sources: Malaysia, Ministry of Primary Industries (1990); author's estimates.

manufacturing in particular is a designated target in Malaysia's plans (Malaysia, Ministry of Finance, 1990).

There are nevertheless constraints on the development of a wood-based products industry which have been recognized by the government. For example, firms are often of inefficient size and use outdated equipment and techniques. Wastage is considerable and the quality of the finished product is poor. The training of workers is inadequate and the supervision poor. The future supply of logs depends on Sabah and Sarawak, but sawmills, plywood/veneer and moulding firms are found mainly in Peninsular Malaysia, while transportation, energy and labour resources are scarce on Sarawak and Sabah. Moreover, international competition is increasing, including that from newly industrializing economies such as Indonesia and the Philippines. Tariff and non-tariff barriers are being erected in consuming countries. Finally, 'downstream' firms will require a reliable, low-cost source of log supply. Thus, the forests in Malaysia needed to be sustained to ensure the sustainability of the wood-based products industry.

The current revenues from timber harvesting and the wood-based products industry are significant, as Table 10.6 details. In terms of total economic activity, timber and the wood-based products industry represent about 10 per cent of GDP and about 2.5 per cent of total employment.

Table 10.6   Forest and wood-based revenues, 1989
(millions of Malaysian ringgits)

Total value	9,600
Total government revenues	
Peninsular Malaysia	1,794
Sabah	226
Sarawak	878
	614
Exports	
Total value	8,714
Log value	5,228
Log volume (millions of cubic metres)	20.9

Sources: Malaysia, Ministry of Primary Industries (n.d.); idem (1990).

Government timber royalties, premiums and cess charges represent 1.7 billion Malaysian ringgits. This is almost 20 per cent of all federal revenues. Alternatively, it represents more than 50 per cent of the state revenues of Sarawak and Sabah. Exports of all timber and processed products represent 13.2 per cent of all exports from Malaysia. Of this amount, about 60 per cent of export revenue comes from logs.

Malaysia's forestry policies and practices

Policy

In Malaysia, responsibility for land use, conversion and management rests with the state governments, not the federal government. Thus, Malaysia has 13 separate forest policies coordinated by but independent of the federal government. The Federal Forestry Department has a sophisticated research unit, the Forest Research Institute of Malaysia (FRIM).

The National Forestry Policy of 1978 referred to above, and the concomitant PFEs, were reinforced in the Fifth Five-Year Plan (1986–90), which devotes an entire chapter to environmental protection, not only of forest products and resources but also of air and water quality. In forestry, the Fifth Plan calls for further forest 'set asides' of unique areas and the gazetting of more state lands into PFE. In addition, it calls for the enforcement of environmental impact assessments (EIA) for development projects affecting the forests including land use conversions, drainage of wetlands and logging of forest areas greater than 500 hectares. EIA are required under the Environmental Quality Act of 1974; in practice they are often not done.

The National Forest Policy of 1978 and succeeding Five-Year Plans provide for the promotion of a wood-based industry in Malaysia (National Forest Policy, 1978). The basic objectives are to transform the wood-based products industry to a major source of employment and export revenue by developing primary, secondary and tertiary processing of timber. This would discourage log exports and encourage use in domestic and foreign value-added activities. Implementation, however, has not always gone smoothly. All states have established PFEs, but the gazetting of lands is often time consuming, and logging activities continue. The individual states often ignore federal advice and guidelines in forestry management, and EIA studies on land development schemes are still haphazard. While progress has been made in developing the wood-based products industry, especially in Peninsular Malaysia, it seems improbable that the goal of processing 20 million cubic metres of logs (out of a total production of 29 million) will be met by 1995, particularly as fewer than 8 million cubic metres were processed in 1990.

Forest management

Over the years, Malaysia's forest management system has evolved from the Malaysia Uniform System (MUS) to the Selective Management System (SMS), reflecting changes in terrain and species diversity, marketability of new species and optimal rotation concerns. The MUS involves the removal of all trees in one felling, with poison girdling of the remaining non-economic tree species. In lowland forests, the result was the creation of an 'open canopy' where sunlight would penetrate and seedlings would grow quickly. The MUS dominated Malaysia's forest practices from 1949 to 1980, and is still practised in some areas today, particularly Sabah and Sarawak.

The MUS was replaced by the SMS in the late 1970s and early 1980s. The SMS involves a use of a pre-harvest inventory to identify, by diameter, those trees selected for harvesting. A post-harvest inventory is made to assess the remaining stock to determine the amount of silviculture treatment necessary. The advantages and potential difficulties of the SMS have been identified (Nor, 1983). The advantages over the MUS include: (a) the regeneration cycle of seedlings is reduced by 50 per cent or more (to 25–30 years); (b) the hill forest is much more heterogenous than the lowland forest – 'open canopy' felling would thus increase waste; and (c) silviculture costs are reduced (less replanting, poison girdling, thin-

Development and Employment: Forestry in Malaysia 269

ning): in fact, the SMS relies on natural regeneration to a great extent. There are potential problems with the SMS, however: proper pre-harvest inventories and proper logging practices are essential, and the monitoring of harvest practices and appropriate post-harvest silviculture treatment are necessary. In practice, the lack of Forestry Department manpower prevents careful pre- and post-harvest inventories and the monitoring of logging practices.

Ideally, the actual annual coupe (harvest) within the area set aside each year (71,200 hectares) is determined by the pre-harvest inventory of the state forestry department. Trees are measured and marked for felling. Once the annual harvest is determined, the new concessions are licensed. There are two broad classes: (a) long-term permits for large sawmills, state corporations and timber complexes which pay a premium (similar to a stumpage fee) for harvest rights; and (b) short-term permits to individuals, based on a bid or tender system. The duration of the permit depends on the size of the lease and the capacity of the logger. Shorter leases increase state revenues at the cost of careful logging practices and sustainability criteria. Abuses of the permit system have been noted in Malaysia's newspapers.

Silviculture treatment of logged-over forests is a key to estimates of sustainability. As noted above, the SMS relies more on natural regeneration of the remaining stand of the forest, but it does require more careful logging practices and post-silviculture treatment. Such treatment includes selective poison girdling, vine clearance or liberation thinning and replanting of species if the volume and variety of the residual stand is inadequate. What constitutes adequate silviculture treatment is a source of interpretation and dispute. Regardless of the quality of treatment, there is a big backlog of land area to be treated (approximately 291,000 hectares in 1990); moreover, the rate of forest rehabilitation on Sabah and Sarawak is far less than for forest harvesting.

A final management strategy practised on some state lands, primarily in Peninsular Malaysia, is the development of plantation forests of fast-growing species such as *Acacia mangium*, *Gmelina arborea* and *Batai*. By 1990, some 53,000 hectares had been planted. The quality of these species is recognized as inferior to that of natural forest species, however.

Supply and demand concerns

Peninsular Malaysia, the focus of the wood-based products industry, faces an imbalance in natural log supply versus anticipated log demand requirements during the latter part of the 1990s, as Table 10.7 suggests. A sharp

Table 10.7 Sources of log supply: Peninsular Malaysia 1988–2010 (millions of cubic metres)

Period	Natural forest[1]	Forest	Rubber	Total
1988–90	8.20	–	1.60	9.80
1991–95	8.20	–	2.25	10.45
1996–2000	4.80	–	1.20	6.00
2001–05	4.80	1.75	1.50	8.05
2006–10	4.80	2.00	1.30	8.10

1. After 1994, harvest is from PFE.
Source: Malaysia, Ministry of Primary Industries (n.d.).

fall in natural log production is expected in 1994 or 1995 as a result of the clear cutting of state lands and the reliance on the PFE for remaining supplies. The logical source of log supply for Peninsular Malaysia would be Sabah and Sarawak, which have only limited processing capabilities. However, both states are intent on developing their own wood-based products industry through downstream processing. These states have introduced export levies to encourage domestic processing, which apply to Peninsular Malaysia as well as to other countries. At the very least, Peninsular Malaysia will suffer a reduction in logging employment, since by 1995–96 employment will be no more than 10,500 in logging (if output/employment ratios hold constant). If logs from Sabah and Sarawak are not forthcoming, employment growth (perhaps employment levels) in sawmills, veneer/plywood, mouldings and furniture will also be reduced. In any case, the transportation costs and export levies will give Peninsular Malaysia an additional cost disadvantage for downstream products.

The case of Sarawak

Controversy over the harvesting of Malaysia's tropical rainforest has been highlighted by recent events in Sarawak. Deforestation coupled with indigenous land claims have led to protests, blockages, arrests and NGO intervention in the state. Threats of timber boycotts followed. Several study missions (United States House of Representatives, 1989; ITTO,

1990) examined harvest practices in Sarawak as a result, prompted further by a request from both Malaysia and Sarawak itself to undertake research concerning sustainable forest management in Sarawak. The International Tropical Timber Organization (ITTO), representing countries possessing more than 75 per cent of the tropical forests in the world and with 95 per cent of the international trade, conducted the Malaysia study. Its objectives were to 'assess the sustainable utilization and conservation of tropical forests and their genetic resources as well as the maintenance of the ecological balance in Sarawak ... to make recommendations ...' Its findings included the following points: (a) in a sentence, the mission's overall assessment is that sustainable management of the forests of Sarawak is being partly achieved ...; (b) admirable features include ... PFE, a Forest Policy which sets watershed protection as a primary objective, the reservation of areas for biodiversity, ... a system of tracing and controlling movement of logs ...; (c) a major impediment to sustainability is overcutting of hill forests and inadequate catchment management and control of felling operations; and (d) the network of totally protective areas (TPA) is insufficient and procedures for gazetting TPA and PFE excessively slow.

The ITTO made three recommendations: hire more forestry staff, improve standards of water catchment in hill forests, and phase down annual harvests to 9.2 million cubic metres (including state forests), i.e. a 30 per cent reduction. In the event, log production in 1989–90 increased to 18 million cubic metres.

The ITTO report was highly criticized, as were logging practices and the treatment of indigenous peoples' land claims. Another mission to Sarawak (United States House of Representatives, 1989) reached the following conclusions: (a) current logging practices are unsustainable and have degraded the environment, e.g. by erosion, flooding, siltation and loss of habitat and wild game and fish; (b) the government's response to native peoples' land claims is inadequate, with competing land claims remaining unresolved; (c) up to one-third of Sarawak's 1.5 million people may be adversely affected by logging activities; (d) timber revenues benefit the local population very little, except for concessionaires; (e) a recruitment freeze in Sarawak's Forestry Department limits the supervision and management of logging practices; (f) the cultural diversity of native peoples is threatened, and there is no local control of the rate of logging and/or economic development; and (g) there is tremendous waste in the timber industry: for example, poor harvest techniques damage the remaining stand of trees, and poor sawmilling practices, outdated technology and poor supervision of workers result in a low recovery rate of wood at sawmills.

Both study missions found Sarawak's forestry practices to be nonsustainable. Sarawak, like Peninsular Malaysia, is attempting to develop its own wood-based products industry but, unlike Peninsular Malaysia, Sarawak suffers from competitive disadvantages in skilled labour, capital and energy supplies and transportation difficulties. With a very small local market, Sarawak would be at the mercy of international market prices and quality standards for almost its entire wood-based products output. It appears that, for Malaysia as a whole, it would be economically more efficient to continue to develop the wood-based products industry on Peninsular Malaysia on account of its existing infrastructure, processing facilities and skilled and semi-skilled labour. This would mean that Sabah and Sarawak would eventually have to export logs to Peninsular Malaysia, either with or without the export levies currently paid by other importers.

MALAYSIAN FOREST POLICY ALTERNATIVES

There is no doubt that current timber harvesting levels in Peninsular Malaysia, Sabah and Sarawak are not sustainable in biological or economic terms. In part this reflects planned changes on state lands from forestry uses to plantation agriculture or urban conversion. But annual harvesting on PFEs is not sustainable either. In addition, the demand by indigenous populations for rights to their customary lands (in both permanent forest and state lands) has not been resolved, even though these rights are protected by law. Settling this issue would further reduce any annual sustainable yield.

The problem is not policy; Malaysia's national policies and implementing institutions are in place. Rather, practices fall short of policy pronouncements. Management plans for forestry management are required: in practice, they are limited to noting activities that have occurred. The Forestry Department of each state provides estimates of sustainable yield on a biological basis, but these are advisory only and state government officials are more concerned with short-term revenue returns from timber harvesting than sustainability. Often, forestry laws governing timber extraction are not enforced owing to lack of qualified personnel. Fines and penalties are too low to be effective.

The consequences of continuing current practices are well known. Government projections indicate that total annual log production in

Peninsular Malaysia will decline after 1993; the decline is more pronounced if natural forest sources only are considered. Sabah's annual production has already declined. Sarawak's forests will be completely harvested in 11 to 15 years at current levels of production (approximately 340,000 hectares per year). After this period, logging would cease until land logged in the early 1970s is ready for reharvesting. With continued deforestation, the loss of amenity values will be exacerbated. This, and the need to reduce annual yields, is being recognized within Malaysia. The question is how, and by how much.

Sustainable annual yield

Sarawak, on which this subsection focuses, has attracted worldwide attention in recent years owing to protests by its indigenous peoples and the support of environmental NGOs. Under pressure, the government of Sarawak made a decision to invite a study mission of its forestry practices (ITTO, 1990). As mentioned above, the study results have been widely criticized.

Definitions of sustainability

At an abstract and non-operational level, sustainability generally means the attainment and maintenance of a natural resource stock yielding a continuing and non-declining flow of products and amenity values, at levels each of which reflect the values of society. One of the non-trivial questions reflecting society's values, it involves the relationship of humans with nature. Does nature exist for humans to exploit and is this compatible with a harmonious relationship with the natural environment? An extreme ecological view puts man subservient to nature, with a belief in the equality of biospecies and the promotion of biological diversity, espoused by the Convention on International Trade of Endangered Species (CITES). Complete biological diversity would occur only when the forest is undisturbed.

Within the broad definition of sustainability, troublesome policy issues are found in Malaysia as elsewhere. These include the following.

(a) *Policy versus practice.* Does the government have a forest policy promoting the sustainability of timber and non-timber objectives? From the subsection on 'Malaysian forestry policies and practices' above, we saw that the answer appears to be 'yes' with regard to policy and 'no' with respect to practice. Sustainability criteria, when achieved in practice, are done so through the advice, monitoring, collection and research activities of the state forest departments.

(b) *Sustainable policy for non-timber values.* Even within the policy context, timber values dominate non-timber amenities. Policy emphasis is on the sustainable yield of timber benefits; non-timber benefits receive less attention. With non-timber values neglected, the income and employment consequences of these values are also neglected. These include, *inter alia*, tourist-related activities, non-timber products related to the forests' biodiversity (cosmetics, drugs, etc.), and the maintenance of the indigenous peoples' lifestyle. Yet some investigators have suggested that state and government officials sometimes consider indigenous peoples a nuisance or as responsible for Sarawak's deforestation (Lau, 1979; ITTO, 1990; Malaysia, Ministry of Primary Industries, 1990). This seems to be a view accepted by several representatives of the ITTO mission to Sarawak as well (ITTO, 1990).

(c) *Sustainability of PFEs and state lands.* Another issue relates to which forest areas are being considered for sustainable management. For this purpose, the major distinction is between PFEs, which are to be managed for forest timber and non-timber uses, and state lands, representing all other lands which are to be used for a multiplicity of purposes. Some state lands are cleared and converted to plantation or other agricultural uses. Other lands are used for urban and industrial purposes. Finally, some state lands are regenerated for subsequent forest production. Sustainability thus depends upon a distinction between current land use and planned land use and on the desirability of converting existing forest lands to agricultural, industrial or residential purposes.

Sustainable employment in Malaysia's forestry sector

The 1990 ITTO mission to Sarawak explored in considerable detail data and estimates relating to the determination of sustainable yield for that state (ITTO, 1990). Unfortunately, lack of space precludes their reproduction here, and interested readers are referred to the ITTO study itself. Estimates for sustainable yield from the PFEs for Sabah and Peninsular Malaysia have been used to derive sustainable employment implications in the timber and wood-based products industry in Malaysia.

The information in Table 10.8 examines current (1990) production of raw logs in the three major producing areas and the associated logging and wood-based products employment. Column four of Table 10.8 represents 'employment units' supported in logging by the annual coupe in each region and for Malaysia as a whole. In general for Malaysia, each

Table 10.8 Current (1990) annual harvests and employment

Area	Total harvest (millions of cubic metres)	Total employment logging	Logging employment	Employment units	(L/Q) Total
Peninsular Malaysia	12.4	79,930	21,000	1.69	6.44
Sabah	9.0	30,283	16,462	1.83	3.36
Sarawak	18.4	39,939	31,520	1.71	2.17
Malaysia	39.8	150,152	68,982	1.73	3.77

Source: Jonish (1992).

1,000 cubic metres cut supports 1.73 logging employment units. The figures for Peninsular Malaysia, Sabah and Sarawak reveal a very similar relationship between annual harvest and logging employment.

Column five represents employment units supported in the entire wood-based products industry, including semi-finished and finished products (L = total employment; Q = total harvest expressed in thousands of cubic metres). The figure for the whole of Malaysia (3.77 employment units per 1,000 cubic metres of annual cut) hides considerable regional disparities. In Peninsular Malaysia, the development of the wood-based products industry has proceeded further into semi-finished and finished product lines than in Sabah and especially Sarawak. The result is greater employment supported by volume of logs cut in Peninsular Malaysia than in Sabah or Sarawak.

For this purpose the following assumptions are maintained: (a) Malaysia relies only on its own timber resources for raw materials, and does not become a net importer of logs; (b) existing technology is maintained, as is the apparent productivity implicit in the employment unit concept (L/Q); (c) no further downstream processing of wood products occurs beyond the level exhibited currently by Peninsular Malaysia; and (d) sustainable yields of timber are from the PFEs only.

Table 10.9 provides some benchmark estimates of sustainable employment. Column one provides the total and regional distribution of the PFEs, while column two represents estimates of sustainable annual yield from the PFEs. The total annual coupe for Malaysia is 14.8 million cubic metres from the PFEs. This contrasts with the actual coupe of 39.8 million cubic metres in 1990 from all sources: PFEs, state lands and plantations.

Table 10.9 Sustainable annual harvests and employment

Area	PFE (millions)	Estimated sustainable yield (millions of cubic metres)	Logging[1]	Employment Total[2]	Total II[3]
Peninsular Malaysia	4.75	4.9	8,325	31,556	31,556
Sabah	3.35	3.9	7,200	13,104	25,116
Sarawak	4.64	6.0	10,345	13,020	38,640
Malaysia	12.74	14.8	25,604	57,680	95,312[4]

1. Employment in logging is Logging Employment Unit (L/Q) × Yield.
2. Total Employment I is Total Employment Unit (L/Q) × Yield.
3. Total Employment II is Peninsular Malaysia Total Employment Unit (L/Q) × Yield.
4. Ignores Peninsular Malaysia plantation yields and associated employment (3.2 million cubic metres per year × 6.44 = 20,600 employees).

The next column provides estimates of logging employment, given the sustainable yield figures. Logging employment would fall from nearly 69,000 persons in 1990 to 25,600 persons under sustainable yield conditions. Sarawak alone would lose 21,000 logging jobs.

The figures in column four reveal Malaysia's forest and wood products employment to be approximately one-third of the current levels under sustainable yield conditions and with the current structure and extent of processing of wood products. All regions suffer in the downturn. Total employment is reduced from 150,000 to less than 58,000 (see Tables 10.8 and 10.9).

The impact of downstream processing is immediately apparent in column five, where the assumption is that processing on Sabah and Sarawak proceeds only to the point where Peninsular Malaysia is at present. Sustainable employment increases from 57,000 to 95,000. Although this is a significant improvement, it still represents a loss of one-third of industry employment from current levels.

While the sustainable employment estimates were based upon conservative assumptions, in the short term these assumptions appear realistic. The adoption and diffusion of new technology, investment in downstream processing facilities and the introduction of new commercial species are not

Development and Employment: Forestry in Malaysia 277

easy commitments that bring immediate returns. It is further noted that Peninsular Malaysia has made an effort to develop and harvest plantation forests (oil palm, rubber) for wood products. Official estimates place this additional source of timber at 3.2 million cubic metres per year – perhaps 20,000 additional wood-based industry jobs.

Sustainable employment, based on current productivity and the structure of the industry, may be only 58,000, including 25,000 logging jobs. Current employment is 150,000, with 69,000 in logging positions. Total employment opportunities would be maintained if downstream processing activities were increased and new sources of supply (oil palm and rubber) were included. Even with these developments, employment losses of 35,000–55,000 could be anticipated.

Clearly, there would be significant transitional costs in any movement towards sustainable yields in Malaysia, and such a policy change would require transitional assistance for displaced workers. However, since in any event current annual coupes are not sustainable in the long run, the downturn in logging employment and associated wood-based products employment will occur either with or without an explicit policy decision respecting sustainable yield. Conceivably, an unplanned downturn may prove more abrupt than a gradual policy directed towards downsizing the sector to levels that are sustainable.

Monitoring and principal/agent concerns

One of the non-controversial recommendations of the ITTO was the need to strengthen comprehensively the State Forestry Department, whose staff levels had been frozen since 1982. Since then, annual harvests have increased by 50 per cent. The result is that most of the staff's time is spent on revenue collection and administration.

Under current conditions in timber harvesting, the State Forestry Department acts as the principal; their agents are the concessionaires. The familiar problem is, how do principals ensure that agents act in a manner consistent with the principals' objectives? The mechanisms are: (a) the monitoring of forest harvest practices; (b) the control of terms and conditions of licences, including time-limits, fees and royalties; and (c) the imposition of penalties for non-compliance or inappropriate behaviour. However, monitoring practices, pre-harvest inventories, post-harvest inspection and rehabilitation efforts seem weak, at best. As a result, in addition to inadvertent environmental consequences resulting from the lack of state supervision, there are reports of illegal re-entry into cut PFE and state lands, game poaching and illegal timber harvesting in national parks and wildlife areas, such as Taman Negara in Peninsular Malaysia (*New Straits Times*, February 1991).

Again, licence and concession practices are coming under increased criticism. The terms of the concessions vary from several years to 25 years, and renewals are not guaranteed. Suggestions have been made to increase the term of the concession so that they match the cutting cycles (at least 35 years), with conditional renewals possible after post-harvest inspections. Policy changes along these lines would attempt to bring the behaviour of concessionaires (as agents) more in line with state (as principal) objectives.

Finally, penalties for non-compliance or inappropriate behaviour need to be strengthened. Violations are often not punished, in part owing to insufficient monitoring, and in any case the small penalties imposed can easily be absorbed by concessionaires whose main interest is in short-term profit. For large or even small operations, the penalties are unlikely to be a significant deterrent.

Reduced demand for log exports

A number of policies have been recommended both within and outside Malaysia to reduce the demand for tropical log exports and thus presumably come closer to sustainability in timber harvesting practices.

Increased production of finished products

A policy option supported by the Malaysian government is to encourage downstream activities in the wood-based products industries. While the potential to move into higher-valued products in these industries exists, the problems facing Malaysia are significant in spite of attendant employment gains. Firms are small (87 per cent have fewer than 50 employees), productivity is low and product quality is variable, with low log recovery rates. Further, the overwhelming bulk of sawmill and wood manufacturing capacity is in Peninsular Malaysia which, in fact, processes most of its own timber harvest. Log exports are perceived problems on Sarawak and Sabah, which have no serious wood-based products industry. The development of such an industry would require closer coordination between Peninsular Malaysia, Sabah and Sarawak. The latter states have resisted cooperation, arguing for a need for wood-based products industries in their own states. But this strategy would be complicated by a lack of skilled labour, small domestic markets and poor transportation facilities.

Aggravating the situation are the import policies of Malaysia's major customers: Japan, the Republic of Korea and Taiwan, China. Logs imported from Malaysia bear low tariffs while finished wood-based products are discouraged by much higher trade barriers (FAO, 1990). The ability to expand Malaysia's wood-based products industry depends on some factors beyond Malaysia's control, including consumer country trade policies and existing and new trade competitors such as Indonesia.

The states of Malaysia are endeavouring to encourage local processing. Sarawak provides a rebate of 80 per cent on the royalty rate of logs processed (within the state only). Log exports to other countries or to other Malay states bear the full royalty rate. In 1988, the revenue forgone by Sarawak was estimated at M$20 million on 1.4 million cubic metres of locally processed logs. Total timber harvests were approximately 13 million cubic metres in 1988 and 18 million cubic metres in 1989 and 1990.

Increased royalties, cess charges, fines

A familiar Pigovian response to externalities would be to internalize the externalities through tax levies and thereby increase the rotation period. To the extent that log product prices were also increased, some reduction in demand would also occur. Alternatively, at constant export prices, concessionaire profits would be reduced.

In 1988, Sarawak derived 480 million ringgits from the forestry sector revenues, including premiums, royalties, cess and fines. This was 53 per cent of state revenues. As noted previously, fines are low and not an effective deterrent to undesirable behaviour. More meaningful penalties, perhaps on a sliding scale for continuing violations, would improve compliance and perhaps enhance revenue collections as well.

Timber export quotas

Timber export quotas have been proposed or adopted by a number of tropical timber countries, including the Philippines and Thailand. Recognition of non-sustainable timber practices is in part responsible, but efforts to encourage downstream wood-based industry production also contributes to these policies. For example, Sarawak and Sabah are making efforts to curb log exports to stimulate domestic downstream activities. They also offer royalty rebates on domestically processed timber.

Export quotas can accomplish what sustainable yield calculations and practices did not: the limiting of annual harvesting and of the export of tropical timber. An additional benefit is that the reduced export supplies by a number of countries might raise market prices, so industry and government revenues would not fall much.

The downside is that export agreements by countries to raise prices have generally been unsuccessful, since individual country goals, resources and population pressures all differ. Furthermore, smuggling contraband tropical timber is not uncommon in the region.

Consumer boycott of tropical timber

A number of environmental groups have suggested that a boycott of tropical timber by consumers would reduce demand and unsustainable harvests. However, there are difficulties with this proposal: (a) a consumer boycott would have to depend on joint cooperation to be effective. As timber processing capacities built up in Far Eastern countries rely heavily on foreign imports of timber, these processing interests are unlikely to be ready to operate against their own self-interest; (b) concern need not abate if a boycott were successful: if tropical forests lost their timber value because of the boycott, this might encourage more rapid clear cutting and conversion to non-timber uses that would accordingly become more attractive; and (c) a consumer boycott requires the 'labelling' or identification of sustainable and non-sustainable sources of timber. Past efforts to convey such information reliably do not inspire confidence.

Property rights allocation

The Sarawak Land Code recognizes native customary land rights. Section 5(2) recognizes rights created prior to 1958. It does not require formal filing of applications, deeds, etc. In practice, the definition of customary land is disputed while logging continues. The state recognizes the right to native dwellings (longhouses) and immediately adjacent areas, but not the adjacent forests. Maps of customary lands do exist, but not officially. The first need is thus to introduce formal maps of customary lands into the official record or to develop a map of customary lands. Unofficial estimates have put claims at about 500,000 hectares. Resolving indigenous land claims would go a long way towards defusing the socioeconomic debate over forestry management.

The indigenous population might also be given property rights in the production of timber and amenity values. More training and employment opportunities would also enable these people to receive more of the direct benefits of timber production and processing activities. Furthermore, employment in parks and reserves and as forest department personnel would involve them in amenity management.

Development and Employment: Forestry in Malaysia 281

Land can be converted to PFE use from state land use and vice versa under the National Forestry Policy and Act. The allocation of more lands to PFEs or to totally protective areas (TPA) would provide greater assurances of long-term sustainability of forests for timber and non-timber uses. But the gazetting process to PFEs is lengthy, and exploitation and conversion proceed while the legal process continues.

Privatization has been recommended as an alternative to state-owned and controlled forests. It is argued that the interests in maximizing net returns over all time-periods by private owners would ensure the sustainability of timber harvests. However, it ignores amenity and social values in the forest which are not captured by the private owners and it ignores the shortness of the private time horizon as governed by the interest rate. Thus, public monitoring and control would remain necessary with private forest management. The most practical mechanism for the privatization of forest lands would be to lengthen the licence/concession period. As mentioned earlier, this varies from several years to 25 years, too short to interest concessionaires in the maintenance and regeneration of the forest resources.

Increased yields and alternative products

Research, testing and commercialization of new ideas point to directions Malaysia might take to increase yields on PFE and state lands. Liberation thinning can increase harvest volume by 40–60 per cent on a 35-year cycle, and 16–50 per cent on a 50-year cycle. While liberation thinning is costly and labour intensive, yield increases are large enough to suggest such practices to be cost-effective. The employment gains would offer an additional social benefit.

Even on clear-cut state lands, subsequent timber production and amenity values are non-zero. Much of the converted agricultural land is in plantation agriculture, oil palm and rubber. Although at reduced levels, these plantations provide timber, water catchment and soil protection. In recent years, ageing rubberwood has been utilized in Peninsular Malaysia as a timber product. Other yield-increasing efforts include the establishment of plantations of fast-growing tropical varieties such as *Acacia mangium*. To the extent that these practices on state lands increase annual harvest yields and provide employment, this may help to reduce pressures on native rain forests.

Another promising strategy is the development of new commercial products within Malaysia's forests. Rubberwood and *Acacia mangium*, mentioned above, are two illustrations. Similarly, the ITTO estimates of sustainability for Sarawak implicitly assume that a commercial market for smaller logs of a poorer quality can be developed.

The potential of new products is sometimes considered enormous, particularly for non-timber forest products (NTFP). A number of products have been identified for commercial feasibility (Dixon, 1988). These include products for the cosmetics and perfume industry, spices, herbal teas, mushrooms, honey and natural foods for which, it is maintained, there is a growing 'green' customer base.

Finally, as indicated earlier, another way to increase net yield is to improve harvest, transport and sawmill practices so that waste is reduced, since even waste by-products (inferior quality logs, waste wood, sawdust) have product uses in the pulp, paper and wood chip industries.

Small-scale swidden agriculture

Swidden agriculture is 'slash and burn' agriculture, a traditional farming method practised by transitory subsistence populations. In areas with high population densities, the extensive use of this farming method can be destructive, even though in Sarawak low population and abundant land make it less of a problem.

State and federal government sources view swidden or shifting cultivation as the main cause of Sarawak's deforestation (United States House of Representatives, 1989). Others view logging practices as mainly responsible. The same data may even be used to support opposing views. According to a state forestry official (Lau, 1979), 'with an average family of six persons, the number of households practising shifting cultivation would be 36,000. Estimates of forested area cleared annually per family unit vary widely, but a conservative average of 7 acres has been estimated. This puts the annual area of forests cut for cultivation at 250,000 acres (100,800 hectares). The percentage of mature high forest cut is around 60 per cent.' The author concluded that the shifting cultivation system is a disaster to sustained forest management, whilst 11 years later an American study group found that:

> An estimated 36,000 families in Sarawak are swiddeners. Studies have shown that swidden households with an average of six people rarely exceed 6.2 acres (2.5 hectares) of land, which represents a total of 180,000 acres (73,000 hectares). Less than half of swidden land is cultivated in primary forest. The amount of primary forest used is not even 15 per cent of the 270,000 hectares logged by the timber industry in Sarawak during 1985.

The authors concluded that it is difficult to blame shifting cultivation for deforestation.

Both views have elements of truth: shifting cultivation does contribute to deforestation (42,000–60,000 hectares per year) but not as much as is claimed by the government. Timber harvesting contributes to deforestation if sustainable practices are not followed, which they are not. In 1989, 340,000 hectares were logged in Sarawak. With a larger population and a smaller land area, shifting cultivation with movement every two to three years could become a serious problem contributing to deforestation. At present in Sarawak, logging practices appear to be a much more important contributor to deforestation.

None the less, there are proposals to reduce the impact of shifting cultivation. Permanent settlements with alternative employment, including plantation agriculture, have been recommended. Improved soil rehabilitation, fertilizers, etc., are also argued to reduce swidden agricultural practices by maintaining productivity on current settlements.

Such proposals are strongly resisted by activists within the various Indian tribes in Malaysia and particularly Sarawak (ITTO, 1990, Appendices). Various NGOs, both within and outside Malaysia, have come to their assistance in providing political, legal and research advice in resolving their problems. As a first major step, the recognition and resolution of indigenous populations' land claims (estimated at 500,000 hectares) would provide a basis for the solution to their predicament.

National and international revenue sharing

It has been argued that the heavy reliance of Sarawak and Sabah on revenues generated from forest harvesting has accelerated unsustainable harvest practices. In Sarawak, timber royalties represent 53 per cent of government revenues. The problem is exacerbated by the fact that forest lands and policy are the domain of the individual states. Each state has its own forest department and coordination among states is voluntary. This suggests two strategies: national and international revenue sharing.

National revenue sharing

While each state has its own forestry department, policies and practices, this is not the case for all natural resources and their exploitation. Oil and gas are under control of the federal government, not individual states. The reason is perhaps that most, if not all, of Malaysia's major oil and non-associated gas fields are offshore. A federal/state revenue-sharing scheme offers Malaysia two possibilities: (a) reducing pressures on Sarawak and Sabah to rely on timber royalties for the bulk of their revenues; and (b) enabling the federal government to exert leverage on state governments to bring timber annual harvests and practices into line with the national policy.

International revenue sharing

Worldwide media and environmental groups have paid considerable attention to Malaysia's forestry practices. Concerns about deforestation, indigenous peoples and global warming have led to significant pressures and a threat of a tropical timber boycott if forest harvesting practices and policies in Malaysia, and particularly Sarawak, are not altered.

Such pressures, however well intentioned, offend nationalistic feelings. It is equally appropriate for countries such as Malaysia to argue that the energy consumption policies of developed countries are responsible for international environmental degradation, including global warming. Both policies, one to reduce pollution from hydrocarbon sources and the other to improve the sustainability of forest management, are required. The economic solution would evaluate the incremental costs and benefits of each strategy and equalize these choices at the margin.

But participant countries differ as regards their total wealth, income per capita and economic development policies. 'Debt-for-conservation' swaps have been recommended to encourage debtor nations to engage in more environmentally sound, conservation-oriented practices. Internationally owned debts are in part forgiven or used to purchase lands for parks, reserves or conservation-enhancing practices. This reduces a country's international repayment obligations while enhancing environmentally sound practices. Malaysia's international debt in 1988 was US$25 billion. In the same year, the United States' international debt was US$400 billion.

In Malaysia's context, such debt-for-conservation swaps might be used to increase the size of the PFE, to purchase for reservation purposes the indigenous land claims, and to reduce annual harvests through revenue supplements to states such as Sabah and Sarawak.

CONCLUSIONS

Land use and sustainable forestry

While 20.5 million hectares (62.4 per cent) of Malaysia remains forested, government policy incorporated in the creation of the PFEs and parks and wildlife areas has devoted 14.1 million hectares (42.9 per cent) to continued forest use. The remaining 6.4 million hectares of state forest lands (including 4.6 million on Sarawak) are available for land use conversion to plantation and related agriculture.

It is estimated that 1,714,000 people are employed in Malaysia's agricultural sector as against 150,152 persons in forestry and the wood-based products industry. It is argued that such land use conversion has generated significant employment opportunities in plantation agriculture, although at least two-thirds of Malaysia's agricultural acreage is in smallholdings which rely on family employment, often unpaid. Sabah and Sarawak have very limited populations, and land use conversion pressures should therefore be minimal. In recent years, with Malaysia's rapid industrialization, labour vacancies in the plantation and timber-based sectors have risen, leading to the immigration of Indonesian and Philippine labour contracted for plantation work. Rural to urban migration should reduce pressures for land use conversion in Peninsular Malaysia. Accordingly, it would appear that the conversion of forest lands to plantation agriculture is not warranted on the ground of employment creation, given the short-term labour market in Malaysia.

Forest management practices are not sustainable. Annual log production is approaching 40 million cubic metres in Malaysia, with Sarawak representing 18 million cubic metres per year. Sabah's log production is already declining, and Peninsular Malaysia's production will fall dramatically after 1994 and it may become an importer of natural logs after 1996 if it is to maintain its current sawmill and plywood/veneer capacities. Sarawak's log production rate suggests a complete harvesting of state and PFE forests in 12–14 years. Re-entry and the initiation of a new cutting cycle will depend on areas originally cut in the 1970s. Without an adequate source of quality logs, the wood-based products industry and its associated employment is at risk in Peninsular Malaysia.

If current harvesting practices do not meet sustainability criteria for long-term timber supplies and the wood-based products industries, it is difficult to see how amenity values can be maintained except in the totally protected forests and national parks and wildlife refuges. Unresolved indigenous land claims further complicate the situation.

The employment implications of a sustainable yield of 37 per cent of current yield are dramatic. Logging employment would fall to 25,000 from current levels of 69,000 and total timber and wood-based products industry employment would decline to 57,000 from current levels of 150,000. A movement into downstream production and the use of supplementary sources of wood supply (plantation forests) would reduce the employment impact, but 35,000–55,000 jobs would still be potentially at risk in the near future.

Concluding comments: policy options

The review of proposed policy options for Malaysia (and particularly Sarawak) reveals a number of strategies to alleviate non-sustainable timber harvesting practices and associated issues. These include the following.

(a) The state forestry departments must be strengthened. Bearing as they do the main responsibility for monitoring the harvest practices of concessionaires and of determining sustainable yield, an increase in the quantity and quality of their personnel is essential. An increase in the employment of indigenous peoples would be welcomed.

(b) The resolution of indigenous populations' land claims and the official acceptance of a map delineating such lands (or the preparation of a new map) seem imperative to defuse this socioeconomic issue in Malaysia's forestry management.

(c) There is a need to increase rehabilitation efforts on state lands and PFEs to increase sustainability volume on once-cut land. Fast-growing species in plantation forestry, the use of ageing rubberwood to supplement traditional forestry supplies and liberation thinning are all technologies which would augment traditional first-cut volumes per hectare. Rehabilitation would reduce pressures on further overcutting of PFEs.

(d) Preliminary research suggests that non-timber forest products (NTP) may have significant economic potential. A comprehensive study by the Malaysia Industrial Development Authority is warranted, as such NTPs would also reduce pressures on the forest products of PFEs. Most of the NTPs identified appear to require small capital resources for entry and are relatively labour intensive.

(e) Several policies to reduce the demand for raw export product were examined. Timber export quotas or consumer boycotts were considered to be ineffectual in improving sustainability practices. Malaysia's own efforts are to encourage the downstream production of wood-based products. This effort is being pursued vigorously but several problems exist. These include external concerns: trade barriers installed by consuming nations and a competitive position in world markets. Domestic issues play a part too, especially the competition between Peninsular Malaysia, Sarawak and Sabah for the location of the downstream industry and the generally inefficient size of firms and variable quality of the finished product.

Development and Employment: Forestry in Malaysia 287

(f) To reduce state government reliance on timber revenues from unsustainable exploitation rates, federal revenue sharing – providing funds but insisting on improved practices, or international debt-for-conservation swaps – would alleviate the pressures to keep up unsustainable rates in the short term.
(g) Small-scale swidden agriculture has been viewed as 'good' (and sustainable) or 'bad' (and leading to deforestation) by different evaluators. Within the present context in Sarawak it is a less important source of deforestation than timber harvesting because of the limited number of swidden agriculturalists and the low population density. However, these conditions will not necessarily be maintained.
(h) An official determination of annual sustainable yields needs to be established a priori and agreed upon in Sarawak and Sabah. Current determinations are advisory, at best, or simply record-keeping exercises of past cuts. Under reasonable assumptions about size and species, liberation thinning and the allocation of lands to indigenous peoples, annual sustainable yield in Sarawak (state lands and PFEs) is approximately 6.0–8.0 million cubic metres on a 35-year cycle. On PFE lands only, the figure is 4.5–6.0 million cubic metres. Current annual harvests in Sarawak are 18 million cubic metres. For Malaysia as a whole, sustainable yields on PFEs alone appear to be approximately 14.8 million cubic metres.
(i) Finally, a re-evaluation of the policy of land use conversion from forest lands to agriculture for employment generation purposes seems warranted. Following recent industrialization efforts, the plantation and logging/timber industry throughout Malaysia has experienced labour shortages, leading to the influx of immigrant labour, often illegal, from nearby nations. While agriculture remains a dominant employer with 1.7 million persons, more than two-thirds of its acreage is in family smallholdings. Much employment in this sector is family labour, often unpaid.

11 Agrarian Structure and Sustainable Livelihoods of Tribal People in Indian Forestry

Vijay Shankar Vyas

POVERTY AND ENVIRONMENTAL DECAY

Two well-recognized and well-documented phenomena pertaining to the developing countries are, first, the high and persistent number of households below the poverty line, and second, the continuing degradation of natural resources. It is easy to establish a correlation between the two, and tempting to advance one as the cause of the other. However, if the objective is to move towards the eradication of poverty without damaging the environment, a deeper probe into the processes of impoverishment and environmental decay is necessary. In this chapter an attempt is made to investigate the phenomena of rural poverty and environmental degradation with a view to examining whether there is a causal relationship between the two, or whether both are manifestations of the prevalent agrarian structure.

To understand this relationship we have selected a region in India with a concentration of poor households where the natural resource base is deteriorating. The vast hilly tracts in the northern, north-eastern, central and southern zones of the country inhabited by the tribal people reflect both these conditions. The tribal belt of Rajasthan, which is a part of the central or the middle zone of tribal concentration in India, extends from the northeast of the state to its southern boundaries. This region, accounting for the largest tribal population in the state, is the part of the tribal belt extending to the adjoining districts of the states of Madhya Pradesh and Gujarat. There is a good deal of similarity between the tribal populations of Rajasthan and the two adjoining states in their ethnological characteristics, natural resource base, economy, physical and institutional infrastructure, and most importantly, in the economic and social handicaps they encounter (Wali, 1992). The findings of this chapter are relevant, to some extent, to the entire tribal belt of central India.

There is a widespread belief that poverty is responsible for environmental degradation and that, once such degradation sets in, a cumulative process is set off which reinforces the trends in both these areas. The reasons given for this are principally the following: the poor depend to a large extent on natural resources to eke out their livelihood; there is a higher rate of population growth in poor households and to that extent the exploitation of the natural resources also progressively increases; and the poor have a high subjective discounting rate and therefore overexploit natural resources (Vyas, 1991). These reasons have been challenged by many observers.

The thesis propounded in this chapter is that, given the natural resource endowment at a point in time, it is the defective agrarian structure which contributes to both impoverishment and environmental degradation. *The agrarian structure is understood in terms of access to natural resources by different sections of rural society, and the institutions that influence the effective use of these resources.* Agrarian structure as so understood is itself conditioned by important macro variables, i.e. the availability of natural resources, technology, demographic pressures, markets, and regulatory and promotional institutions. These external variables impact on the agrarian structure, which, in turn, influences the process of impoverishment as well as environmental degradation. The model can be conceptualized as in Figure 11.1.

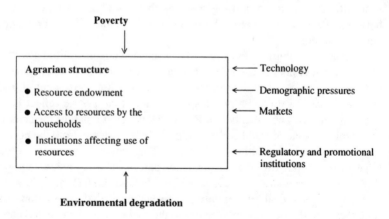

Figure 11.1 Relationships between poverty, environmental degradation and agrarian structure.

Agrarian Structure and Livelihood: Indian Forestry

POOR PEOPLE IN A POOR REGION

The conceptual framework suggested above is examined in the context of one of the poorest regions of Rajasthan, the southern, tribal, part of the state. Two districts, Banswara and Dungarpur, are taken up for intensive study.

Natural resources

The selected districts lie in the southern Aravalli zone of Rajasthan. The Aravalli Hills stretch about 700 kilometres from the Banaskantha and Sabarkantha districts of north Gujarat to Delhi. The southern part of the Aravallis is the main habitat of the adivasies (scheduled tribes) in Rajasthan.

The natural resources of this area may be discussed in terms of water, land and forest resources.

Water

Southern Rajasthan receives abundant rainfall, especially when compared with the amount of rainfall in the state as a whole. It is also spread more evenly, with more rainy days. The region is also blessed with a number of rivers and rivulets. Though not perennial, each of these rivers carries plenty of water. There are several *artificial* water bodies. Surface water percolates through a number of artesian wells. In terms of the overall availability of water the region can be considered as well watered. Although the quality of drinking water is poor (and this is responsible for a number of water-borne diseases), water is not a limiting factor in this part of the state.

Soils

These range from thin, infertile mountain soils typical of hilly areas to brown alluvial soils, with some parts identified as deep medium black soils. These last are the most fertile. Most of the area is characterized by medium to high fertility soils. The soil profile of the region, so far, is not discouraging although soil degradation is becoming serious (Bhattacharya, 1987).

Forests

The region was known until very recently as the jungle area, with trees and shrubs growing lushly and a rich and varied wildlife. The situation has changed now because of large-scale and indiscriminate deforestation.

Satellite data over the past two decades suggest a continuous degradation of the resource base, particularly in the tree-covered area, pasture land, soil cover, etc. Deforestation in the Aravalli gaps has made this hill range ineffective in checking the onward march of the Thar desert to the west. On the hillsides indiscriminate deforestation has transformed forested hills into barren ones.

The environmental situation is thus quite serious. For example, the region is now experiencing a growing frequency and intensity of drought, of dust storms in the Aravalli gaps and east of these gaps, and of biotic interference due to overgrazing and overcultivation on marginal lands; a continuous expansion of the rocky, sandy and barren wastelands; depletion of forest products such as timber, fuelwood, fodder, etc.; a continuous decline in groundwater tables; a critical transformation in the wildlife habitat; and the continuous siltation of rivers and reservoirs (Dhabariya, 1988). In short, the forest-based lifestyle of the tribal population is being threatened, and the task of rehabilitating their economy has become all the more difficult. The question one would naturally ask is: why has this happened? To answer this we shall briefly examine the social, demographic and economic characteristics of the region, more particularly its tribal population.

Characteristics of population in the selected districts

Southern Rajasthan is economically the most backward region in the state, and the two districts chosen for study are the poorest in the state. Nearly 70.5 per cent of the population in Banswara and 58.4 per cent in Dungarpur are below the poverty line, as against 26.8 per cent in the state as a whole. They account for about 5 per cent of the state's population, but their share in its poor population is about 11 per cent.

In Rajasthan the tribal population does not represent an overwhelming proportion of the total (12.4 per cent in 1991). However, in the six southern districts of the state, the proportion of the tribal population ranges from 73.5 per cent in Banswara and 65.8 per cent in Dungarpur to 9.0 per cent in Bhilwara. The two selected districts, which account for less than 5 per cent of the state's total population, house more than 25 per cent of its tribal population, and the social and economic features of this part of the state are largely determined by its predominantly tribal character.

Demographic features

The rate of population growth in Rajasthan is one of the fastest in the country (2.8 per cent per annum in Rajasthan as against 2.3 per cent in

Agrarian Structure and Livelihood: Indian Forestry

India as a whole from 1981 to 1991). In the selected districts the population density has significantly increased (from 176 to 229 per square kilometre in Banswara and 181 to 232 in Dungarpur) over the past decade. These districts are among the most densely populated in the state. The population is predominantly rural (92 per cent in Banswara and 93 per cent in Dungarpur). Comparative population data are shown in Table 11.1.

Livelihood pattern

The tribal population in these districts is predominantly rural. Agriculture is the mainstay of their economy. It is generally presumed that tribals living in the forest have forest-related activities as their principal source of livelihood. This, however, is no longer true in the case of the tribals of

Table 11.1 Demographic features of the selected districts and the state

Demographic features	Banswara		Dungarpur		Rajasthan	
	1981	1991	1981	1991	1981	1991
Total population	886,600	1,155,600	682,845	874,549	34,261,862	44,005,990
Rural population	831,413	1,066,406	638,719	810,732	27,051,354	33,938,877
Urban population	55,187	89,194	44,126	63,817	7,210,508	10,067,113
Tribal population	643,966	849,050	440,026	575,805	4,183,000	547,881
Decennial rate of growth of total population	35.44	30.34	28.78	28.07	32.97	28.44
Decennial rate of growth of rural population	33.80	28.26	28.00	26.96	21.55	25.46
Rate of growth of urban population	66.20	61.62	41.17	44.62	36.98	39.62
Rate of growth of tribal population	34.90	31 185	30.39	30.86	25.29	30.88
Sex ratio	984	969	1.045	995	919	910
Density of population	176	229	181	232	100	129

Source: India, Government of Rajasthan, Directorate of Census Operations (1991).

southern Rajasthan. The 1981 Census divided the workforce between main workers and marginal workers, depending on the time spent by the worker on the economic activities of the household, and indicates that the proportion of marginal workers in the tribal households is much higher compared with that of the state as a whole: correspondingly, the proportion of main workers is lower. Hardly 3 per cent of the main workers in Dungarpur report forestry and other allied activities as their main occupation, and in Banswara the proportion of workers in these activities was negligible. Thus, the tribals in this region, though living in the midst of forests, are not greatly dependent on forest-related activities for their livelihood.

Seventy-five per cent of the people in Banswara and 72 per cent in Dungarpur depend on agriculture as a source of livelihood. The size of the holdings is small (1.76 ha in Banswara and 1.37 ha in Dungarpur, on average) and the plots are scattered. On most of these holdings low-value crops dominate. Cereals and pulses account for nearly 83 per cent and 89 per cent of the total cropped area in Banswara and Dungarpur respectively. Maize is the principal cereal in the area (India, Government of Rajasthan, Directorate of Agriculture, 1992). Productivity is low for most of the crops, although the yield per ha in the winter crops is better in Banswara than in the state as a whole. The same is true for the summer crops in Dungarpur.

The causes of low productivity are not hard to find. Apart from the small size of the holdings, irrigation is limited, especially on tribal holdings: 28.7 per cent of the net sown area is irrigated in Banswara, as against only 16.2 in Dungarpur. The use of fertilizers and improved seeds is virtually limited to the towns.

The handicaps imposed by the poor resource base are compounded by the poor infrastructure support in credit, input supplies and marketing. As a result, the income earned from agriculture, which is the principal source of livelihood, is meagre. Gross return per ha is around Rs 3,600 in Banswara and Rs 2,900 in Dungarpur. A disturbing factor is the rising proportion of agricultural labourers. As we shall see later, a large part of this labour is bonded labour in one form or another.

Social and political deprivations

To these economic handicaps is added social deprivation. Two of the principal social services, namely primary education and primary health, were not available to the tribal population to any measurable extent until very recently. This is reflected in the fact that the level of literacy for the males in these two districts was 38 per cent (Banswara) and 46 per cent (Dungarpur), and that for females was less than 15 per cent in both the districts. The situation is worse in the case of the tribal population. Female literacy in the tribal population in Banswara is around 3 per cent, that in Dungarpur is less than 5 per cent. The situation should improve as one of these two districts, Dungarpur, has been selected as a 'full literacy' district with the objective of achieving 80+ per cent literacy by the end of the current Five-Year Plan.

Similarly, in terms of the availability of and access to primary health facilities these districts are at a great disadvantage. The high infant mortality rate (111 in Dungarpur and 108 in Banswara) is a significant indicator. There has been some progress in protection against water-borne diseases. The incidence of guinea-worm, one of the most debilitating diseases in the area, has gone down, but that of malaria rose over the decade of the 1980s (Mathur and Yadav, 1993).

As might be expected, the political clout of the tribals is very weak. They do not have proportionate representation in any of the decision-making bodies. Along with a remarkable change in their economy and society, the tribal polity has also undergone a decisive change (Pamecha, 1985). The modern state, which is not attuned to controlling dispersed settlements or a nomadic mode of living, encouraged more sedentary and conglomerative settlement patterns. In that process the tribal society lost its autonomy. The dependence on the state for a livelihood as well as for social services has, as with other sections of society, increased. Some of the prerogatives, such as the reservation of jobs for scheduled tribes, enshrined in the Constitution of India, and the legislative measures, such as the prohibition of the sale of tribal lands to non-tribals, have only helped to aggravate the differentials within the tribal community. An élite has emerged in the tribal society, which is pre-empting most of the benefits that are supposed to reach the tribal masses. With their co-option into the 'Sanskritized', non-tribal society, they have also acquired social hegemony over the mass of tribals and consolidated their superior status.

In brief, a number of suppositions regarding tribal people, e.g. that they depend on forests, are not constrained by scarcity of land, have a less differentiated social organization, and are autonomous and self-governing, are no longer valid. The agrarian structure in the southern, tribal part of Rajasthan suggests that the inhabitants, particularly the tribals, have limited access to cultivable land and forests. The productivity-enhancing technology (e.g. irrigation) and institutions (e.g. credit) are weak and or confined to small areas. The investment in human capital is inadequate. These facts have to be juxtaposed with the natural resource degradation as exemplified in deforestation. How these two processes have worked simultaneously is discussed in the next section.

PROCESS OF IMPOVERISHMENT AND ENVIRONMENTAL DEGRADATION

Impoverishment of the tribals

From plains to hills

It appears that some 500 or 600 years ago the tribals of southern Rajasthan (mainly the Bhils) were driven out of the fertile lands they inhabited in the foothills of the Aravalli Hills into the barren and rough uplands, with little arable land and infertile soils. The Rajput princes who were responsible for this displacement maintained their hegemony over the tribals, who gradually adapted their economic, social and political institutions to their new habitat, where they survived by hunting and gathering forest by-products.

At the same time a small number of *adivasies*, beginning with the more powerful of them, were assimilated into the Rajput-dominated Hindu society. This process of assimilation gained momentum, boosted by the demand for tribal labour for agricultural work in the valleys.

By the end of the nineteenth century the political, social and economic factors were all moving towards the establishment of an integrated economy with two unequal partners: impoverished and suppressed *adivasies* on the one hand and the higher castes of Rajputs, Brahmins, Patels and Mahajans on the other. The latter were resourceful and had economic and political clout. The hegemony of the upper caste was firmly established and the circle of the assimilated *adivasies* progressively widened. Thus the isolation of the tribals and their exclusive dependence on forests was ended, and the period of settled agriculture in the former forest areas began.

From forest to cultivation

Since the beginning of this century a remarkable change has occurred in the livelihood pattern of the tribals. They were forced to obtain the major part of their income from cultivation rather than from forest-related activities. Three main reasons contributed to this transition: (a) legislation relating to the ownership and management of forests; (b) growing commercialization; and (c) demographic pressures.

The introduction of modern legislation converted the former common property resources of forests into state property, controlled and managed by the forest bureaucracy. More important than the legislation, however, was the growing commercialization of the forest property. The rise in the economic value of timber and forest products on the one hand and the resource requirements of the state government on the other changed the conservationist role of the Forestry Department into that of an important revenue-earning department. This led to the entry of timber merchants and contractors. With the connivance of the forest officials the contractors organized the large-scale felling of trees. Though the objective of the forestry legislation was the conservation of the forest wealth, exactly the opposite took place. Vast areas were felled in no time, with hardly any effort being made towards the regeneration of the forests.

The tribals now had only two options: (a) to migrate from the hilly areas to the nearby plains; or (b) to extend cultivation on fragile lands. Both the alternatives were adopted. A number of tribals found their way from the hills to the farms, factories and mines in the plains. The jobs that they got in these places were menial, unskilled and low paid.

The main pressure was borne by agriculture. Experts maintain that most of the hill areas of the Aravalli should be under a well-maintained forest cover for sustained productivity and for soil and water conservation (Dhir, 1987). But, because of the population pressure, progressively larger areas of unfertile and marginal lands were brought under the plough, with disastrous consequences.

Growing dependence on the livestock, mainly to supplement the meagre income from agriculture, has led to an increase in the biotic pressure on land. While the number of livestock remained more or less steady over the decade from 1977 to 1988, the grazing area declined significantly. The productivity from the livestock naturally declined also. This, coupled with the stagnant yield from agriculture, contributed to the decline in the living standards of the tribals. In more recent years there has been a reversal of this trend, at least in the valleys and plains of the districts, where irrigation and herds of buffalo have been introduced. However, these changes do not greatly affect the hilly areas where most of the *adivasies* live.

From cultivators to agricultural labourers

As a result of the shift to infertile land and the decline in the income from agriculture and livestock enterprises, the *adivasies* were forced to overexploit the forests, at least to connive in rapid deforestation. However, not only was the current income of the tribals in the hilly areas declining, but landlessness also set in, with landownership being concentrated in a few élite tribal households, and with a growing differentiation within the tribal society (Vyas, 1983).

The most powerful factor in land alienation is neither growing population nor a fall in the income from the traditional sources of livelihood, but pressure for 'social' consumption coupled with malfunctioning of the credit market. The bulk of the demand for credit by the tribal households is for meeting social obligations at the time of a birth, death or marriage in the household. Since land alienation to non-tribals (who are the major suppliers of credit) is illegal, the repayment is asked for, and given, in terms of human labour. By different types of informal labour contracts, the *adivasies* are reduced to the status of bonded labour (Vyas, 1980). The proportion of households recording agricultural labour as their main occupation has increased from an insignificant figure to nearly 6 per cent in Banswara and 8 per cent in Dungarpur. These figures do not tell the whole story. A large number of small and marginal farmers, whose number is progressively increasing, have mortgaged their land to moneylenders and are reduced to the status of sharecroppers, if not bonded labour.

The tribals who do enjoy a comfortable living are generally those who, with some educational attainment, left for the urban areas and took advantage of the constitutional provision of the 'reservation' of a certain percentage of jobs for the scheduled tribes. The tribals remaining in the hillside, with their infertile lands, restricted access to forests and lack of diversification in occupational structure, are doomed to a low standard of living.

DETERMINANTS OF CHANGE

The resource endowment

Agrarian structures do not evolve in a vacuum. Initial conditions and the subsequent development of the macro variables determine the set of institutions and practices which affect peoples' access to, and use of, natural resources. As regards the 'initial conditions', the most relevant factor in the context of our discussion on poverty and environment is the natural factor endowment of a region.

In economic literature natural factors did not receive thorough analytical treatment until very recently. These factors were considered as 'given'. Their impact on economic growth was considered by the classical economists, particularly Ricardo, who gave great importance to rent from land in explaining his model of the stationary state. In more recent years resources such as natural gas and oil have received some attention. However, the problem of missing or incomplete markets has led to a more cautious treatment of these resources through purely economic concepts and tools. The absence or incompleteness of markets in natural resources means that there are problems in the valuation of these resources, as well as in the implementation of policies, since the markets are not there to guide the decisions (Karshenas, 1992).

The difficulties of identifying norms for valuation or guidelines for policy implementation have not, however, deterred development researchers from recognizing the importance of natural factors in influencing the pace and pattern of economic growth. Climate, rainfall, soil characteristics and water availability, for example, are considered important constraining or facilitating factors, and natural resources, both renewable and non-renewable, are analysed to explain the economic performance of pre-commercial societies to the modern market economies. The more popular exercises in 'carrying capacity' are tried out at the global as well as regional levels (see, for example, Parikh and Rabbas, 1981).

The impact of the environmental and natural factors is clearly evident in the *adivasi* areas of the southern Aravallis. Rugged terrain has made communications difficult and the cost of transportation of man and material expensive. In most parts of the region the soils are thin and infertile. The yield of the main crops is adversely affected. While these are adverse factors, there are also factors which favour rapid growth in this area. Water availability is not a major problem. The climate is more balanced than in other parts of the state. There is a rich diversity of flora and fauna. The region has a wide variety of minerals. The terrain, difficult for certain purposes (such as transport and communications), is quite favourable in other respects. There are numerous small dams to retain water. The soil is not very suitable for crop raising but is quite good for growing a wide variety of trees.

Through science and technology the few natural handicaps can be corrected, and several favourable features can be further accentuated. If this has not happened so far, the role of the other macro variables needs to be examined. In fact, owing to certain changes in macro factors, new environmental constraints have emerged or the existing ones have been exacerbated. For example, commercialization leading to indiscriminate deforestation has adversely affected the soils and has led to a degradation of the soil quality. The water-retaining capacity is adversely affected, resulting in the formation of gullies and ravines. Some forms of environmental pollution, e.g. the greenhouse effect, are not yet serious problems. At the same time, pollution encouraged by open-cast mining is becoming a serious problem (Shankar, 1993). Agricultural technology with intensive use of chemicals is slowly making inroads in the plains of the region. The growth in the animal population is putting a heavy pressure on the natural pastures and grazing lands. It is also responsible for an increase in methane gases in the atmosphere. The composition of the livestock, more than their numbers, is another relevant consideration. The larger proportion of small ruminants, which is partly a reflection of the inability of the households to support large cattle, especially in the hilly areas, has led to the indiscriminate use of grazing lands.

A closer study of the natural resource base of the region suggests that: (a) the quality of development of the region is significantly influenced by the availability of natural resources, although a precise evaluation and the optimal intervention are difficult to decide; (b) the natural resource endowment in the selected region, even with a few handicaps, is generally favourable for sustainable growth; and (c) the prospects of such growth are jeopardized because of human interventions and institutions.

Demographic pressure

The mutually reinforcing relationship between population growth, environmental degradation and the process of impoverishment has now become a part of the conventional wisdom (Bifani, 1992). It is generally assumed that the poor have a higher rate of population growth and that the increasing size of their households has an adverse impact on the resource base of a region. One may question the underlying assumption, i.e. the faster population growth in the poorer households. Even if this is accepted, neither on a priori grounds nor on the basis of experience can one draw 'straightforward' conclusions. Of the two districts we are studying, Banswara, has a higher population growth rate than the state, while in Dungarpur the rate is in fact slightly lower than that of the state. Both districts are equally poor. How is population growth likely to affect the poor and, in turn, the resource use in a region?

The literature identifies two principal ways in which this may happen: (a) with the growth in the population in poor households the supply outpaces the demand for labour which is therefore marginalized; and (b) the growing population leads to a greater demand for food, which would lead to an extension of the area under food crops, bringing marginal land under the plough (Leach and Mearns, 1992).

There is, however, another important though largely neglected way, in that population growth in the poor farmers' households leads to further subdivision of their tiny holdings. Although there is a strong body of theoretical and empirical research which suggests an inverse relationship between the size of holdings and the productivity of land, it has to be granted that there is a floor below which a holding will be unproductive and a positive size/productivity relationship will obtain (Vyas, 1980). In

Agrarian Structure and Livelihood: Indian Forestry 301

the small-farm-dominated agrarian structure, the creation of the tiny holding sector is the most important effect of population growth. All these three effects of growing population can be illustrated and explained in the context of the selected region.

The population in tribal areas, as in other parts of the country, continued to increase at a relatively moderate rate in the first 50 years of this century. Since the 1950s there has been a sudden upsurge, mainly owing to a reduction in child mortality and a rise in life expectancy. The spread of medical facilities in the tribal areas during the past few decades has helped to reduce the incidence of some common diseases, although not to the same extent as in relatively more developed regions.

The growth in population is directly reflected in the reduction of the size of agricultural holdings. Even within a decade or so the average size of holdings in these districts has come down: in Banswara from 2.17 ha (in 1976–77) to 1.76 ha (in 1990–91); in Dungarpur from 1.64 ha to 1.37 ha. More importantly, the proportion of holdings below 1 ha increased from 46.6 per cent to 47.7 per cent in Banswara between 1976–77 and 1990–91, and from 50.0 per cent to 56.9 per cent in Dungarpur.

Two other implications of population growth on agrarian structure, mentioned above, are also exemplified in this region. Population growth generally leads to an extension in the area under cultivation or the intensive use of existing arable land. In the selected districts both these developments have taken place. As was shown earlier, a sizeable expansion in the arable area took place by bringing the fallow land, grazing areas, etc., under the plough. There are indications to suggest that until the 1960s this expansion occurred even at the cost of the forest areas. This process seems to have been halted in recent years. The extension of cultivation to marginal lands, as noted earlier, contributed to soil degradation. Population growth can also lead to intensive cultivation, either with the reduction in the 'current fallows' or by bringing larger land into the 'area sown more than once' category. This, again, can lead to loss of productivity, especially on the light and fragile soils. The selected region did witness a reduction in 'current fallows' and an increase in 'area sown more than once'.

The impact of population growth on the labour supply is reflected in the high level of unemployment and underemployment in these districts. As no substantial diversification in the occupational structure has taken place, the increased labour supply has meant, principally, an increase in the number of workers per hectare and a fall in labour productivity in agriculture.

The other manifestation of the growing population has been the migration of tribal labour from the hills to the towns in the valleys. The tribal population in urban areas, though still small in number, is growing rapidly. As was mentioned earlier, this has resulted in tribals entering low-paid, low-productivity occupations, and also in an increase in the number of tribal agricultural workers in the non-tribal areas.

Technology

Technology enters the discussion of poverty and the environment at several levels. In the first place, how far do available technologies resolve the existing environmental problems, especially those pertaining to air pollution, water pollution, the greenhouse effect, etc.? Secondly, technologies could be considered for their likely contribution to sustainable development. In this respect, attention is focused mainly on the use of technologies to rejuvenate the degraded ecological areas, as far as possible those technologies which also generate employment and income for the poor. The labour-intensive works for soil and water conservation are examples of the technologies which aim at 'employment for environment' (World Bank, 1990a). A third and more basic objective is to use technologies in such a manner that the pressure on resources is reduced and the danger of overexploitation is avoided. An example is the agricultural research aiming to optimize the use of water. Finally, efforts could be made to widen the choices for income generation. Thus, advances in off-farm activities in regions where there is too much pressure on agriculture can be considered environmentally friendly technological interventions. In all these cases the objective is to evolve technologies which are economically sound, namely where the cost-benefit ratio is favourable (Doeleman, 1992).

Pollution-reducing technologies

In the area we have studied, not much attention is paid to pollution-reducing technologies. This does not mean that pollution is absent. Both air and water pollution exist, and their impact on the health of the inhabitants is considerable. Three potentially dangerous pollutants prevail in the area. The excessive emission of methane gas from animal dung etc. could be harmful. Secondly, smoke from burning wood mainly affects the health of the women who are responsible for cooking. The pollution of drinking water is the third particularly serious pollutant as it can cause several water-borne diseases.

Only token efforts to mitigate these dangers have been made. The only exception is a large-scale, rather successful drive for clean drinking water to safeguard the population from water-borne diseases such as guinea-worm, hookworm, etc. No similar efforts have been made to control other forms of pollution. Biogas for cooking and lighting is available, and there are a few *gobar* gas plants installed in the region. These have not become very popular in the selected districts. The same is true of the technologies to avoid pollution from burning wood, such as the smokeless *chulha* woodstoves.

Technologies for sustainable development

Here the main advances in this region, as in the state as a whole, are: (a) water conservation through the watershed development projects; and (b) afforestation efforts to bring larger areas under tree canopies and to ensure a better survival rate for the plants.

The watershed development projects, so far, have been government-sponsored and bureaucratically managed. However, the emphasis in these programmes is gradually shifting to the involvement of the local community in one form or the other. For example, different organizational innovations have been tried out for popularizing social forestry and farm forestry. Even with all the wastage and inefficiencies these efforts have started paying dividends. Not only has deforestation been halted, but also, at least during the past decade, there has been a net expansion in the area under forest.

The benefits of such an expansion in the forest area have, however, been limited as far as the local *adivasi* population is concerned. The afforestation programme on the government lands has not directly benefited the poor. Also, most of the efforts in farm forestry have been confined to the larger holdings. The afforestation programme on the village commons or on the smallholdings has not yet been tried on a measurable scale. Thus there has been no fundamental change in entitlement or access to the forest resources by the local tribal population. In this respect the agrarian structure, interpreted in the broader sense, i.e. to include entitlement to all natural resources, has not changed significantly and, here again, the benefits have not percolated to the poorer sections.

Higher-yielding technologies

The major weaknesses are in the third level of technology application, namely in evolving less (cash) input-intensive yet higher-yielding technologies in crop-raising and animal husbandry. More concerted research on the crops and animals of the poor, and determined efforts to increase 'value added', are missing from the agricultural research agenda.

Undoubtedly, some progress is being made in all the three directions of technological upgrading, yet the efforts made so far do not touch even the fringe of the problem. What are most lacking are evolving technologies which would offer economically superior and environmentally friendly alternatives to the poor households which depend mainly on agricultural production. In fact, the main alternative available to a large number of people in this area is employment in opencast mining. This is not necessarily a better income-earning proposition for the smallest cultivators or agricultural labourers, and is definitely inferior in terms of its impact on the environment (Shankar, 1993). Technological (and organizational) efforts to develop sustainable forest-based enterprises have not been given priority in the R & D policy for the tribal region.

If technologies can be evolved and extended in the directions mentioned above, some of the constraints imposed by the agrarian structure can be significantly relaxed, thus ensuring better access by the poor to the productive resources of the area and more effective utilization of the region's resources in alleviating poverty without degrading these resources.

Macroeconomic policies

The effectiveness of economic measures to influence the allocation and use of environmental resources is constrained to a large extent by the phenomenon of 'missing markets' or 'incomplete markets'. The World Bank, suggested four principal reasons for the inadequacy of the markets to reflect social priorities in this area: (a) it is difficult to establish and enforce property rights in the ownership and use of some natural resources, e.g. air; (b) in many cases the markets are incomplete because all benefits (and costs) cannot be taken into account – for example, while the forest wealth may be evaluated for its timber value, the contribution of the forests to watershed protection may not be recognized; (c) because of the well-known problems of externalities, it is difficult to establish the share and responsibility of individuals; and (d) the lack of information makes it difficult to assess the environmental impact of various measures (World Bank, 1992).

The other, and less appreciated, factor in this market failure is the lack of assets and purchasing power which keeps a section of the population at the periphery of the markets. The relevance of the markets becomes restricted to those who can respond to the market signals, either as producers or as consumers. Not that any section of the population, including *adivasies*, is totally outside the markets, but to them many market transactions are largely irrelevant. To bring these households into the orbit of the market becomes an objective in itself. In any event, it is necessary to differentiate clearly between the impact of policy measures on resource-rich households and the impact on resource-poor households.

Yet there are several policy measures which do affect the value of natural resources and influence the decisions of a large section of the poor. A common example is the use of subsidies. Many resources (e.g. forest products, water or energy) are overexploited because they come cheap to the users. The subsidized inputs used in the complementary or competitive enterprises, such as agriculture, animal husbandry or forestry, may tilt the balance in favour of one or the other and may have repercussions on other enterprises. Subsidized agriculture may compete with forest land, and may lead to the extension of the area under the plough at the expense of forests. The policy on subsidies and taxation will thus have a major impact, direct as well as indirect, on the use of natural resources (World Bank, 1992). The same is true about the trade policies. If excessive importance is attached to exports at a particular stage of macro policy reforms, this can contribute to the overexploitation of the resources and may have deleterious effects on the consumption and welfare of domestic consumers.

Another example of the importance of policy initiatives in the context of the optimal use of natural resources is the credit policy. The malfunctioning of credit markets, i.e. discriminatory credit delivery, high rates of interest, short gestation loans, requirement for large collateral, etc., affects the economic calculus of the poor in a manner which is injurious to sustainable growth.

The other government decisions affecting the environmental resources in the context of a planned economy such as that of India are the decisions pertaining to public investment. By setting and executing certain investment priorities, the government can not only contribute directly to the development of the natural resources but also give signals to private parties which influence their investment decisions. The complementarity of public and private investment is more pertinent in a situation where the government has to take the lead in long-gestation and lumpy investments.

The natural resource management in southern Rajasthan illustrates the importance as well as the limitations of macro policy intervention. First, the markets in natural resources in this area are incomplete. The valuation of even the main product, e.g. timber, is not necessarily based on market demand and supply. Indirect benefits of the forests, such as soil or water conservation, are hardly taken into account. The subsidies on agricultural inputs, making agricultural production more profitable, have changed the relative profitability of agricultural and non-agricultural enterprises, and have contributed, together with population growth, to the expansion of agriculture in the forest areas. The subsidies on fertilizers and water have led to the overexploitation of these inputs. Although the use of these inputs has not reached the level where it will start having injurious effects on soil or water quality, the dangers of soil and water degradation are imminent. The effect of the growing trade in timber and forest by-products, coupled with the monopolistic role of the state agencies as the sole owner of the forests, have led to the indiscriminate felling of trees and neglect of rejuvenating the forests. The 'cheap' prices of timber and charcoal will have similar effects. This should not, however, lead one to conclude that high prices will automatically conserve the natural resources.

The investment priorities for the development of natural resources cannot be calculated at this stage of our knowledge because of the difficulties in evaluating these resources. Insights from other disciplines (ethics, sociology, anthropology, political science) have to be brought in to determine priorities which more clearly reflect social values as well as the norms of sustainability. This does not mean that there is no scope for the markets to play their role or that all the decisions have to be made intuitively, or by the best judgement of the bureaucrats or political leaders. Even with present imperfections the market signals can be, and should be, used at least for indicating the directions to be taken.

Some government policies further weaken the functioning of the markets in natural resources. Land legislation provides an example of such policies. With the restrictions placed on the sale of the land owned by the tribals to the non-tribals and by virtually abolishing the lease market in land, the land markets are distorted, with the poor suffering the most. The lessons are clear. We should recognize both the limitations and the strength of the markets in natural resources; we should examine all the policy measures which distort commodity as well as asset markets; and we should have a strategy to entitle the poor to have access to resources so that they may respond to market signals.

Institutional interventions

The access to resources and the efficacy of the delivery systems are expressions of the institutional framework. Both these are critical to the well-being of the poor. In the context of natural resources, the issue of access has to be looked at in terms of the nature of property rights or entitlements, on the one hand, and the security of such entitlements, on the other. The criteria for efficient delivery systems in relation to the management of resources by the poor would include, first and foremost, a congruence between the delivery system and the recipient system. The other criterion could be the extent of peoples' participation in management and decision-making functions. It may also involve the integration of various layers of organizations without a concomitant loss of transparency and accountability. An effective monitoring and information system should be able to contribute towards these objectives.

Property rights in natural resources could take the form of private ownership, communal ownership, state ownership or any combination of the three. In many instances, private ownership of natural resources is not possible as it is difficult to assign proprietary rights to individuals or households. The usual pattern in such cases is communal or state ownership. Different facets of communal ownership have been widely discussed in India and elsewhere in the context of Common Property Resources (CPR). It is now widely recognized that the poor depend much more on CPR than do the rich (Jodha, 1990). However, the nature and extent of inequality of access to CPR is much less researched (Leach and Mearns, 1992). The attrition of CPR with the increase in population, the hegemony of powerful groups or the ideological orientation of the state is quite common. The basic reason for such attrition is that the systems of control of CPR under communal ownership are not resilient enough to adapt to changes in the size and composition of the participating groups.

From a weakened communal control of CPR to the state ownership of these resources is not necessarily a progressive move. The resources owned by the state are managed, basically, by bureaucrats, whose lack of training in resource management and short time horizons make their role even weaker. At the same time, state ownership generally leads to 'open access' to these resources to an extent that proves injurious to the maintenance of the natural resource base.

The delivery systems designed to make the institutions more poor-friendly face similar problems when, with the laudable objectives of providing social infrastructure to the poor, the state enters into this area in a major way. Even with the private enterprises the task of adapting these institutions (e.g. the institutions to deliver credit or input supplies and those for output marketing) to the reality of small-scale agriculture demanding small quantities of inputs, including credit, and also offering a small marketable surplus, is not an easy one. The high transaction costs, on the one hand, and the problem of 'moral hazards' on the other, lead private enterprises to exclude the small clients from their purview. The ingenuity lies in adapting different institutional structures which combine characteristics of the collective as well as private enterprises. In many circumstances cooperative enterprises may fulfil these requirements.

The experience of the institutions governing the access to natural resources and organizing the delivery of services in the selected region substantiates the above propositions. The following discussion, however, is confined mainly to property and user rights in forest land and products.

For the past 50 years or more the government has played a major role in the management of forest resources, thus influencing the level of income and welfare of the *adivasies*. As was mentioned earlier, as the feudal lords had a hegemony over most of the region there were stringent and arbitrary rules for the management of the forest resources. They had been taxing the *adivasies* on the forest by-products collected by them, even on the fruits plucked from their own trees. In the pre-Independence period forests were treated, basically, as the game reserves of the native rulers. In a few progressive princely states, however, legislation was enacted to regularize various practices and specific rights were granted to the people living in the forest areas. However, such legislation was the exception rather than the rule. The real change occurred only after the enactment of the Rajasthan Forest Act in 1953, modelled on the Indian Forest Act of 1927.

The objective of this legislation and the subsequent activities of the state in relation to the forests was protective rather than promotional. This is reflected even in the terminology used in the legislation and the rules derived from it. For example, timber is defined as 'major forest produce', while other forest produce, which is more important for the *adivasies*, is 'minor forest produce'. The intention of the law-makers seemed to be that the forests should be protected from the people who at best might be given certain 'concessions' rather than rights.

The legislation did recognize some rights of the people living in the forests. These included right of way, right to water courses or to use water, grazing rights and rights to forest produce. But for all practical purposes the state usurped what had been since time immemorial a community resource. In other words, the common property became the state property. The denial to *adivasies* of their customary rights only led to their indulging in petty theft or conniving with the timber merchants and contractors in large-scale deforestation. The state property degenerated into 'open access' property.

The major problem for the *adivasies* arose from the non-recognition of their rights to land cultivated by them when an area was declared a forest area. They were then considered encroachers and were subjected to harsh punishment. Thus, while the state was enacting legislation for the cultivators in the non-forest areas to provide 'land for the tiller', it was denying even the occupancy rights to the people who had been cultivating land for generations in the designated forest areas (Singh and Srivastava, 1993).

Further, the problem of landlessness was accentuated as in these parts of the state the land of the *adivasies* was acquired by the government to build large dams, e.g. the Mahi and Kadana dams. The resettlement of the dispossessed was given lowest priority. For example, the *adivasies* whose land was submerged following the construction of the Mahi dam were given meagre resources and infertile lands in the distant Chittor area. Most of these *adivasies* soon lost their land, together with whatever amount they had been given as compensation.

The reduction in the rights of the *adivasies*, in other words the restrictions on their access to forest resources, is a result of a variety of factors. In the first place the Forestry Department's role is ill defined. The entire land belongs to the Revenue Department of the state and, in theory at least, the forest land is leased to the Forestry Department. Again, the jurisdiction of the state and the central government in this regard is ill defined. But in practice the central government has the upper hand, if not the final say, since diverting even a small area of forest for non-forest uses requires its agreement. It would be fair to assume that people on the spot would probably be more aware of the issues than those in the state capital, and much less in New Delhi.

The functioning of the Forestry Department is highly centralized, with the lower echelons having little power and less understanding of the purposes of the legislation or the correct ways for its implementation. Coupled with that is the inadequacy of the staff, who are supposed to implement the Department's highly detailed work plans in addition to their routine regulatory functions.

These defects in the functioning of the principal regulatory agency are now fully recognized. Several attempts have been made to involve the beneficiaries in the forest activities. An important programme of this nature is the Joint Forest Management (JFM) scheme, introduced by the Forestry Department in April 1992. Under this arrangement a partnership is forged between the Forestry Department and the people. The JFM schemes take account of ecological and social considerations. When forests are cut down the denuded land still has some root-stocks which, if protected, can regenerate and produce new forests. It is on such denuded lands with some remaining root-stocks that a JFM partnership is built up, the principal objective being that the forests be regenerated and protected by the people rather than by forest guards and other officials of the Forestry Department. Such a partnership with the local people in the protection of the forests is possible only if the ownership, management and benefits of the forest land are shared.

The programme in Rajasthan was inspired by the successful experiment on these lines in a few other states, such as Haryana and West Bengal. In

Rajasthan about 20 village-level joint forest committees were set up (in 1992) which aim at protecting 2,000 ha of land. In some of the selected areas this experiment has proved quite successful. The experience in the states which pioneered in JFM schemes, such as West Bengal, Orissa and Haryana, suggests that these schemes require sufficient time for learning and experimentation and a very sympathetic understanding of the local communities. For example, in Haryana where one of the JFM schemes has acquired nationwide recognition it was pointed out by one of the important leaders of the project that it took them a year and a half of learning and experimentation to draft an appropriate state-level policy. This was done after consultation between Forestry Department officers and the villagers. It took another two years before the rules to implement this policy could be framed. The task was to organize the villagers and to ensure their working in a democratic way. A committee was constituted with one member from each of the households of the village. Care was taken to see that women were not neglected.

Apart from this partnership between the state and the people, NGOs can also play an important role in devising the right policy framework and in implementing various policies in an effective manner. The NGOs perform, basically, three roles. The first and foremost role is that of advocacy. The *adivasies* are amongst the poorest sections of society and they need resourceful persons to strive on their behalf. This important role is being performed by several NGOs in Rajasthan: securing the *adivasies* basic rights as forest dwellers, legitimizing the so-called 'encroachments', protecting their grazing rights, ensuring minimum wages on forest-related works, and so on. The second area of concern for the NGOs is that of protecting the forest with the people's cooperation. The felling of trees by timber merchants and contractors with the connivance of the local people, indiscriminate cutting by the forest dwellers themselves, or overgrazing of the forest land, have all to be regulated if forests are to be restored on denuded lands. The NGOs can help to organize the local people to safeguard their own interests through regulated forest operations. Thirdly, the NGOs can assist local people in promoting projects and programmes, similar to Joint Forest Management schemes, for the development of their areas.

Fortunately, in Rajasthan a large number of NGOs are working in all three areas and, what is more encouraging, they are finding a responsive forest administration to help them in their task. The rejuvenation of the forests with an equitable access to these resources by the *adivasies* is the first step in alleviating their poverty and in preventing environmental degradation. The next step is systematic afforestation and the regulated use of forest products and forest spaces.

Agrarian Structure and Livelihood: Indian Forestry 311

With the combined efforts of the state and the NGOs, there has been some improvement in the *adivasies'* access to forest resources. The same cannot, unfortunately, be said of other institutions and programmes in the selected area. The government's Integrated Rural Development Programme has helped the 'non-poor' by transferring assets, mainly livestock and other businesses. However, the poorest of the poor have not gained significantly from the programme (Mehta and Joshi, 1993).

The reason for bypassing the bulk of the poor lies in the nature of the delivery systems and the way they function. The limitations of the delivery systems that are not geared specifically to the need of the 'recipients', as listed above, provide an explanation for their ineffectiveness in ameliorating the lot of the poor *adivasies*. The lack of any effective organization of the poor to help them make use of scale-conscious, urban-biased delivery systems (for credit, input supplies, marketing) exacerbates their difficulties. In the absence of support from these institutions, the options open to the *adivasies*, both as producers and as consumers, remain limited.

CONCLUSION: THE LESSONS DRAWN

We have maintained in this chapter that in our selected region the economic well-being of a large number of households depends, largely, on the agrarian structure, this being understood as easy access to natural resources and institutional support for the effective use of these resources. Appropriate government policies and programmes, as well as autonomous forces such as technology, markets and institutions, could contribute to make the agrarian structure favourable to the poor, the environment, or both. These factors could also have adverse implications for both the poor and the environment. The institutional change, policy initiatives and organizational interventions which may help in making the agrarian structure favourable for the poor *and* the environment are briefly indicated below.

(a) There are three important forms of proprietary rights in the selected area in relation to land and forest resources: state ownership, communal ownership and private ownership. All three forms of proprietary rights are likely to continue. With regard to state property, the lesson of the past few decades suggests that, without a genuine partnership with the poor and their involvement in conserving natural resources, state-owned property lapses into 'open access' property. The objective of conserving natural resources and enhancing their value can only be realized if the beneficiaries are involved in their management. Joint Forest Management (JFM) is an example of such a partnership. Normally, participation is possible only at the local level. In the context of state ownership of natural resources, it is imperative that decision-making be decentralized, especially in regard to the formation and implementation of local programmes of resource conservation. Decentralization in decision-making will facilitate the peoples' participation.

(b) As far as communally owned resources are concerned, the general experience is that they are being usurped by the powerful households, or are victims of indifference and neglect by the community. The main reason for the erosion of the community's authority in enforcing the equitable and sustainable use of these resources is the remarkable change in the size and composition of the groups who are co-partners. Unless these groups reorganize and adapt to the changing circumstances, the future for Common Property Resources (CPR) is bleak. There has been a significant development in our understanding of the formation of groups and the viability of collective action. One of the major findings of the recent studies in this area is that the groups can be effective only to the extent that they are small in size (to ensure mutuality) and homogeneous in character (to minimize moral hazards). These insights should help in organizing effective communal action on CPR.

(c) In the tribal areas studied by us, serious distortions have taken place in the markets for privately held land resources because of the restrictions imposed on the land markets (i.e. prohibition of the sale of tribal lands to non-tribals) and the abolition of the lease market. The restrictions on the land market have not served the purpose of helping the poor tribals to retain their land but have meant that the market for land has been restricted: the richer tribals have cornered the land offered for sale by their poor brethren and the poor sellers are faced with an artificially low price of land. These reforms have aggravated the differences within tribal society and encouraged the emergence of an élite group. These restrictions on land and lease markets should be removed to help households who wish to move out of unprofitable agriculture. Activating the lease market after safeguarding the interests of the landowners as well as the tenants can assist in the creation of larger numbers of viable holdings. Given the present weak economic conditions of most tribals, the prevailing ceilings on the owned as well as the leased holdings may be retained. This will prevent a more skewed distribution of the size of holdings or the emergence of 'reverse tenancy' (i.e. when

Agrarian Structure and Livelihood: Indian Forestry 313

larger farmers contract leases from small landholders), both of which in the present circumstances go against the interests of the poor.

(d) Technology can play an important role in enabling small farmers to increase their incomes from their main resource, namely land, and in preventing the degradation of this resource by overexploitation. We examined earlier the role of technology at various levels, i.e. (i) to prevent the pollution of air and water; (ii) to contribute to sustainable development; (iii) to contribute to 'high value added' agriculture; and (iv) to give the poor farmers a wider choice by opening up alternative employment opportunities. More importantly, there are various measures of environment protection, e.g. soil and water conservation, afforestation, etc., which are also labour intensive. The technologies which can promote 'environment for employment' are extremely important in areas of high unemployment and underemployment such as prevail in the tribal regions of southern Rajasthan. Technology generation and extension in these areas are neglected because such developments will mainly benefit the poor resourceless and 'voiceless' households. In fact, the accumulated knowledge of these very households can be a major source of advancement of the relevant technologies.

(e) As well as the land and lease markets, the credit market plays an important role in the economic decision-making of the poor households. It was noted above that, because of the problems of high transaction costs and 'moral hazards', formal credit institutions are reluctant to deal with a large number of the poor households. As a result these households are left at the mercy of the moneylenders who also combine the roles of traders and landowners. The interlocked markets over which they preside take full advantage of the poor tribals' helplessness. The formal credit institutions will be able to assist these households more effectively if they can deal with them in groups, rather than as numerous individuals. The formal or informal groups of the *adivasies* can have a meaningful dialogue with credit institutions.

(f) Reforms in proprietary rights, or in the credit market, or even the desired thrusts in technology have to be instituted in the context of the prevailing macro policy framework. Macro policy reforms on pricing, subsidies and taxation should be designed in such a manner that they do not encourage the overexploitation of natural resources, or lead to a further strengthening of the hegemony of the richer sections. Care should be taken that macro policies which have a strong export orientation do not result in the unsustainable exploitation of the natural resources. In a situation where large numbers of households are poor and depend significantly on natural resources, the impact of macro economic policies on the poor or on the exploitation of natural resources cannot be ignored. While one can always compromise between the rate of economic growth and the rate of resource depletion, the choice is much harder if there is a trade-off between poverty alleviation and resource conservation. Fortunately, with appropriate institutional and market reforms and the application of the right technologies, the strategies to meet the two objectives can be made complementary.

(g) Throughout this discussion we have emphasized economic norms, markets and policies and have dwelt on economic consequences. However, it should be recognized that in a situation of 'missing' or 'incomplete' markets, which is widespread in the case of environmental resources, economic principles cannot provide reliable guidelines for a number of decisions – all the more so, when a large number of resourceless households are only at the periphery of the markets. In such circumstances other considerations, ethical, social or political, become equally, if not more, important.

Part Three
Conclusion and Policy Synthesis

12 Sustainable Livelihood and Employment: Pragmatic Approaches

Iftikhar Ahmed

In this concluding chapter we attempt to highlight and synthesize the major themes emerging from the conceptual and case study chapters. For a sharper focus on the central theme of sustainable livelihood, the discussion here is limited to the following major interrelated subjects: (a) the relationship between environment and growth; (b) issues of sustainable livelihood, including linkages between environmental and socioeconomic sustainability; (c) the quantitative and qualitative employment dimensions of sustainable development, including the important issue of environment–employment trade-offs; and (d) policy instruments and institutional reform.

Being an overview chapter, it synthesizes the major findings of the earlier chapters. However, it goes far beyond these. It furnishes additional empirical evidence and introduces more rigorous analytical arguments so as to set the various issues against a much wider background. It builds on and refines the evidence and concepts presented earlier, thus bringing out clearly the volume's overall contribution to theory, to the identification of data gaps and to policy analysis.

IS DEVELOPMENT POSSIBLE WITHOUT ENVIRONMENTAL DEGRADATION?

Cross-country evidence in Chapter 2 demonstrated that in the course of economic development environmental degradation at first increases and then begins to fall. This inverted U-shape relationship between per capita income and the rate of environmental degradation has been described as an environmental Kuznets curve. The existence of the curve has also been confirmed by time series data from country case studies (Chapter 2; Pezzey, 1992; Dixon, 1993; Pearce, 1993) and country case studies from Côte d'Ivoire, Mexico and Thailand (Reed, 1992). The confirmation of the environmental Kuznets phenomenon through time series data is important, as findings based exclusively on cross-sectional data would be critically distorted because of the overall problem of simultaneity between economy, environment, economic structure and trade (Stern *et al.*, 1994). This only confirms the need for further time series analysis of the environmental Kuznets curve for different countries and country groupings.

Therefore, if the trend illustrated by the environmental Kuznets curve is indeed inescapable at the initial stages of the development process, can the curve be flattened through policies such as eliminating price distortions, internalizing environmental costs and defining and enforcing property rights over the use of natural resources? If not, there is a danger of creating a 'grow first, clean up later' mentality. Even worse, it may spread the simplistic impression that development is necessary in order to clean up the effects of growth (Pezzey, 1989). These issues are taken up in detail later. In the next section we deal with several important aspects of sustainable livelihood.

SUSTAINABLE LIVELIHOOD

The discussion on sustainable livelihood is presented in three parts: (a) the causes and consequences of the destruction of sustainable livelihoods; (b) feasible options for the promotion of sustainable livelihoods; and (c) conditions under which sustainable livelihood is possible and socioeconomic and environmental sustainability in general can be achieved.

Destruction of sustainable livelihood: causes and consequences

Environmental destruction in the tropics is invariably linked with deforestation. It is important to realize that it is not merely a question of stopping the indiscriminate felling of trees. Indigenous peoples have lived in these forests for thousands of years and developed sustainable resource utilization systems. For instance, 1.5 million indigenous people in the Brazilian Amazon depend on forests for their livelihood. There are as many as 300,000 rubber tappers. Their economic activities, cropping patterns and sources of income are diversified out of an awareness of the links between their livelihood and the maintenance of economic diversity. They cultivate and manage fruit trees, palms and forest species. They are not just subsistence producers. Their average cash income is twice the Brazilian minimum wage (Redclift, 1992).

Swidden agriculture as practised by the indigenous population of Malaysia (Chapter 10), although dubbed as environmentally destructive, is insignificant when compared with the gigantic scale of commercial logging.[1] Just a few bulldozers could easily wipe out more tropical forests than many thousands of indigenous people. Moreover, the recognition and resolution of Malaysia's indigenous people's land claims (500,000 hectares) would pave the way for sustainable practices.

There is now overwhelming evidence to prove that the massive conversion of eastern Amazonian forest areas to cattle pastures had little to do with ranching and more to do with establishing land claims. A single tax incentive – a corporate tax incentive to beef-cattle ranchers – was responsible for 30 per cent of forest conversion in Amazonia until 1983 (Lopez, 1992), and as many as six major policy incentives contributing to deforestation have been identified (Binswanger, 1991). More than 20 million hectares of Amazonian land have shifted from public to private hands.

The plundering of forests and the destruction of the sustainable livelihoods of the indigenous population are not justified on economic grounds. The amount of pasture land per head of cattle is very high – 0.73 cattle per hectare – which is equivalent to the production of 22 kg of meat per hectare. The sheer magnitude of economic waste is conveyed by the finding that the production of one quarter-pound hamburger requires the clearing of half a ton of wood! (Nugent, 1994).

Moreover, ranching simply encroaches on land with high opportunity cost. If the same forestry land were cleared for agriculture instead of ranching, crop yields could be from 20 to 70 times higher (Nugent, 1994). The net present value of indigenous utilization of Amazonian rainforest involving fruits, latex and selective cutting (7 per cent of timber logged) is calculated at US$6,820, compared with US$3,184 for timber extraction of pulpwood and the lowest net present value of US$2,960 for cattle pasture, even assuming fully stocked herd size (Munasinghe, 1993).

Cattle ranching contributes little to employment creation. Extensive ranching creates one job for 30 hectares of pasture, while agriculture creates a job for only 3 hectares brought under cultivation (Nugent, 1994). The state hardly benefits from tax revenue collected from ranching when compared with agriculture. For instance, tax revenue collected by the state from the cultivation of black pepper is US$18 per hectare as compared with only US$0.50 collected as tax revenue from extensive ranching. Tax revenue per unit of land from cassava is five times higher than that from ranching (Nugent, 1994).

It is scarcely surprising that many pastures were abandoned in just 10 years, as extensive ranching could not be profitable without subsidies (Mattos and Uhl, 1994). Moreover, big ranches depended on the sale of logging rights to subsidize their operations. A survey of beef-cattle ranchers revealed that 93 per cent of them were involved in logging (Nugent, 1994). The ecological cost of such a large-scale conversion of tropical forests into unprofitable pastures has been colossal. Several hundred tons of biomass containing thousands of species are reduced to 10 tons of biomass with negligible biodiversity when 1 hectare of forest land is cleared for ranching (Mattos and Uhl, 1994). The conversion also affects regional rainfall patterns. Despite the colossal ecological devastation and the massive destruction of the sustainable livelihood of indigenous people caused by clearing forest land for unprofitable ranching, the debate is not over as to whether cattle ranching should be totally abandoned. This would certainly be difficult as long as ranching remains a convenient way of claiming landownership in frontier areas.

However, policy rationalization is needed immediately, so that uneconomic, non-specialized extensive ranching (characterized by low capital investments, low animal stocking densities and low returns) may be completely discouraged. Instead of this form of unviable ranching, the following sustainable ranching practices should be supported (Mattos and Uhl, 1994): (a) specialization by small operators (calf-fattening/dairy); (b) intensification (restoration of degraded pastures and increased production); and (c) livestock raising by small producers (diversification and security).

This strategy is also attractive as regards employment generation. Assuming that the magnitude of labour absorption can be derived from proportionate labour costs, labour costs represent 72 per cent of annual costs for small-scale dairy farmers as against 35 per cent to 37 per cent for the extensive and semi-extensive ranchers (Nugent, 1994).

Incentives for deforestation and the consequent destruction of sustainable livelihood are not entirely confined to forest land conversion to ranching. Thirty-four projects based on charcoal-consuming industries (mainly pig iron) in the Amazon Basin could bring down trees representing 14 million tons of wood per year, and are supported by federal government fiscal incentives such as tax exemption for a ten-year period (Anderson, 1990).

In Bangladesh, the application of subsidized chemical fertilizers and pesticides in rice cultivation affects the nutritional levels of the whole population, particularly the poor, as fish production in paddy fields has been reduced by 60 to 75 per cent (Chapter 9).

The challenge of sustainable livelihood

Increasing unemployment accelerated upland migration in Asia, which in turn intensified environmental degradation and served to redistribute

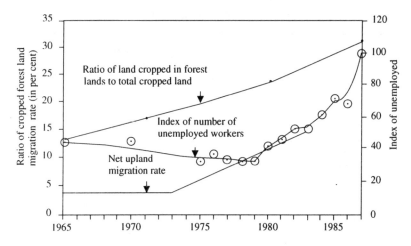

Figure 12.1 Unemployment, upland migration and environmental destruction: Philippines, 1965–87. (*Source*: Cruz and Repetto, 1992, Figure 1.2.)

poverty spatially. For instance, there appears to be a clear correlation between the trends in the unemployment rate, the net upland migration rate and the rate of deforestation in the Philippine uplands (Figure 12.1). Upland migration could be further accelerated by increases in poverty. It is clear that the incidence of poverty is a net upland migration push factor in the Philippines, while higher income measured by regional GDP per capita attracts immigration or discourages out-migration (Table 12.1).

While poverty drives landless agricultural workers from crowded farmlands onto the uplands, poverty also awaits them in such ecologically fragile lands. One-third of all Filipinos are packed into the uplands, which constitute over one-half of the country's area. The population of the uplands of both Java (Indonesia) and the Philippines are among the poorest (Cruz and Cruz, 1990). An average per capita income of 2,168 pesos in the Philippine uplands is well below the official poverty line (Magrath and Doolette, 1990).

Owing to the lack of secure claims on the land, there is little incentive for the growing ranks of the poor, described as squatters in the Philippines, to engage in environmental protection in the uplands. To stem the environmental degradation of the uplands, soil conservation is an essential investment in sustaining or increasing the future productivity of land. With no secure claims on future income flows from the resource, users extract what they can in the present with little regard for the consequences in terms of reduced income and livelihoods in the future.

Table 12.1 Incomes, poverty and upland migration: Philippines, 1980–85

Region	Per capita GDP (pesos at 1972 prices) 1980	Poverty incidence 1980 (%)	Net upland migration rate (%)[1] 1980–85
Ilocos	989.00	40.30	16.8
Cagayan Valley	981.00	43.10	10.4
Central Luzon	1,466.00	27.40	8.5
Southern Tagalog	1,820.00	31.30	18.5
Bicol	783.00	42.72	–0.1
Western Visayas	1,422.00	50.50	–1.1
Central Visayas	1,509.00	48.10	–1.1
Eastern Visayas	718.00	33.00	–2.9
Western Mindanao	1,130.00	40.10	24.0
Northern Mindanao	1,368.00	38.60	10.8
Southern Mindanao	1,605.00	33.30	18.3
Central Mindanao	1,395.00	28.40	23.7
National capital region	3,893.00	11.20	7.2[2]
All regions	1,655.00	14.30	14.5

1. A negative figure indicates net outflow of population from the uplands.
2. Upland urban areas.
Source: Cruz and Repetto (1992), Tables 4.6 and 4.7.

Employment intensity and sustainable livelihood

Assuming that the provision of land rights provides inducements for investment in soil conservation by the producers in the uplands, the question of the economic viability and job creation potential of alternative environmental conservation techniques becomes important. Data from the Philippines would permit the following broad conclusions (Table 12.2):

(a) Environmental protection is compatible with employment.
(b) Labour intensity and average costs of soil conservation are inversely related to the erosion potential (slope of upland serving as an indicator).
(c) Labour intensity as well as average costs vary with the environmental protection technique applied.
(d) A divergence in the social and private profitability from erosion abatement is noted. Assuming that the quantity (value) of on-site nutrient loss due to soil erosion prevented by environmental

Table 12.2 Employment created in environmental conservation in the Philippines uplands (per hectare)

Erosion potential (slopes)	Environmental conservation techniques	Employment in establishment			Employment for maintenance		
		Workdays	Workdays + workdays with work animals	Costs (US$)	Workdays	Workdays + workdays with work animals	Costs (US$)
≤ 25%	*Biological or vegetative*						
	Contour strip cropping	34	6 + 7	53 or 31	42	14 + 7	66
	Buffer strip cropping	14	7 + 2	22 or 17	20	31	
	Mulching	38 per year		42		66	
≤ 12%	*Mechanical or structural Conservation tillage*						
	Minimum tillage and mulch	42 per year	1 + 5	33 or 17	40	0 + 5	31 or 16
	Precision tillage and strip zone	21		3	20		31
	Zero tillage	10		16	10		16
	Contour	60 per year	2 + 7	47 or 25	56		44 or 22
≥ 50%	Terraces						
	Bench	500		786	26		39
	Orchard	112		176	6		9
	Individual basin	12		19	6		8
≥ 50 %	Ditches						
	Contour	31		49	14		22
	Hillside	100		16	5		8

Sources: Cruz et al. (1988), Table 42, p. 58; Cruz and M. Cruz (1990).

protection represents potential gains or benefits from erosion abatement, the potential private benefit is not always greater than the various conservation methods.

(e) When shadow prices of fertilizers are used to measure the social cost of nutrient losses (US$12 worth of fertilizer lost per ton of affected land), the potential benefit is clearly larger than the private benefit (US$51 valued at nominal fertilizer prices). This is also greater than the cost of most of the soil conservation methods (Cruz and Cruz, 1990).

(f) The differences between private and social conservation benefits should form the basis for government subsidies to upland farmers as incentives for adopting soil conservation measures (Cruz and Cruz, 1990).

While the search for employment-intensive approaches to environmental conservation has been the right strategy in labour-surplus economies in Asia, labour shortages have been a major source of environmental degradation in the Latin American highlands (e.g. the Andes mountains), as labour-intensive soil conservation practices had to be abandoned owing to a high rate of rural–urban migration in Bolivia, Mexico and Peru (Garcia Barrios and Garcia Barrios, 1990; Lopez, 1993; Zimmerer, 1993). For instance, in Bolivia nearly half the peasant households surveyed abandoned at least one of the soil conservation techniques on account of the growing shortage of labour as workers took up employment in non-farm work in extra-regional markets (Table 12.3).[2]

Does environmental sustainability depend on socioeconomic sustainability?

Having looked at the causes and consequences of the destruction of sustainable livelihood as well as strategies of maintaining sustainability under conditions of surplus labour and labour scarcity, we should now ask whether socioeconomic sustainability depends on environmental sustainability or whether the converse is true. Furthermore, this section attempts to verify empirically the hypothesis (Chapter 3) that the sustainable livelihood of target groups requires the economic and environmental sustainability of the community as a whole to which the target group belongs.

Data from a survey carried out in Kenya (Oniang'o and Mukudi, 1993) most powerfully demonstrate that socioeconomic sustainability and environmental sustainability are not only interrelated but also mutually

Table 12.3 Labour scarcity and soil conservation: Bolivia (numbers of households)

Household age	Sample size	Labour scarcity	Ceased soil conservation techinques	Labour scarcity and ceased conservation techinques	Increased farm work by women (1980–91)
Young (< 25 years)	19	17	11	11	na
Middle (25–40 years)	20	11	9	7	18
Elderly (> 49 years)	11	5	3	3	5
Total	50	33	23	21	23

Source: Zimmerer (1993), Table 3.

supportive and inseparable phenomena. The basic question is, why are some poor communities more environment-friendly than others? The principal attributes of a Kenyan household's socioeconomic sustainability noted from the survey (Table 12.4) were as follows: (a) dependency on natural resources for their livelihoods; (b) ownership of land

Table 12.4 Linkages among technology, socioeconomic and environmental sustainability: rural Kenya (proportion of households)[1]

Indicator	Proportion of households (percentages)
SOCIOECONOMIC SUSTAINABILITY	
Average size of household	6 members
Level of education: no formal education	32
Household ownership of land	98
Housing owned	100
Farm size (average)	3.7 acres
Farm size distribution (majority)	3–5 acres
Sources of income	
Agriculture/livestock	95
Off-farm employment	28
Women members as a proportion of those seeking off-farm employment	83
Cropping pattern	
Tea	100
Coffee	85
Food crops	90
Sources of farm labour	
Female	93
Male	75
Food security	
Enough cash income to purchase food	68
Famine preparedness: well-preparedness	68
Coping mechanism: purchase food	88
AGRICULTURAL TECHNOLOGY	
Fertilizer use	93
Pesticide use	42
Knowledge of adverse health effects of chemicals	48
Use of protective gear: none	53

Table 12.4 (continued)

Indicator	Proportion of household (percentages)
ENVIRONMENTAL SUSTAINABILITY	
Energy:	
Type of fuel used	Majority: firewood
Sources of firewood	
Own	67
Collected	18
Purchased	15
Distance to source of firewood (1/2 km)	73
Collectors of firewood: female	90
Frequency of collection	
Daily	52
3–4 times a week	12
No. of meals cooked a day	
1–2 meals	17
3 meals	75
4 meals	8
Cooking fuel pollution (smoke) reported	87
Adverse effect on health	78
Fuel-saving technology:	
Knowledge of technology	85
Sources of information	
Indigenous	68
Extension agencies	5
Adoption of technology	33
Intend adoption	23
Reasons for adoption	
Energy-saving	80
Cleaner	8
Time-saving	5
Reasons for non-adoption	
Costs	7
Water:	
Source: tap water	94
Responsibility: women	100
Distance	
Within home	83
1/2 km	15

Table 12.4 (continued)

Indicator	Proportion of household (percentages)
Quality	
Believe treated	87
Believe not treated	7
Water treatment technology	
Knowledge of boiling as a method	87
Believe already treated	77
Practice of boiling	23
Sources of knowledge on technology	
MCH/hospital	47
School	25
Extension agencies	5
Sanitation:	
Access to toilet facility	98
Ventilated improved pit (VIP) technology	
Knowledge	67
Adoption	40
Reasons for non-adoption: costs	40
Sources of information	
Neighbours/friends/display	50
MCH/hospital	10
Extension services	2
Forestry resources:	
Proximity to forest: close	97
Resources extracted	
Firewood	27
Timber	18
Employment in forestry	
None	87
Timber harvesting	11
Firewood collection	2
Knowledge of restrictions (type)	
Permits needed	93
Selected harvesting	37
Enforcement of restrictions: supporters	82
Reasons for restrictions on access to forests	
Save forest	72
Avoid drought	50
Ensure future fuel supply	5

Table 12.4 (continued)

Indicator	Proportion of household (percentages)		
Tree planting:			
Planting on farms	97		
Ownership			
Males	65		
Females	5		
Household	27		
Planting by			
Males	72		
Females	35		
Household	18		
Use of trees			
Fuel	98		
Building	82		
Charcoal making	42		
Soil conservation:			
Value of soil conservation: positive opinion	93		

Soil conservation technology	Knowledge	Adoption	Practice by community
Types of technology			
Terracing/benching	97	87	88
Nappier grass	65	58	45
Tree planting	37	17	27
Maize over contours	17	2	8
None	2	7	
Source of information			
Indigenous channels		26	
Extension services		65	

Table 12.4 (continued)

Indicator	Proportion of household (percentages)		

Waste management:

Household waste disposal	Household rubbish	Kitchen solid waste	Water waste
In the farm	87	68	80
Garbage pit	8	10	8
Front/back yard	–	7	8
Livestock feed	2	7	–
Made into manure	3	7	2

Livestock and poultry waste			
Converted into manure		67	
Into the garden		33	

Disposal of tea/coffee processing waste (industrial effluent)	Solid waste	Liquid waste	
Treated in pond	7	38	
Drain into river	2	10	
Collected by people	48	2	
Do not know	23	28	

1. Based on 60 respondents (95 per cent women) obtained by a two-stage stratified cluster (200 households) sample selection procedure on the basis of probability proportional to size.

Source: Compiled from Oniang'o and Mukudi (1993), numerous tables.

(property rights); (c) more equitable farm size distribution; (d) stability of income and employment; (e) diversification of cropping patterns; (f) greater food security; (g) satisfaction of basic needs (access to water, sanitation and housing); and (h) application of productivity-enhancing agricultural technologies.

It is scarcely surprising that hardly anyone from such a community in such a socioeconomically sustainable setting depends on forestry for employment or livelihood, despite their living very close to the forests.

Table 12.5 Technology, socioeconomic and environmental sustainability linkages: Kenya – rural Embu and Kibera urban slums, 1993 (percentage of households)[1]

Indicator	Rural Kenya	Urban Kibera slums
SOCIOECONOMIC SUSTAINABILITY		
Average size of households	6 members	5 members
Level of education:		
no formal education	32	21
Own land	98	–
Housing		
Owned	100	22
Rented	0	78
Number of rooms		
1	8	74
2–3	62	14
4–5	17	3.5
Sources of livelihood		
Agriculture/livestock	95	–
Male salary/wages/business	4	60
Female small business/salary	2	25
ENVIRONMENTAL SUSTAINABILITY		
Energy:		
Type of energy		
Firewood	Majority	–
Paraffin		79
Charcoal		21
Distance to sources of fuel (1/2 km)	73	48
Frequency of collection		
Daily (firewood)	52	–
When need arises (paraffin)	–	56
Number of meals cooked a day		
1–2	17	7
3	75	78
4	8	14
Cooking fuel pollution (smoke) reported	87	40
Effect on health	78	57

Table 12.5 (continued)

Indicator	Rural Kenya	Urban Kibera slums
Fuel saving technology		
Knowledge	85	88
Adoption	33	42
Advantages: energy-saving	80	77
Reasons for non-adoption: costs	7	37
Water:		
Source		
Own tap	94	0
Purchased	0	97
Public tank/tap	0	86
Distance		
Within home	83	0
1/2 km	15	90
Responsibility: female	100	83
Means of transportation: Back/head/hand	–	93
Water treatment technology		
Knowledge of boiling as a method	87	90
Believe already treated	77	81
Practice of boiling	23	16
Sanitation:		
Access to toilet facility	98	85
Ventilated improved pit (VIP) technology		
Knowledge	67	64
Adoption	40	36
Reasons for non-adoption		
Costs	40	5
Not own house	–	59
Sources of information		
Indigenous	50	88
MCH/hospital	10	3
Extension services	2	5

Sustainable Livelihood and Employment

Table 12.5 (continued)

Indicator	Kenya slums					
Waste management (household):	Household rubbish		Kitchen solid waste		Water waste	
	Rural	Urban	Rural	Urban	Rural	Urban
In the farm	87	–	68	–	80	–
Garbage pit	8	69	10	64	8	19
Unmanaged drainage trench	–	7	–	–	–	9
Livestock feed	2	–	7	–	–	–
Made into manure	3	–	7	–	2	–

1. Based on 60 respondents (95 per cent and 70 per cent women respectively in the rural and urban samples) obtained by a two-stage stratified cluster (200 households) sample selection procedure on the basis of probability proportional to size.

Source: Compiled from Oniang'o and Mukudi (1993), numerous tables.

They simply wish the forests to be saved and view the value of forests as a way of avoiding drought. Tree planting is practised by the entire community. They are aware of alternative soil conservation technologies and a vast majority of them are engaged in soil conservation as they recognize the positive ecological value of such techniques.

Similar linkages between environmental and economic sustainability have been observed from a survey of woodlot-growing households in Kenya, which have diversified economic activities, receive remittances from the wages of migrant family members and enjoy adequate incomes (Dewees, 1993).

The impact of proprietary rights and a community's dependence on natural resources is most dramatically brought out by two contrasting populations in Kenya (Table 12.5). In rural Kenya, the majority of households convert animal waste into manure and use household waste in the farm. In contrast, most urban slum-dwellers dump their household waste into garbage pits. In rural Kenya, hardly any industrial effluent from tea/coffee processing plants drains into the river.

The above finding clearly confirms the hypothesis postulated in Chapter 3 that sustainable livelihood of any given target group, e.g. female-headed rural households or the rural poor, can be achieved if the entire community to which that target group belongs is, as a whole, both economically and environmentally sustainable, and that the distribution of income within that community is such that the needs of all its members are met.

ARE ENVIRONMENT, DEVELOPMENT AND EMPLOYMENT INTERRELATED?

Having briefly surveyed the various dimensions of sustainable livelihood within the conceptual framework developed in Chapter 3, we should investigate whether environment, development and employment are interrelated. There is no simple straightforward answer to this question. Two approaches are followed in the search for clues and possible scenarios. Firstly, some evidence is examined on the trilateral linkages between environment, growth and employment, and secondly, the statistics on the employment dimensions of sustainable development, including projections, are examined, with conclusions from findings with respect to industrialized countries being often drawn on. Where property rights are undefined and the distribution of the environmental assets is unequal, employment creation constitutes a major instrument for promoting sustainable livelihoods.

Trade-offs

The Zambian case study (Chapter 6) clearly demonstrated the trade-offs between employment, environmental quality, food security and aggregate income. For instance, if the dual objectives of full employment and aggregate income maximization are simultaneously pursued, only half of the available forestry land can be preserved. Total income maximization is achieved through a trade-off between tourism and agriculture as sources of livelihood for the fully employed workforce (Table 12.6). However, more country case studies are needed before the wider validity of this finding is established.

This conclusion reflects the direct trade-offs between employment, income and the environment. If the indirect (backward/forward linkages) effects are taken into account, the outcomes on the trade-offs could well be different. If the shadow (present) value of the 'preservation of rare species' were taken into account, the size of the tourism revenue could be much larger to capture the totality of benefits.

The study of Malaysian forestry (Chapter 10) showed that sustainability requires current sacrifices in both output and employment. To ensure sus-

Table 12.6 Trade-offs between employment, income, food security and environmental quality: Zambia

Policy objectives	Maximum possible extent of forest land kwacha) (percentage)	Tourism revenue (millions of kwacha)	Agricultural income (millions of kwacha)	Total income (millions of kwacha)
Environmental quality	100	1,230	0	1,230
Full employment	48	590	1,269	1,859
Food security	62	762	690	1,452
Full employment and food security	35	430	1,325	1,755

Source: Chapter 6, Table 6.9.

tainability of future flows of incomes and employment, a reduction in yield of 37 per cent over current yield levels would lead to a tremendous (62 per cent) loss in employment over current levels of employment (Table 12.7).[3] Over 90,000 forestry workers out of the present total of 150,000 employed workers would lose their jobs.

The above two illustrations, though limited, indicate the existence of such trade-offs in many more instances. Evidence on the nature, magnitude and intensity of Chilean environmental problems appears to be highly sector-specific, and, owing to the regional specialization in the type of economic activity, the environmental problems are also highly region-specific (Malman, 1994). By matching the sectoral environmental problems (Tables 12.8 and 12.9) with the corresponding sectoral growth rates in value added and employment, certain connections between growth, employment and environmental problems are deduced below for a range of economic sectors (Figures 12.2 and 12.3). These are based on a survey of a total of 12,874 national press reports on environmental problems (Tables 12.8 and 12.9).

(a) One piece of evidence on the type of environmental issue raising national concern in the fishing industry provides a surprise. There is much greater national media awareness of the pollution problems of *fish-processing* industries than there is of those of over-fishing. However, as expected, conservation issues dominate the environmental problems in forestry, as is revealed by the substantial statistics of press reports. While this was happening (from 1986 to

Table 12.7 Current and sustainable levels of forestry output and employment: Malaysia

Region	Yields (millions of cubic metres)		Employment (number of workers)			
			Logging		Total	
	Current	Sustainable	Current	Sustainable	Current	Sustainable
Peninsular Malaysia	12.9	4.9	21,000	8,325	79,930	31,556
Sabah	9.0	3.0	16,462	7,200	30,283	13,104
Sarawak	18.4	6.0	31,520	10,345	39,939	13,020
Malaysia	39.8	14.8	68,982	25,604	150,152	57,680

Source: Chapter 10, Tables 10.8 and 10.9.

Table 12.8 Types and intensity of environmental problems faced by economic sectors: Chile, 1992

Sector of economic activity	Type of pollution						Other[2]	Total	(Numbers in parentheses)
	Air	Water	Soil	Other[1]	Infrastructure	Conservation			
Trade	56.1	0.8	1.2	30.3	4.1	2.5	4.9	100.0	(244)
Fishing	20.3	35.6	0.5	20.0	3.3	5.1	15.1	100.0	(390)
Electricity, gas and water	31.3	1.4	0.0	36.3	0.8	24.9	5.5	100.0	(659)
Forestry	12.0	1.0	0.6	1.6	1.0	48.0	35.8	100.0	(500)
Agriculture	23.6	5.8	7.5	17.0	17.5	5.7	23.0	100.0	(670)
Construction	30.5	9.6	0.8	3.8	48.7	3.7	2.9	100.0	(863)
Mining	67.8	16.8	3.1	1.1	3.2	5.4	2.7	100.0	(1,414)
Other services	45.6	11.1	2.3	12.1	10.9	9.1	8.9	100.0	(1,498)
Industry	64.7	6.9	1.0	10.7	7.7	3.4	5.6	100.0	(2,303)
Transport and communications	38.6	3.3	0.1	21.9	33.8	0.9	1.4	100.0	(4,333)
Total	4.3	7.6	1.3	15.0	18.8	6.4	6.4	100.0	(12,874)

1. Indoor, radioactive, acoustic, visual, food contamination, odour, hazardous waste.
2. Environmental degradation, natural disasters, ecosystems, global issues.
Source: Malman (1994), Table 9.

Table 12.9 Sectoral distribution of specific environmental problems: Chile, 1992

Sector of economic activity	Type of pollution						Total
	Air	Water	Soil	Other[1]	Infrastructure	Conservation	Other[2]	
Trade	2.4	0.2	1.7	3.8	0.4	0.7	1.5	1.9
Fishing	1.4	14.1	1.2	4.0	0.5	2.4	7.2	3.0
Electricity, gas and water	3.6	0.9	0.0	12.3	0.2	19.8	4.4	5.1
Forestry	1.1	0.5	1.7	0.4	0.2	28.9	21.7	3.9
Agriculture	2.8	4.0	29.1	5.9	4.8	4.6	18.7	5.2
Construction	4.6	8.4	4.1	1.7	17.3	3.9	3.0	6.7
Mining	16.8	24.1	25.6	0.8	1.9	9.3	4.6	11.0
Other services	12.0	16.9	19.8	9.4	6.7	16.4	16.2	11.6
Industry	26.1	16.3	13.4	12.7	7.3	9.4	15.5	17.9
Transport and communications	29.3	14.5	3.5	48.9	60.6	4.7	7.3	33.7
Total	100.0	100.0	100.0	100.0	100.0	100.0	100.0	100.0
(Numbers in parentheses)	(5,706)	(983)	(172)	(1,937)	(2,421)	(830)	(825)	(12,874)

1. Indoor, radioactive, acoustic, visual, food contamination, odour, hazardous wastes.
2. Environmental degradation, natural disasters, ecosystems, global issues.
Source: Malman (1994), Table 9.

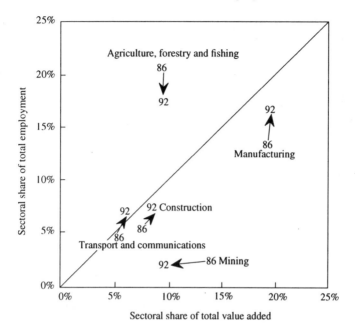

Figure 12.2 Change in sectoral value added and employment: Chile, 1986–92. (*Source*: Malman, 1994.)

1992), the sectoral share of *employment* in agriculture, forestry and fishery declined sharply while the sectoral share in value added remained unchanged. However, many more jobs are created by these three sectors, as agriculture has both a high backward and a high forward linkage to the rest of the economy. In contrast, many more additional jobs will be indirectly created by forestry and fishery activities as they both have forward linkages to the rest of the economy. The relevant policy question is: will the adoption of environmental measures in these sectors contribute to a further slide in employment? This issue is discussed in the section on the economy-wide employment implications of sectoral environmental action.

(b) Most of the environmental problems in the *mining* sector related to air pollution, while gaseous emissions received far more national attention than solid or liquid wastes in the industrial sector. At about the same time (from 1986 to 1992) as this was happening, there was a large relative decline in the share of value added in

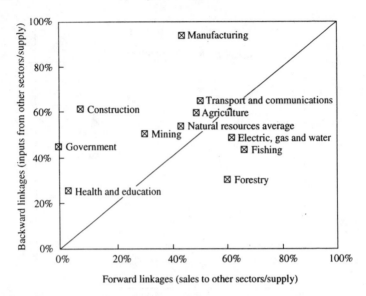

Figure 12.3 Direct forward and backward linkage of economic sectors: Chile. (*Source*: Malman, 1994, Graph 2.)

mining with a rather modest decline in the share of employment in this sector. However, the decline in direct employment may be more than compensated by jobs created indirectly elsewhere in the Chilean economy, given mining's high forward and backward linkages to the rest of the economy. In contrast, the *manufacturing* sector's share of employment grew substantially while remaining a constant proportion of value added. However, the manufacturing sector's exceptionally high backward and forward linkages to the rest of the economy will create additional jobs throughout the Chilean economy. Here again the policy calling for further investigation is: will the proposed environmental protective measures lead to a further increase in unemployment in mining or to a brake on or even a decline in employment gains in manufacturing?

(c) The data (12,874 observations) on the national press coverage of environmental issues in the *transport* and *construction* sectors paid as much attention to air pollution problems as to congestion and infrastructural issues. At the same time it is remarkable that both these sectors have tremendous growth and employment potential, as their sectoral shares of both employment and value added grew significantly between 1986 and 1992. The transport and construction sector also creates many more jobs indirectly, given the very high forward and backward linkages to the rest of the economy, although their number cannot be given accurately. The additional jobs created indirectly by the construction sector arise primarily through its backward linkages to the rest of the economy (Figure 12.3). Once again, we need to know whether an attack on the environmental problems in these two sectors will retard the current high rates of growth in output and employment experienced by both.

Therefore, the types of policy intervention will clearly vary from sector to sector, depending on the nature of the environmental or pollution problem and keeping in view the significance of the corresponding contribution being made to further sectoral growth and sectoral employment expansion. A single, unique economy-wide policy prescription for environmental protection will not be appropriate if growth and job creation are jeopardized when the policy measure has to be tuned to tackle the particular sector-specific environmental problem.

Job potential in the environmental industry

Since no quantitative estimate on the overall job-creating potential of environmental activities is available for Third World countries, some general indications can be obtained from statistics for advanced countries. In 1988, a pollution control expenditure of US$100 billion created 3 million jobs directly and indirectly in the United States. In the European Union countries an estimated 1.2 to 1.5 million jobs were created directly in pollution control (Renner, 1991). Such jobs are on the increase. Nearly 4 million environmental protection-related jobs were created in the United States in 1992, and in Germany the number of jobs directly and indirectly created by environmental policies increased from 433,000 in 1984 to 546,000 in 1990 (Sprenger, 1994).

The estimated total expenditure of 63,000 million ECU in the European Union supported over 1 million jobs in those countries in 1992 (Table 12.10). Over three-quarters of the total environmental expenditure was on the environmental protection industry, with over one-third of the total 1992 environmental employment in the European Union countries indicating substantial backward linkages.

It is expected that both the environmental protection industry and employment generated by it will grow by over 70 per cent between 1992 and 2000 in the European Union countries (ECOTEC, 1993).

Table 12.10 Employment created by environmental expenditure by type of environmental activity: European Union, 1992

Environmental activity	Environmental employment	
	(000s)	(%)
Environmental protection industry[1]	370	34
Water supply industry	226	21
Public authority environmental management	333	30
Industry environmental management	167	15
Total	1,096	100

1. Excludes employment generated by exports.
Source: ECOTEC (1993), Table 5.1b.

Job creation in pollution havens: facts and myths

The most polluting industries are the fastest-growing industries in Third World countries (Table 12.11). These include pulp and paper, chemical products, petroleum refineries, non-ferrous metals, leather tanning and textiles. The developing countries' relative share of industrial value added in the world total for some of the most highly polluting industries has increased dramatically.

It is hardly surprising that these very dirty industries have been the engines of employment growth in Third World countries (Table 12.12). It is quite clear that, while the share of value added is shrinking rapidly and employment is declining in absolute terms in the advanced countries, these industries are a major source of growth and jobs in developing countries.

Some have attributed the growth of this phenomenon to two basic factors: (a) the developing countries' less rigorous environmental regulations encourage a huge southward flight of 'dirty' industries, creating jobs in the South and costing jobs in the North, and making Third World countries dirtier still; and (b) plant closures have resulted from high environmental compliance costs in advanced countries, which have even been described as 'job-killers' (Sprenger, 1994). The validity of these two reasonings is verified by empirical evidence below.

Table 12.11 Average annual growth of selected industries with a higher impact on the environment: advanced and developing countries, 1985–92

Branch of industry	Advanced countries		Developing countries	
	1975–85	1985–92	1975–85	1985–92
Textiles	0.2	0.0	2.8	3.9
Leather and fur products	−0.3	−0.2	4.4	5.3
Pulp and paper products	1.7	3.4	5.0	5.1
Industrial chemicals	1.6	3.5	6.7	7.4
Petroleum refineries	0.7	1.2	7.8	5.3
Miscellaneous petroleum and coal products	2.0	1.7	8.1	4.1
Non-ferrous metals	−1.5	1.0	6.4	4.2
	0.9	3.2	7.2	5.4

Source: UNIDO (1992b).

Environmental compliance costs

Environmental compliance costs account for only 1.2 per cent of manufacturing value added in European industry and less than 2 per cent in the

Table 12.12 Index numbers[1] of employment in industries with a high impact on the environment: Advanced[2] and developing[3] countries, 1978–90 (1980 = 100)

Branch of industry/region	1978	1981	1985	1990
Food, beverages and tobacco				
Developed	101	98	95	97
Developing	93	102	107	115
Textiles				
Developed	107	91	80	78
Developing	97	95	91	86
Wearing apparel, leather and footwear				
Developed	103	95	86	82
Developing	98	103	123	145
Wood products and furniture				
Developed	102	93	84	87
Developing	92	101	131	183
Paper, printing and publishing				
Developed	98	98	97	104
Developing	94	104	108	118
Paper and paper products				
Developed	101	97	89	90
Developing	93	106	111	121
Chemical, petroleum and plastic products				
Developed	100	98	96	103
Developing	92	102	115	124

Table 12.12 (continued)

Branch of industry/region	1978	1981	1985	1990
Basic metals				
Developed	103	96	74	68
Developing	92	100	112	118
Metal products, machinery and equipment				
Developed	99	98	95	96
Developing	92	101	103	112

1. The indices are designed to measure trends in the average number of persons engaged. Estimates of number of persons engaged, including working proprietors and active business partners, unpaid family workers as well as employees are used as weighting coefficients.
2. Includes North America (Canada and the US), Western Europe, Australia, Israel, Japan and New Zealand.
3. Includes Caribbean, Central and South America, Africa (excluding South Africa), Asia (excluding China, North Korea, Israel, Japan and Viet Nam), Oceania (excluding Australia and New Zealand).

Source: UNIDO (1992a).

Table 12.13 Environmental compliance costs in manufacturing: European Union countries and the United States

Country/regional bloc	Manufacturing environmental spending/manufacturing added value (%)
Belgium/Luxembourg	0.7
Denmark	0.7
France	1.2
Germany	1.2
Greece	1.1
Ireland	0.3
Italy	0.6
Netherlands	1.0
Portugal	0.5
Spain	0.7
United Kingdom	1.2
European Union	1.2
United States	1.7

Sources: Economic Research and Advisory Consortium (ERECO), quoted in ECOTEC (1993), Table 4.5; United States Congress (1993).

United States (Table 12.13). Therefore, it is inconceivable that they would create pressures for plant closures or the flight of 'dirty' industries to Third World countries. The labelling of environmental compliance costs as 'job-killers' is more a myth than a reality. Environmental regulations have generally been one of the less important reasons for investing abroad (Sprenger, 1994). Only the pollution abatement and control expenditures of about 5 per cent of the value added in four polluting process industries (chemicals, petroleum, pulp and paper, primary metals) are higher than the average of 1.72 per cent of United States manufacturing as a whole (United States Congress, 1993).[4]

Despite the relatively higher compliance costs, the chemicals and wood pulp industries in the United States are highly competitive internationally, with significant trade surpluses (United States Congress, 1993).

Profitability, returns and industrial competitiveness

A survey of 500 United States enterprises revealed that the adoption of environmentally sound technologies could reduce the production of hazardous wastes by between 85 and 100 per cent. These industries also

Table 12.14 Returns from industrial pollution prevention in the United States

Industry	Method	% reduction of wastes	Payback
Pharmaceutical production	Water-based solvent replaced organic solvent	100	< 1 year
Equipment manufacture	Ultrafiltration	100 of solvent and oil, 98 of paint	2 years
Farm equipment	Proprietary process manufacture	80 of sludge	2.5 years
Motor vehicle manufacture	Pneumatic cleaning replaced caustic process	100 of sludge	2 years
Microelectronics	Vibratory cleaning replaced caustic process	95 of cumene	3 years
Organic chemical production	Absorption, scrap condenser, conservation vent, floating roof	100 of sludge	1 month
Photographic film processing	Electrolytic recovery ion exchange	85 of developer, 95 of fixer, silver and solvent	< 1 year

Source: Huising (1989), p. 6.

Table 12.15 Returns from energy conservation by Tata Iron and Steel Company (TISCO), India, 1981–87

Energy conservation areas	Total investment (Rs million)	Cumulative savings (Rs million)
Steam leaks, rationalization and insulation	5.50	6.60
Cold blast main insulation at blast furnaces	1.00	1.20
Insulation of LSHS storage tanks	0.15	0.12
Improvement in insulation in soaking pits; reheating furnaces and forge furnaces	2.00	1.80
Improvement in oil-firing equipment and high efficiency burners in reheating furnaces and forge furnaces	2.00	3.00
Improvement in blast furnace stoves for higher hot-blast temperature	10.50	12.60
Soaking-pit recuperator design changes	1.50	1.00
Improvement in furnace and energy distribution; instrumentation for energy conservation	38.00	45.60
LD gas recovery system and concast	140.00	390.10
Utilization of waste LP nitrogen for steam savings	7.00	5.60
Solar heating system for LSHS heating	0.45	0.41
Portable energy monitoring device	1.50	1.35
Energy promotion activities, including interdepartmental competition	0.65	0.78
Total	210.25	470.16

Source: Pachauri (1990), p. 22.

amortized their investments in pollution abatement over periods ranging from one month to three years (Table 12.14). At the plant level, environmentally sound technologies can be profitable in the medium term (Table 12.15). Returns from pollution prevention and environmental clean-up in the dirty industries of a range of developing countries are encouraging for their diffusion prospects (Table 12.16).

Table 12.16 Returns from pollution prevention and environmental clean up: Brazil, India, China and Zambia

Country	Technology/approach	Industry	Benefits
Brazil	1. Technologies that promote better use of the laminator 2. Water and gas emissions treatment 3. Coal (from trees) management system	Steel	Reduction of emissions on water, air, decreased consumption of coal Annual savings US$1.5 million emissions treatment Prevented exploitation of 1,000 hectares of forest a year
China	Utilization of organic wastes	Small straw pulp mills	Production of agricultural fertilizers
India	1. Technology which allows recycling of emissions 2. Reuse of chlorohydric acid and caustic soda	Polyfibres	Considerable savings
	Utilization of solid waste called willow dust	Textile	Production of biogas Savings in energy consumption
Zambia	Technologies to filter dust from the stove	Cement	Net annual savings US$40,000 Pollution inside the factory avoided. Productivity raised Reduction of investment in equipment

Sources: Compiled by Almeida (1993) from UNIDO (1992c) and Schmidheiny (1992).

Demise of manufacturing

The declining trend in dirty industries in the advanced countries has to be viewed in the light of the overall declining trend in manufacturing as a whole. The service sector accounts for nearly three-quarters of United States GDP and of its jobs. In the United Kingdom and Canada manufacturing accounts for less than 20 per cent of total output. In the leading Japanese and German economies, manufacturing does not account for even one-third of GDP (*The Economist* (London), 20 February 1993).

Employment-intensive options

In order to avoid generalizations arising from too broad a discussion of employment-intensive options for environmental action, it is better to focus in depth on the employment implications of the alternatives of (a) environmentally sound technologies; and (b) waste management.

Environmentally sound technologies and jobs

It is estimated that cleaning technologies, i.e. 'end-of-pipe' solutions removing pollutants from waste streams, currently account for 80 per cent of total investment in environmentally sound technologies (EST). Clean technologies relevant for pollution prevention account for the remaining 20 per cent (United Nations, ECLAC, 1991).

In the section on profitability, returns and competitiveness, we saw that ESTs can be profitable (Tables 12.14, 12.15 and 12.16). However, not much is known about their employment intensity. The adoption of cleaner production technologies in leather tanning simply involves the substitution of one type of chemical for another, e.g. the use of no-chrome technologies instead of low-chrome technologies to reduce pollution (OECD, 1992). This should not lead to labour displacement.

Evidence from the United States shows that the application of ESTs in two of the most polluting industries (chemicals and pulp and paper) does not reduce the competitiveness of these industries, although their impact on employment has not been assessed. For instance, chemical process technologies cut waste generation significantly and at the same time reduce capital and operating costs. Similarly, ESTs in pulp and paper, which help to recover one-third to two-thirds of organic substances (some of which are not biodegradable), require capital installation with lower operating costs (United States Congress, 1993).

Employment in one 'end-of-pipe' EST in Mexico City could create US$75 million of business. Expenditure on catalytic converters, mandatory since 1991, at US$500 each for 600,000 motor cars in Mexico City amounts to US$300 million annually. If the 3 million vehicles in Mexico City were to pass an inspection test (costing US$12.50 each) to check for emission of gases twice a year, this would generate business worth US$75 million annually with a corresponding degree of job creation in this activity (Kate and Joost, 1994).

Waste management

Waste management appears to be the most employment intensive of all environmental protection measures. For instance, waste management

Table 12.17 Market and employment in the European Union environmental protection industry, by environmental media, 1992

Environmental media	Employment[1]		Market size	
	No.	%	Millions of ECU	%
Air pollution control	49,700	13	7,428	15.5
Contaminated land remediation	13,400	4	1,419	3.0
Waste management	177,700	48	23,405	48.9
Waste-water treatment	129,400	35	15,647	32.7
All media	370,000	100	47,899	100

1. Adjusted for imports to the European Union. Excludes employment, associated with the production of exports, of 92,000 jobs.
Source: ECOTEC (1993), Tables 3.1a and 3.3

alone accounts for 80 per cent of the 370,000 jobs created by environmental protection in the European Union with an equally large (80 per cent) market demand (Table 12.17).

Evidence on waste management alternatives from the United States shows that recycling creates relatively more employment than either landfills or incineration. For every million tons of waste processed, recycling facilities in Vermont generate 550 to 2,000 jobs as compared with 150 to 1,100 jobs by incinerators and only 50 to 360 jobs by landfills. This pattern is also confirmed by evidence from New York (Table 12.18).

As regards relative costs, on account of economies of scale in the larger New York facility, more waste is handled per employee in New York than in Vermont (Renner, 1991). Recycling as a waste-management alternative is not only the most employment intensive but also the least costly option. The cost (US$500 million) of an incinerator is three times that of a recycling facility that can deal with the same amount of waste (Renner, 1991). Therefore, recycling stands out as the best option for waste processing in terms of both employment maximization and cost minimization.

Data on the employment gains from environment-friendly waste management for Third World countries are not available, but when 20 river basins (out of a total of 320) receive 90 per cent of the Mexican urban-industrial waste-water (Kate and Joost, 1994), the prospects of cleaner job

Table 12.18   Jobs created by type of pollution abatement, Vermont and New York City (number of jobs per 1 million tons of waste processed)

Type of pollution abatement	Vermont	New York City[1]
Landfills	50 to 360	40 to 60
Incinerators	150 to 1,100	90 to 300
Mixed solid waste composting	–	210 to 310
Recycling	550 to 2000	390 to 590

1. The numerical figures for the range of jobs created were approximated from the graphical presentation by Renner (1991) based on statistics obtained from the New York City Department of Sanitation.
Source: Renner (1991), pp. 34–35

creation are quite evident from the illustrations on employment-intensive waste management provided for the European Union and the United States.

Similar contributions could be made to job creation from pollution control in Argentina if the industrial waste and untreated sewage could be processed in waste management facilities and thereby bring an end to the human assault on the environment. Indeed, factories, slaughterhouses and tanneries dump waste ranging from animal blood to heavy metals into the river running through the industrial belt of Buenos Aires (*Financial Times* (London), 27 October 1993).

Economy-wide job creation

The above discussions on the quantitative impact of environmental measures have largely focused on direct or immediate, often short-term, employment, often at the sectoral level. However, macroeconomic intersectoral or inter-industry models for European Union countries suggest that the *net* employment effect of environmental expenditures for a number of advanced countries is positive (Sprenger, 1994).

Owing to data constraints for Third World countries it is not possible to trace the intersectoral repercussions of environmental action on employment. However, as noted earlier, reconciling the information available for Chile on the intersectoral input-output linkages (Figure 12.3), the sectoral employment and output growth potential (Figure 12.2) and the type of environmental problems (Tables 12.8 and 12.9) could provide some clue

Sustainable Livelihood and Employment

on the aggregate or net employment effect of a given environmental action in Chile. For instance, agriculture, manufacturing, transport and communications have high forward and backward linkages to the rest of the Chilean economy (Figure 12.3), clearly holding out the promise of extra job creation through environmental action taking into account the direct employment and growth potential (Figure 12.2) in these sectors individually.

Qualitative aspects of employment

While quantitative data on the employment impact of environmental measures in Third World countries are scarce, information on the qualitative aspects is even scarcer. However, some indications of the qualitative impact on employment if environmental policies were adopted can be deduced for the Third World from a review of some evidence available for advanced countries. This deals with two aspects: skills and regional impact. It is useful to have some idea of the relative role of the public and private sectors in the creation of 'green' jobs.

It is estimated that, if environmental policies were applied to the energy sector, the employment of low-skilled workers would increase dramatically, while high-skilled employment would decline sharply in the European Union countries (Majocchi, 1994). The extra low-skilled jobs would probably often arise in economically depressed areas with high levels of European Union unemployment (ECOTEC, 1993). In contrast, environmental action in the transport industry leading to job losses in the car industry could affect certain regions of the European Union, including the Ruhr region of Germany, the West Midlands of the United Kingdom, France, Italy, Spain and Belgium (ECOTEC, 1994a). It is expected that action on specific environmental problems linked to regional specialization of economic activity in Chile will have a range of varied employment effects in several regions (Malman, 1994).

Public versus private

The public sector appears to make an important contribution to job creation through environmental action. For instance, of the 500,000 jobs created by environmental activities in the European Union in 1992, two-thirds were created by public authorities with environmental responsibilities and only one-third by industry (ECOTEC, 1993).

In developing countries a similar prominent role may be expected. For instance, in Chile an active government role in environmental matters is noted in virtually every sector of the economy except mining and trade, where the business community appears to play a more dominant part (Malman, 1994). This, of course, does not provide indications on the relative contribution to job creation by private industry following the worldwide trends in privatization.

POLICY INSTRUMENTS

The case studies in this volume as well as the section on sustainable livelihood in this chapter clearly demonstrate that biases and distortions in incentive structures have largely been responsible for the overuse of natural resources (Chapters 4 and 5) and the destruction of tropical forests. There also appears to be a policy dilemma. Low prices for agricultural products reduce farm incomes, dampening incentives for soil conservation. On the other hand, high farm prices could contribute to extensive cultivation of fragile lands (Barbier and Burgess, 1992). Incentives for agricultural input use (e.g. irrigation, fertilizers and pesticides) could also contribute to land degradation. However, a critical review of subsidies on fertilizers leads to the conclusion that the benefits of increased productivity and reduced poverty outweigh the adverse effects (e.g. fertilizer run-off causing water pollution and land degradation) on the environment (Barbier and Burgess, 1992).

Subsidies in agriculture could also be used for promoting environmental goals and sustainable livelihoods. Instead of offering incentives for the wider use of environmentally degrading chemical inputs through subsidies in the Philippines (Chapter 8), the same resources could be used for subsidizing job creation through environmental conservation in the Philippine uplands (Table 12.2). Subsidies could make up for the divergence between the social and the private cost of benefit of the most employment-intensive options, as discussed in the section on sustainable livelihood in this chapter.

Under efficiency, feasibility and cost-effectiveness criteria, different types of environmental problems require different policy approaches (Table 12.19). For instance, government investment is neither needed nor effective as long as 'command and control' (CAC) approaches are most effective in controlling lead pollution through the installation of catalytic converters in cars. As noted earlier, the mandatory installation of such catalytic converters in Mexico City since 1991 creates US$300 million annual sales of converters and their inspection for compliance generates additional business of US$75 million per year (Kate and Joost, 1994).

Table 12.19 Effectiveness of policy approaches by type of pollution

Type of pollution	Policy approach		
	Government investment	Command and control approaches (CAC)	Market-based incentives
Air			
Reduce lead	–	Quite effective	Effective
Reduce TSP	Effective	Effective	Effective
Reduce SO_2	–	Effective	Effective
Water			
Increase potable water supply	Effective	–	Effective
Treat sewage	Quite effective	Effective	–
Reduce waste water discharge	–	Effective	Quite effective

Source: Dixon (1993).

However, CAC approaches would be unsuitable for controlling certain specific types of environmental pollution in developing countries. For instance, in Philippine agriculture, the control of nitrate leaching by CAC methods raises technical difficulties, as setting standards, weak institutions and the lack of administrative skills would lead to high enforcement and monitoring costs (Chapter 8). Market-based incentives are effective for industrial emissions (e.g. tradable emission permits). Government investment is both necessary and effective for the treatment of domestic sewage from considerations of equity and economies of scale, while market-based incentives are of limited use. As noted earlier from examples of waste management in the United States and the European Union, employment-intensive waste management alternatives are available. Waste-water discharges from industries are not effectively handled by government investment, e.g. to tackle the exceptionally high pollution of rivers in Argentina and Mexico. Market-based incentives and CAC methods (e.g. pollution rollbacks) would be more effective (Dixon, 1993).

Policy approaches could again be classified in respect of direct and indirect instruments, as illustrated in Table 12.20.

Table 12.20 Policy approaches and instruments

	Policy instrument	
Policy approaches	Direct	Indirect
Market-based incentives	Effluent charges, tradable permits and deposit refund systems	Input/output taxes and subsidies, subsidies for substitutes and abatement inputs
Government and control approaches	Emission regulations (source-specific, non-transferable quotas)	Regulation of equipment processes, inputs and outputs
Government investments	Regulatory agency expenditures for purification, clean-up, waste disposal and enforcement	Development of 'clean' technologies

Source: Eskeland and Jiminez (1992).

INSTITUTIONAL DIMENSIONS

The various policy instruments discussed above are more relevant for urban industrial pollution issues. From the perspective of sustainable livelihood, institutional dimensions of natural resource management are more important. This section purposely focuses on three institutional issues of importance to sustainable livelihood (property rights, free access to common property and land reform), in order to avoid less useful generalities from too broad a treatment.

Property rights

Defining property rights is important because those who use natural resources have an interest in conserving them for the future. It removes uncertainty in resource ownership, which deters conservation and leads to overexploitation of natural resources. Property rights should be created or defined more clearly where such rights are vague or incomplete. The contribution of property rights to environmental sustainability was demonstrated dramatically by evidence from Kenya.

As noted earlier, the establishment of the property rights of the indigenous population in Malaysian forestry could contribute to greater conservation through the practice of swidden agriculture (Chapter 10), while the legislation of similar rights for the indigenous people in the Amazon forests could prevent the economically unjustified, colossal destruction of forests in further areas simply by establishing land claims by outsiders.

Free access to common property

Hardin's 'tragedy of the commons' (Hardin, 1968) professed that, wherever many individuals freely use a common property resource, it is doomed to be degraded through overuse and will bring ruin to all. However, this assertion has been challenged by evidence that grassroots-level communal management of such resources could indeed be shared and collectively benefit all users, and could also permit resource conservation. This view is supported by analytical arguments in Chapter 3. Nevertheless, the fact remains that the highest proportion of poor workers in the Philippines are landless or dependent on open access resources (Cruz and Repetto, 1992).

As regards free access to common property, the case studies provide a number of scenarios:

(a) Environmentally harmful swidden agricultural practices are the consequence of free access to forestry land by the indigenous population who have no legal property rights (Chapter 10).
(b) In Bangladesh, owing to the absence of land rights, the indigenous population practise shifting cultivation (*Jhum*). At the same time, because of the intensification of commercial agricultural production induced by the Green Revolution, free access to common property has drastically diminished. This has worsened poverty because the portion of the income of the poor obtained from non-exchange sources, e.g. food obtained by gathering, fishing, collecting of thatch and fuel and the availability of free fodder and sustenance for livestock, has simply disappeared (Chapter 9).
(c) Legislation in India converted common property forestry resources into state property controlled and managed by the forestry bureaucracy, leaving the indigenous tribal people with two options: either migrate from the hills to the plains, or extend cultivation to fragile lands (Chapter 11).

Land reform

In parts of Bangladesh (Chapter 9) where *Jhum* cultivation, which contributes to soil erosion, is practised, a greater inequality in the distribution of cultivated land area has been noted.

In India (Chapter 11) the tribal masses lost their access to natural resources as the state forestry bureaucracy took over the management and (over-)exploitation of the forest lands and as a handful of tribal élites secured the ownership of the bulk of the tribal land as a result of legislation preventing the sale of tribal land to non-tribals. The richer tribals simply took advantage of the artificially low prices offered by the poor tribals in this imperfect land market. An acute differentiation in the tribal population emerged and vast numbers of landless tribal people were driven to cultivating small parcels of fragile marginal lands or to working as casual agricultural workers, both for the tribal élites and for landowners outside the tribal belt. Many of the tribal people were transformed into bonded labour when these landless and assetless workers were compelled to repay loans in terms of human labour to the non-tribal population who were the major suppliers of credit.

Discriminatory legislation in pre-independence Zimbabwe created a dualistic land tenure which has protected the large well-maintained commercial farming areas from degradation, thus conserving the natural resource base for future generations. In contrast, a tremendous pressure on natural resources is created by packing most of the Zimbabwean population into marginal areas (commercial lands) cultivated by peasant agriculturists (Davies and Rattsø, 1994). This poses a dilemma for the policy-makers who wish to pursue the dual objectives of equity and sustainable development. However, at the very least, property rights which encourage conservation would require the localizing (privatization) of ownership and the management of natural resources at the village level.

In addition to defining property rights for the indigenous population for environmental sustainability, lack of tenurial security discourages improvements such as the plantation of fruit trees or plantations, e.g. in Honduras (Dorner and Thiesenhusen, 1992).

CONCLUDING REMARKS

While the environmental Kuznets phenomenon is inescapable, it is clear that the *inverted* U-shape relationship between per capita income and the rate of environmental degradation can be flattened by eliminating price distortions and institutional biases, promoting clean industrial production and employment-intensive waste management, internalizing environmental costs, and defining and enforcing property rights over natural resources. A more egalitarian and economically diversified society having access to environmentally sound technologies will achieve growth with lower levels of environmental degradation.

There is, by now, overwhelming evidence that the systematic destruction of the sustainable livelihood of indigenous populations was the direct consequence of environmental destruction deliberately fostered by precise fiscal incentives and land legislation enabling individuals to acquire land rights automatically by clearing tropical forests. The relentless elimination of indigenous populations engaged in sustainable natural resource management practices and their replacement by gigantic unprofitable cattle ranches commanding huge areas of rapidly degrading pastures has to be immediately stopped. Subsidies to cattle ranching and logging rights which directly contribute to environmental destruction have to be discontinued. Since the cattle lobby will not permit the closure of already established ranches, only the small-scale specialized employment-intensive ranches with sustainable ranching practices should be allowed to remain, as they are likely to be profitable following the withdrawal of subsidies and discontinuation of distortion in policy incentives.

There is fairly strong statistical evidence to demonstrate that socio-economic sustainability and environmental sustainability are not only interrelated but also mutually supportive and inseparable phenomena. As hypothesized in Chapter 3, the basic precondition for the sustainable livelihood of a given target group depends not on the sustainability of the few (the target group) but on the economic and environmental sustainability of the community as a *whole*. Another important criterion for the achievement of sustainable livelihood is that the distribution of income within that community should be such that the needs of all its members are met.

Trade-offs are observed between employment, environmental quality, food security and aggregate income in the use of natural resources in Zambia. Lowering current levels of output and employment may be necessary to sustain a consistent level of future flows of incomes and employment in the natural resource sector of Malaysia. Instances are not rare where considerable commercial justifications (the basis of imputed net present value or derived revenue) exist for natural resource conservation in both Africa and Latin America.

Increasing unemployment and the intensification of poverty accelerated migration to ecologically fragile areas (e.g. the Philippine uplands). Such migration contributed to environmental degradation and merely served to redistribute poverty spatially. Even in such contexts, employment creation can be complementary to environmental protection through the use of employment-intensive soil conservation methods.

Employment intensity varies with the environmental protection techniques which are currently available for ecologically fragile uplands inhabited by the poor of Asia. The obvious pro-poor environment-friendly strategy would be to promote the application of the most employment-intensive option for environmental protection. However, in other parts of the developing world, e.g. the Latin American highlands, environmental protection had to be abandoned because of labour scarcity.

The considerable extra employment created by the environmental industry in the United States and the European Union countries should encourage developing countries in promoting their own environmental industry.

Environmental and employment problems are often sector- and region-specific. Therefore, a single economy-wide policy prescription which groups together all types of environmental problems will not be appropriate if growth and job creation, as well as the prospects of sustainable development, are not to be jeopardized.

The polluting industries being the fastest growing, they are engines of employment growth in Third World countries in the early stages of industrialization. The environmental compliance costs are too insignificant to cause a flight of dirty industries from the North to pollution havens in the South. There is little empirical proof of plant closures on account of high costs of compliance and, therefore, little justification to consider them as 'job-killers' in the advanced countries. An autonomous change in sectoral structure, involving a shift in employment away from manufacturing to services, was a trend the beginnings of which preceded the rigorous enforcement of environmental regulations in the advanced countries.

Since non-farm job creation in Third World countries critically hinges on the growth of polluting industries, the rapid introduction of clean production technologies is crucial. There is already some evidence of their economic attractiveness, and to adopt them would involve little risk of loss of international competitiveness; however, hardly any data are available on their impact on employment.

Evidence on the high employment-creating capacity in industrial and household waste management from the United States and European Union countries should inspire Third World countries vigorously to promote such measures, particularly in countries such as Argentina and Mexico, to spare the highly polluted rivers from further dumping of wastes. Some of the most employment-intensive waste management alternatives are often also the economically most attractive.

Sustainable development strategies in some sectors, e.g. energy, will enhance employment relatively more for unskilled workers than for skilled workers. This is again a welcome trend for Third World countries, where the large numbers of jobless unskilled workers could find employment. Sectoral environmental policies also have regional implications. As has been noted, although sustainable development strategies in the transport sector raise overall European Union employment, the incidence of unemployment will be high in regions where car manufacturing is concentrated. Furthermore, the significant amount of environmental job creation by public authorities in the European Union raises hopes of similar job creation by the public authorities in developing countries, especially when the process of privatization of industry and services is virtually universal. Indeed, the Government of Chile is seen as a very active agent in environmental matters.

Further information on the sectoral breakdown of growth in employment and value added, intersectoral backward and forward linkages and the type of environmental pollution for Chile could provide the basis for extending further quantitative work on employment multiplier and economy-wide net employment effects. At present data constraints do not permit the application of sophisticated tools of statistical analysis, e.g. input-output and econometric models used in advanced countries to quantify the aggregate employment and income effects.

There is, in general, support for the withdrawal of subsidies on environmentally degrading agricultural inputs, and on incentive structures which lead to overuse of natural resources, and a recognition of the lack of fiscal disincentives to the use of industrial inputs causing environmental pollution. While, clearly, a change in policy is needed to protect the environment, there is little knowledge of the precise effect on employment and income distribution of such changes. On the other hand, the most employment-intensive environmental protection method in the Philippines may require a state subsidy to bridge the divergence between the private and social cost/benefit perceptions.

A fairly clear picture emerges about the effectiveness of direct and indirect environmental policy instruments in tackling specific environmental problems. Here again, there is a need for systematic analysis of the impact of the use of a given instrument simultaneously on employment and the environment, particularly for market-based incentives which are considered as the most efficient.

Clearly, the provision of property rights for indigenous populations engaged in swidden agriculture in Malaysia, shifting cultivation (*Jhum*) in Bangladesh and peasant agriculture on communal lands in Zimbabwe will lead to more sustainable agricultural practices. In the Amazon frontier areas, forest clearing as a means of automatic acquisition of land titles could have been prevented if only the indigenous population engaged in sustainable natural resource use were given the property rights. In Bangladesh the disappearance of free access common property has led to the intensification of poverty. In India, the forestry bureaucracy took over free access to common property forest areas, forcing the indigenous tribal people either to migrate to the plains to serve as low-paid workers or to be pushed to ecologically fragile areas to eke out a living from cultivating marginal lands. Well-intended land legislation in India which banned the sale of tribal land to non-tribals enabled a handful of tribal élites to corner the land market, offering artificially low prices to the poor tribals selling land in times of distress. What is worse, the landless tribals are transformed into bonded labour when they offer their future labour services as collateral for credit obtained from non-tribal moneylenders.

Notes

1. Swidden agriculture is 'slash and burn' agriculture, a traditional farming method practised by transitory subsistence populations.
2. Actually, environmental protection activities reduced because of labour shortages which directly or indirectly helped soil conservation include diversion ditches, protective rockwalls and agricultural terraces.
3. Revenues for ecotourism in Costa Rican rainforests are calculated at US$35 per household. Annual viewing values for elephants in Kenyan safaris are estimated as US$25 million, creating a consumer surplus for safaris of US$182 million to US$218 million per year (Munasinghe, 1993).
4. These polluting process industries in the United States spend an average of 15 per cent of capital expenditure on pollution control and abatement, compared with a meagre 3.4 per cent for all other manufacturing sectors in the United States. These industries also account for three-quarters of pollution abatement expenditure by manufacturing but only 22 per cent of manufacturing value added (United States Congress, 1993).

Bibliography

Abel, N.O.J.; Flint, M.E.; Hunter, N.D.; Chandler, D.; Maka, G.: *Cattle keeping, ecological change and communal management in Ngwaketse* (Gaborone, Botswana: Integrated Farming Pilot Project, 1987).
Addison, A.; Demery, L.: 'The economics of rural poverty alleviation', in S. Commander (ed.): *Structural adjustment and agriculture: Theory and practice in Africa and Latin America* (London: Overseas Development Institute, 1989).
Adelman, I.; Fetini, H.; Golan, E.H.: *Development strategies and the environment*, paper presented at the WIDER Conference on the Environment, Helsinki, September 1990.
Ahmed, I.: 'Advanced agricultural biotechnologies: Some empirical findings on their social impact', in *International Labour Review* (Geneva, ILO), vol. 128, no. 5 (1985).
Ahmed, M.: 'The use and abuse of pesticides and the protection of environment', in Bangladesh Ministry of Education: *Protection of the environment from degradation*, op. cit. (1986).
Ahmed, M.F.: 'Modern agriculture and its impact on environmental degradation', in Bangladesh Ministry of Education: *Protection of the environment from degradation*, op. cit. (1986).
Akita, S.; Moss, D.N.: 'Photosynthetic responses to CO_2 and light by maize and wheat leaves adjusted for constant stomatal apertures', in *Crop Science* (Madison, Wisconsin: Crop Science Society of America), vol. 13, no. 2 (1973).
Alamgir, M.: *Famine in south Asia* (Cambridge, Massachusetts: Oelgeschlager, Gunn and Hain, 1980).
Alauddin, M.; Tisdell, C.A.: 'Poverty, resource distribution and security: The impact of the new agricultural technology in rural Bangladesh', in *Journal of Development Studies* (London, Frank Cass), vol. 25, no. 4 (1989).
Alauddin, M.; Tisdell, C.A.: 'The "Green Revolution" and labour absorption in Bangladesh agriculture: The relevance of the east Asian experience', in *Pakistan Development Review* (Islamabad, Pakistan Institute of Development Economics), vol. 30, no. 2 (1991a).
Alauddin, M.; Tisdell, C.A.: *The Green Revolution and economic development: The process and its impact in Bangladesh* (London: Macmillan, 1991b).
Allan, W.: *The African husbandman* (Edinburgh: Oliver and Boyd, 1965).
Almeida, C.: *Development and transfer of environmentally sound technologies in manufacturing: A survey*, United Nations Conference on Trade and Development Discussion Paper No. 58 (Geneva: UNCTAD, 1993).
Anderson, A.B.: 'Smokestacks in the rainforest: Industrial development and deforestation in the Amazon Basin', in *World Development* (Oxford, Pergamon Press), vol. 18, no. 9 (1990).
Anderson, J.R.; Thampapillai, D.J.: *Soil conservation in developing countries: Project and policy intervention*, World Bank Policy Research Series No. 8 (Washington, DC, 1990).

Arnold, J.E.M.: 'Community forestry and meeting fuelwood needs', in *Commonwealth Forestry Review* (Oxford, Commonwealth Forestry Association), vol. 62, no. 1, pp. 183–189 (1983).
Arntzen, J.W.; Veenendaal, E.M.: *A profile of environment and development in Botswana* (Gaborone, Botswana: University of Botswana, National Institute of Development, Research and Documentation, 1986).
Arrow, K.J.; Fisher, A.C.: 'Environmental preservation, uncertainty, and irreversibility', in *Quarterly Journal of Economics* (Cambridge, Massachusetts, Harvard University, Department of Economics), vol. 88, no. 2 (May 1974).
Ault, D.E.; Rutman, G.L.: 'The development of individual rights to property in tribal Africa', in *Journal of Law and Economics* (Chicago, University of Chicago Press), vol. 22, no. 1 (1979).
Balisacan, A.M.: 'Goals and consequences of fertiliser policies in the Philippines, 1960–87', in *UPLB Agricultural Policy Research Program* (Laguna, Philippines: University of the Philippines at Los Banos, 1989a).
Balisacan, A.M.: 'Survey of Philippine research on the economics of agriculture', in *Philippine Review of Economics and Business* (Quezon City, Philippines, University of the Philippines), vol. 26, no. 1 (1989b).
Balisacan, A.M.: *Fertilizers and fertilizer policies in Philippine agricultural development*, UPLB Agricultural Policy Research Programme Monograph 90-02 (Laguna, Philippines: University of the Philippines at Los Banos, 1990).
Balisacan, A.M.: *The war against rural poverty: The Philippines* (Quezon City, Philippines: School of Economics, University of the Philippines, 1991).
Bangladesh Bureau of Statistics: *The Bangladesh Census of Agriculture and Livestock, 1983–84*, Volume I: *Structure of agricultural holdings and livestock population* (Dhaka, 1986).
Bangladesh Bureau of Statistics: *Monthly Statistical Bulletin of Bangladesh* (Dhaka), December 1990.
Bangladesh Ministry of Education: *Protection of the environment from degradation*, Proceedings of a South Asian Association for Regional Co-operation (SAARC) Seminar, 1985 (Dhaka: Bangladesh Ministry of Education, Science and Technology Division, 1986).
Barbier, E.B.: 'Cash crops, food crops and sustainability: The case of Indonesia', in *World Development* (Oxford, Pergamon Press), vol. 17, no. 6 (1989).
Barbier, E.B.; Burgess, J.C.: *Agricultural pricing and environmental degradation*, Policy Research Working Paper WPS 960 (Washington, DC: World Bank, 1992).
Barker, R.; Cordover, V.G.: 'Labour utilisation in rice production', in International Rice Research Institute: *Economic consequences of new rice technology* (Los Banos, Philippines, 1978).
Barlow, C.; Jayasuriya, S.; Price, E.C.: *Evaluating technology for new farming systems in case studies from Philippines farms* (Los Banos, Philippines: International Rice Research Institute, 1983).
Beghin, J.: *A game theoretic model of agriculture and food price policies in Senegal*, PhD dissertation (Berkeley, California: University of California, 1988).
Behnke, R.H.: *Open-range management and property rights in pastoral Africa: A case of spontaneous range enclosure in South Darfur, Sudan* (London: Overseas Development Institute, 1990).

Bernus, E.: 'Desertification in the Eghazer and Azawak Region, Niger', in J.A. Mabbutt and C. Floret (eds.): *Case studies in desertification*, Natural Resources Research, vol. XVIII (Paris: UNESCO, 1981).

Berry, A.; Cline, W.: *Agrarian structure and productivity in developing countries* (Baltimore, Maryland: Johns Hopkins University Press, 1979).

Berry, L.: *Assessment of desertification in the Sudano-Sahelian region 1978–1984*, UNEP Governing Council, 12th Session (Nairobi: United Nations Environmental Programme (UNEP), 1984).

Beynon, J.G.: 'Pricism v. structuralism in sub-Saharan African agriculture', in *Journal of Agricultural Economics* (Ashford (Kent), United Kingdom, Agricultural Economics Society), vol. 40, no. 3 (1989).

Bhalla, A.S. (ed.): *Environment, employment and development* (Geneva: ILO, 1992).

Bhalla, A.S.: *Clean, green and all that: Some aspects of cleaner technologies and production*, paper prepared for the Conference on Economic Growth with Clean Production jointly organized by CSIRO Australia and the United Nations Industrial Development Organization (Melbourne, 7–10 February 1994).

Bhalla, A.S.; James, J.: 'New technology revolution: Myths or reality for developing countries?' in P. Hall (ed.): *Technology, innovation and economic policy* (Oxford: Philip Allan, 1986).

Bhattacharya, A.N.: *Southern Aravalli: The habitat of adivasies*, paper presented at the Conference on Aravalli 2001 AD, Udaipur, 31 December 1986 – 2 January 1987 (Ubheshwar Vikas Mandal, Udaipur, 1987).

Bifani, P.: 'Environment degradation in rural areas', in A.S. Bhalla (ed.): *Environment, employment and development* (Geneva: ILO, 1992).

Biggs, S.D.; Clay, E.J.: 'Sources of innovation in agricultural technology', in *World Development* (Oxford, Pergamon Press), vol. 9, no. 4 (1981).

Binswanger, H.: 'Brazilian policies that encourage deforestation in the Amazon', in *World Development* (Oxford, Pergamon Press), vol. 19, no. 7 (July 1991).

Blackmer, A.M.: 'Losses and transport of nitrogen form soils', in F.M. D'Itri and L.G. Wolfson (eds.): *Rural groundwater contamination* (Chelsea, Michigan: Lewis Publishers, 1988).

Bond, M.E.: *Agricultural responses to prices in sub-Saharan African countries*, International Monetary Fund Staff Papers No. 30 (Washington, DC: International Monetary Fund, 1983).

Boserup, E.: *The conditions of agricultural growth* (London: George Allen & Unwin; New York: Aldine Publishing, 1965).

Boserup, E.: *Population and technological change: A study of long-term trends* (Oxford: Basil Blackwell, 1981).

Bowes, M.; Krutilla, J.: 'Multiple use management of public forests', in A.V. Kneese and J.L. Sweeney (eds.): *Handbook of natural resource and energy economics* (Amsterdam, Elsevier Science Publishers, 1985), vol. II.

Bruce, J.: 'A perspective on indigenous land tenure systems and land concentration', in R.E. Downs and S. Reyna (eds.): *Land and society in contemporary Africa* (Hanover, New Hampshire: University Press of New England, 1988).

Bruce, J.; Fortmann, L.: 'Trees and tenure', in L. Fortmann and J. Riddell (eds.): *Trees and tenure: An annotated bibliography for agroforesters and others* (Madison, Wisconsin: University of Wisconsin, Land Tenure Center; Nairobi: International Council for Research in Agroforestry, 1985), pp. vii–xvii.

Cain, M.: 'Landlessness in India and Bangladesh: A critical review of national data sources', in *Economic Development and Cultural Change* (Chicago, University of Chicago Press), vol. 32, no. 1 (1983).
Caldwell, J.C.: 'Towards a restatement of demographic transition theory', in *Population and Development Review* (New York, Population Council), vol. 2, nos. 3 and 4 (1976).
Charney, J.: 'Dynamics of deserts and drought in the Sahel', in *Quarterly Journal of the Royal Meteorological Society* (Bracknell, United Kingdom, Royal Meteorological Society), vol. 101, no. 428 (1975).
Chaudhury, R.H.: 'The seasonality of prices and wages in Bangladesh', in R. Chambers, R. Longhurst and A. Pacey (eds.): *Seasonal dimensions of rural poverty* (London: Frances Pinter, 1981).
Ciriacy-Wantrup, S.V.: *Resources conservation* (Berkeley, California: University of California Press, 1952).
Clark, C.: *Mathematical bioeconomics: The optimal management of renewable resources* (New York: John Wiley, 1976).
Clawson, M. (ed.): *Research in forest economics and forest policy* (Baltimore, Maryland: Johns Hopkins University Press for Resources for the Future, 1978).
Clayton, E.: *Agriculture, poverty and freedom in developing countries* (London: Macmillan, 1983).
Clayton, R.: 'Demand and supply of hydro-electric power from the Kariba Dam', in W.L. Handlos and G.W. Howard (eds.): *Development prospects for the Zambezi Valley in Zambia* (Lusaka: University of Zambia, Kafue Basin Research Committee, 1985).
Cleaver, K.: *The impact of price and exchange rate policies on agriculture in sub-Saharan Africa*, World Bank Staff Working Paper No. 728 (Washington, DC: World Bank, 1985).
Colchester, M.: *Sustaining the forests: The community-based approach in south and south-east Asia*, United Nations Research Institute for Social Development Discussion Paper DP 35 (Geneva: UNRISD, 1992).
Colclough, C.; Fallon, P.: 'Rural poverty in Botswana: Dimensions, causes, constraints', in D. Ghai and S. Radwan (eds.): *Agrarian policies and rural poverty in Africa* (Geneva: ILO, 1983).
Conway, G.R.: 'Agricultural ecology and farming systems research', in J.V. Remenyi (ed.): *Agricultural systems research for developing countries*, ACIAR Proceedings No. 11 (Canberra: Australian Centre for International Agricultural Research, 1985).
Conway, G.R.: 'The properties of agroecosystems', in *Agricultural Systems* (Barking, United Kingdom, Elsevier Applied Science Publishers), vol. 24, no. 2 (1987).
Conway, G.R.; Barbier, E.B.: *After the Green Revolution: Sustainable agriculture for development* (London: Earthscan, 1990).
Cousins, B.: *A survey of current grazing schemes in the communal lands of Zimbabwe* (Harare: University of Zimbabwe, Centre for Applied Social Sciences, 1987).
Cruz, W.; Francisco, H.A.; Tapawan-Conway, Z.: *The on-site and downstream costs of soil erosion*, Working Paper Series No. 88-11 (Los Banos, Philippines: Philippine Institute for Development Studies, 1988).
Cruz, W.D.; Cruz, M.C.J.: 'Population pressure and deforestation in the Philippines', in *Asean Economic Bulletin* (Singapore, Institute of Southeast Asian Studies), vol. 7, no. 2 (1990).

Cruz, W.; Repetto, R.: *The environmental effects of stabilization and structural adjustment programs: The Philippines case* (Washington, DC: World Resources Institute, 1992).

Daly, H.: *Ecological economics and sustainable development: From concept to policy*, World Bank Environment Department, Policy and Research Division, Divisional Working Paper 1991-24 (Washington, DC: World Bank, 1991).

Daly, H.; Cobb, J.B.: *For the common good* (Boston, Massachusetts: Beacon Press, 1989).

Darity, W.A.: 'The Boserup theory of agricultural growth: A model for anthropological economics', in *Journal of Development Economics* (Amsterdam, Elsevier Science Publishers), vol. 7, no. 2 (1980).

Dasgupta, P.S.: 'Population, resources and poverty', in *Ambio* (Stockholm, Royal Swedish Academy of Science), vol. 21, no. 1 (1992).

Dasgupta, P.S.; Heal, G.M.: *Economic theory and exhaustible resources* (Cambridge: Cambridge University Press, 1979).

Davies, R.; Rattsø, J.: *Land reform as a response to environmental degradation in Zimbabwe: Some economic issues*, paper presented at the WIDER Conference on Medium-Term Development Strategy, Helsinki, 15–17 April 1994.

de Janvry, A.; Sadoulet, E.; Wilcox, L.: 'Rural Labour in Latin America', *International Labour Review*, 128, no. 6 (1989).

Delgado, G.L.; Mellor, J.W.: 'A structural view of policy issues in African agricultural development', in *American Journal of Agricultural Economics* (Ames, Iowa, American Agricultural Economics Association), vol. 66, no. 5 (1984).

Dewees, P.A.: *Trees, land and labor*, World Bank Environment Department Working Paper No. 4 (Washington, DC: World Bank, 1993).

Dhabariya, S.S.: *Aravalli mountain region: Issues and perspectives*, paper presented at the Conference on Aravalli 2001 AD, Udaipur, 31 December 1986 – 2 January 1987 (Ubheshwar Vikas Mandal, Udaipur, 1987).

Dhabariya, S.S.: *Eco-crisis in Aravalli hill region* (Jaipur, 1988).

Dhir, R.P.: *Eco-degradation in the Aravallis and the need for control*, paper presented at the Conference on Aravalli 2001 AD, Udaipur, 31 December 1986 – 2 January 1987 (Ubheshwar Vikas Mandal, Udaipur, 1987).

Dixon, J.: *The urban environmental challenge in Latin America*, LATEN Dissemination Note No. 4 (Washington, DC: World Bank, 1993).

Dixon, J.A.; James, D.E.; Sherman, P.B.: *The economics of dryland management* (London: Earthscan, 1989).

Dixon, R.K.: 'Forest biotechnology opportunities in developing countries', in *Journal of Developing Areas* (Macomb, Illinois, Western Illinois University), vol. 22, no. 2 (1988).

Doeleman, J.A.: 'Employment concerns and environmental policy', in A.S. Bhalla (ed.): *Environment, employment and development* (Geneva: ILO, 1992).

Dorfman, R.: 'An economist's view of natural resource and environmental problems', in R. Repetto (ed.): *The global possible: Resources, development and the new century* (New Haven, Connecticut: Yale University Press, 1985).

Dorner, P.: *Land reform and economic development* (Harmondsworth, United Kingdom: Penguin, 1972).

Dorner, P.; Thiesenhusen, W.C.: *Land tenure and deforestation: Interactions and environmental implications*, United Nations Research Institute for Social Development Discussion Paper DP 34 (Geneva: UNRISD, 1992).

Douglass, G.K.: 'The meanings of agricultural sustainability', in G.K. Douglass (ed.): *Agricultural sustainability in a changing world order* (Boulder, Colorado: Westview Press, 1984), pp. 3–29.

Dregne, H.E.: *Desertification of arid lands* (New York: Harwood, 1983).

Drewnowski, J.: 'Poverty: Its meaning and measurement', in *Development and Change* (London, Sage Publications), vol. 8, no. 2 (1977).

Dyson-Hudson, N.: 'Adaptive resource use by African pastoralists', in F. Di Castri, F.W.G. Baker and M. Hadley (eds.): *Ecology in practice* (Dublin: Tycooly International, 1984).

Eckholm, E.P.: *Losing ground: Environmental stress and world food prospects* (New York: Norton, 1976).

ECOTEC: *Sustainability, employment and growth: The employment impact of environmental policies*, Discussion Paper Two, May 1993 (Birmingham/Brussels, ECOTEC Research and Consulting Ltd., 1993).

ECOTEC: *The potential for new employment opportunities from pursuing sustainable development: A brief note on employment opportunities in transport*, paper presented at the Workshop on the Employment Potential of Sustainable Development Policies organized by the European Foundation for the Improvement of Living and Working Conditions (Dublin, 20–21 April 1994a).

ECOTEC: *The potential for new employment opportunities from pursuing sustainable development: A brief note on energy*, paper presented at the Workshop on the Employment Potential of Sustainable Development Policies organized by the European Foundation for the Improvement of Living and Working Conditions (Dublin, 20–21 April 1994b).

Ehrlich, P.R.: 'The limits to substitution: Meta-resource depletion and a new economic-ecological paradigm', in *Ecological Economics* (Amsterdam, Elsevier Science Publishers), vol. 1, no. 1 (1989).

El Serafy, S.: 'The proper calculation of income from depletable natural resources', in Y.J. Ahmad, S. El Serafy and E. Lutz (eds.): *Environmental accounting for sustainable development* (Washington, DC: World Bank, 1989).

Ellis, J.; Swift, D.: 'Stability of African pastoral ecosystems: Alternate paradigms and implications for development', in *Journal of Range Management* (Denver, Colorado, Society for Range Management), vol. 41 (1988).

Eskeland, G.S.; Jiminez, E.: 'Policy instruments for pollution control in developing countries', in *Research Observer* (Washington, DC, World Bank), vol. 7, no. 2 (1992).

Feder, G.; Noronha, R.: 'Land rights systems and agricultural development in sub-Saharan Africa', in *World Bank Research Observer* (Washington, DC, World Bank), vol. 2, no. 2 (1987).

Fei, J.C.H.; Ranis, G.: 'Agrarianism, dualism, and economic development', in S.P. Singh (ed.): *Underdevelopment to developing economies* (Oxford: Oxford University Press, 1978).

Food and Agriculture Organization (FAO): *Map of the fuelwood situation in developing countries* (Rome, 1981a).

Food and Agriculture Organization: *Review of forest management systems of tropical Asia* (Rome, 1981b).

Food and Agriculture Organization: *The State of Food and Agriculture 1984* (Rome, 1985).

Food and Agriculture Organization: *Yearbook of Forest Products 1988*, vol. 23 (Rome, 1990).
Franke, R.; Chasin, B.: 'Peasants, peanuts, profits and pastoralists', in *The Ecologist* (Camelford (Cornwall), United Kingdom, Ecosystems Ltd), vol. 11, July–August 1981.
Friedman, J.; Rangan, H. (eds.): *In defense of livelihoods: Comparative studies on environmental action* (West Harford, Connecticut: Kumarian Press, 1993).
Gammage, S.: *Report on environmental economics in the developing world* (Washington, DC: USAID, 1990).
Garcia-Barrios, R.; Garcia-Barrios, L.: 'Environmental and technological degradation in peasant agriculture: A consequence of development in Mexico', in *World Development* (Oxford, Pergamon Press), vol. 18, no. 11 (1990).
Garrity, D.P.: 'Agronomic research on biofertilisers at IRRI: Overcoming system-level constraints', in *Proceedings of the National Symposium on Bio- and Organic Fertilisers, October 9–12, 1990, University of the Philippines at Los Banos* (Los Banos, Philippines: 1990).
Garrity, D.P.; Flinn, J.C.: 'Farm-level management systems for green manure crops in Asian rice environments', in International Rice Research Institute: *Proceedings of a symposium on sustainable agriculture – The role of green manure crops in rice farming systems* (Los Banos, Philippines, 1987).
Garrity, D.P.; Bantilan, R.T.; Bantilan, C.C.; Tin, P.; Mann, R.: Indigoflora tinctoria: *Farmer-proven green manure for rainfed rice lands*, paper presented at the National Symposium on Bio- and Organic Fertilisers, 9–12 October 1989, University of the Philippines at Los Banos (Laguna, Philippines, 1990).
Gear, A. (ed.): *The organic food guide* (Barking, United Kingdom: Henry Doubleday Research Association, 1983).
Ghai, D.; Radwan, S. (eds.): *Agrarian policies and rural poverty in Africa* (Geneva: ILO, 1983).
Ghai, D.; Smith, L.D.: *Agricultural prices, policy and equity in sub-Saharan Africa* (Boulder, Colorado: Lynne Rienner, 1987).
Ghai, D.; Vivian, J.M. (eds.): *Grassroots environmental action: People's participation in sustainable development* (London: Routledge, 1992).
Ghatak, S.; Ingersent, K.: *Agriculture and economic development* (Baltimore, Maryland: Johns Hopkins University Press, 1984).
Golan, E.H.: *Land tenure reform in Senegal: An economic study from the peanut basin*, Land Tenure Center Research Paper No. 101 (Madison, Wisconsin: University of Wisconsin, 1990).
Gorse, J.-E.: *Desertification in the Sahelian and Sudanian zones of West Africa*, World Bank Technical Paper No. 61 (Washington, DC, 1987).
Grainger, A.: *The threatening desert: Controlling desertification* (London: Earthscan, 1990).
Gregerson, H.M.: *Village forestry in the Republic of Korea: A case study*, Forestry for Local Community Development Programme Series (Rome: FAO, 1982).
Griffin, K.: 'Comments on labor utilisation in rice production', in International Rice Research Institute: *Economic consequences of the new rice technology* (Los Banos, Philippines, 1978).
Griffin, K.: *Alternative strategies for economic development* (Basingstoke, United Kingdom: Macmillan in association with OECD, 1989).

Gritzner, Jeffery: *The West African Sahel: Human agency and environmental change* (Chicago: University of Chicago Press, 1988).

Grossman, G.M.; Krueger, A.B.: *Environmental impacts of a North American Free Trade Agreement*, Working Paper No. 3914 (Cambridge, Massachusetts: National Bureau of Economic Research, 1991).

Grouzis, M.: *Dynamics of Sahelian ecological systems: The case of Oursi Pond, Burkino Faso*, paper presented at the Technical Meeting on Savanna Development and Pasture Production, Woburn, November 1990 (London: Commonwealth Secretariat, 1990).

Gupta, T: 'The economics of tree crops on marginal agricultural lands with special reference to the hot arid region in Rajasthan, India', in *International Tree Crops Journal* (Bicester, United Kingdom, AB Academic Publishers), vol. 2, no. 2, (1983), pp.155–194.

Hamid, M.A.; Saha, S.K.; Rahman, M.A.; Khan, A.J.: *Irrigation technologies in Bangladesh: A study in some selected areas* (Rajshahi, Bangladesh: University of Rajshahi, Department of Economics, 1978).

Haque, B.A.; Hoque, M.M.: 'Faecal pollution of surface water and diseases in Bangladesh', in A.A. Rahman, S. Huq and G.R. Conway (eds.): *Environmental aspects of surface water systems of Bangladesh* (Dhaka: University Press, 1990), pp. 180–200.

Hardaker, J.B.; Troncoso, J.L.: 'The formulation of MOTAD programming models for farm planning using subjectively elicited activity net revenues', in *European Review of Agricultural Economics* (Amsterdam, Mouton de Gruyter), vol. 6, no. 1 (1979).

Hardin, G.J.: 'The tragedy of the commons' in *Science* (Washington, DC, American Association for the Advancement of Science), vol. 162, pp. 1243–1248.

Hardy, E. (a.k.a. E.H. Golan): *An economic analysis of tenure security in West Africa: The case of the Senegalese peanut basin*, PhD dissertation (University of California at Berkeley, 1989).

Hare, F.K.: 'Climate and desertification', in United Nations Conference on Desertification: *Desertification: Its causes and consequences* (Oxford: Pergamon Press, 1977).

Harrison, P.: *The greening of Africa: Breaking through in the battle for land and food* (London: Paladin, 1987).

Harriss, B.: 'Paddy milling: Problems in policy and choice of technology', in B.H. Farmer (ed.): *Green Revolution? Technology and change in rice-growing areas of Tamil Nadu and Sri Lanka* (London: Macmillan, 1977).

Hart, G.: *The mechanisation of Malaysian rice production: Will the petty producers survive?* (Geneva: ILO, 1987).

Hartwick, J.M.: 'Intergenerational equity and the investing of rents from exhaustible resources', in *American Economic Review* (Nashville, Tennessee, American Economic Association), vol. LXVII, no. 5 (1977).

Hartwick, J.M.: *Economic depreciation of mineral stocks and the contribution of El Serafy*, World Bank Environment Department Working Paper No. 4 (Washington, DC, 1991).

Hayami, Y.: 'Assessment of the Green Revolution', in C.K. Eicher and J.M. Staatz (eds.): *Agricultural development in the Third World* (Baltimore, Maryland: Johns Hopkins University Press, 1984).

Hayami, Y.; Ruttan, V.W.: *Agricultural development: An international perspective* (Baltimore, Maryland: Johns Hopkins University Press, 1985).

Hazell, P.B.R.: 'A linear alternative to quadratic and semi-variance programming for farm planning under uncertainty', in *American Journal of Agricultural Economics* (Ames, Iowa, American Agricultural Economics Association), vol. 53, no. 1 (1971).

Hazell, P.B.R.; Norton, R.D.: *Mathematical programming for economic analysis in agriculture* (New York: Macmillan, 1986).

Herath, H.M.G.: 'Economics of salinity control in Sri Lanka: Some exploratory results', in *Agricultural Administration* (Barking, United Kingdom, Elsevier Applied Science Publishers), no. 18 (1985).

Herskovits, M.J.: *The economic life of primitive peoples* (New York: Knopf, 1940).

Hoben, A.: *Land tenure among the Amhara of Ethiopia* (Chicago: University of Chicago Press, 1973).

Holburt, M.: 'International problems on the Colorado River', in *Water Supply and Management*, vol. 6, no. 1 (1982).

Holling, C.S.: 'Resilience and stability of ecological systems', in *Annual Review of Ecology and Systematics* (Palo Alto, California, Annual Reviews Inc.), vol. 4, no. 1 (1973).

Holling, C.S.: 'The resilience of terrestrial ecosystems: Local surprise and global change', in W.C. Clark and R.E. Munn (eds.): *Sustainable development of the biosphere* (Cambridge: Cambridge University Press, 1986).

Holling, C.S.: 'Simplifying the complex: The paradigms of ecological function and structure', in *European Journal of Operational Research* (Amsterdam, Elsevier Science Publishers), vol. 30 (1987).

Holling, C.S.; Bocking, S.: 'Surprise and opportunity: In evolution, in ecosystems, in society', in C. Mungall and D.J. McLaren, D.J. (eds.): *Planet under stress: The challenge of global change* (Oxford: Oxford University Press, 1990).

Hossain, M.: *Agriculture in Bangladesh: Performance, problems and prospects* (Dhaka: University Press, 1991).

Hotelling, H.: Letter quoted in United States Department of the Interior, National Parks Service: *The economics of public recreation: An economic study of the monetary evaluation of recreation in the national parks* (Washington, DC, 1949).

Howard, R.: 'Formation and stratification of the peasantry in colonial Ghana', in *Journal of Peasant Studies* (London, Frank Cass), vol. 8, no. 1 (1980).

Hrabovszky, J.P.: 'Agriculture: The land base', in R. Repetto (ed.): *The global possible* (New Haven, Connecticut: Yale University Press, 1985)

Hudson, N.: *Soil and water conservation in semi-arid areas* (Rome: FAO, 1987).

Huising, D.: 'Cleaner technologies through process modifications, material substitutions and ecologically based ethical values', in *Industry and Environment* (Paris, United Nations Environment Programme, Industry and Environment Office), vol. 12, no. 1 (1989).

India, Government of Rajasthan, Directorate of Agriculture: *Trends in land use statistics* (Jaipur, 1992).

India, Government of Rajasthan, Directorate of Census Operations: *Rajasthan Census of India 1991*, Series – 21 (Jaipur, 1991).

Intergovernmental Panel on Climate Change (IPCC): *Scientific assessment of climate change* (Geneva: World Meteorological Organization; Nairobi: United Nations Environment Programme (UNEP), 1990).

International Labour Office (ILO): *The socio-economic impact of technical cooperation projects concerning rural development*, Advisory Committee on Rural Development, Eleventh Session, Geneva, 1990, doc. ACRD/XI/1990/III (Geneva, 1990).

International Rice Research Institute (IRRI): *Green manure in rice farming*, Proceedings of a symposium on sustainable agriculture, The Role of Green Manure Crops in Rice Farming Systems, 25–29 May 1987 (Los Banos, Philippines, 1988).

International Tropical Timber Organization (ITTO): *The promotion of sustainable forest management: A case study in Sarawak, Malaysia*, mission report to the Eighth Session of the International Tropical Timber Council, Denpasar, Bali, Indonesia, 16–23 May (1990).

Ishikawa, S.: *Labour absorption in Asian agriculture: An issue paper* (Bangkok: ILO/ARTEP, 1978).

Islam, Md A.: 'Consequences of increased pesticide use', in A.A. Rahman, S. Huq and G.R. Conway (eds.): *Environmental aspects of agricultural developments in Bangladesh* (Dhaka: University Press, 1990).

Islam, S.: 'The decline of soil quality', in A.A. Rahman, S. Huq and G.R. Conway (eds.): *Environmental aspects of agricultural developments in Bangladesh* (Dhaka: University Press, 1990).

D'Itri, F.M.; Wolfson, L.G. (eds.): *Rural groundwater contamination* (Chelsea, Michigan: Lewis Publishers, 1988).

Jagannathan, N.V.: *Poverty, public policies and the environment*, World Bank Environment Department Working Paper No. 24 (Washington, DC, 1989).

Jamal, V.: 'Nomads and farmers: Incomes and poverty in rural Somalia', in D. Ghai and S. Radwan (eds.): *Agrarian policies and rural poverty in Africa* (Geneva: ILO, 1983).

James, R.W.: *Land tenure and policy in Tanzania* (Toronto: University of Toronto Press, 1971).

Jodha, N.S.: 'Population growth and the decline of common property resources in Rajasthan, India', in *Population and Development Review* (New York, Population Council), vol. 11, no. 2 (1985).

Jodha, N.S.: 'Common property resources and rural poor in dry regions of India', in *Economic and Political Weekly* (Bombay, Sameeksha Trust), vol. 21, no. 27 (1986).

Jodha, N.S.: *Rural common property resources: Contributions and crisis*, Foundation Day lecture, 16 May 1990 (New Delhi: Society for the Promotion of Wastelands Development, 1990).

Johnson, H.G.: 'Factor market distortions and the shape of the transformation curve', in *Econometrica* (Oxford, Basil Blackwell), vol. 34, no. 3 (1966).

Johnson, O.E.G.: 'The agricultural sector in IMF stand-by arrangements', in S. Commander (ed.): *Structural adjustment and agriculture: Theory and practice in Africa and Latin America* (London: Overseas Development Institute, 1989).

Johnston, B.F.; Cownie, J.: 'The seed-fertiliser revolution and labour force absorption', in *American Economic Review* (Nashville, Tennessee, American Economic Association), vol. LIX, no. 4 (1969).

Jones, S.: 'Agrarian structure and agricultural innovations in Bangladesh: Panimara Village, Dhaka District', in T.P. Bayliss-Smith and S. Wanmali (eds.):

Understanding green revolutions: Agrarian change and development planning in South Asia (Cambridge: Cambridge University Press, 1984).
Jonish, J.: *Sustainable development and employment: Forestry in Malaysia*, mimeographed WEP working paper WEP 2-22/WP. 234 (Geneva: ILO, 1992).
Joshi, P.K.: 'Benefit-cost analysis of alkali land reclamation technology: An expost evaluation', in *Agricultural Situation in India* (New Delhi, Ministry of Agriculture), vol.38, no. 2, pp. 467–470 (1983).
Junankar, P.N.: 'The response of peasant farmers to price incentives: The use and misuse of profit functions', in *Journal of Development Studies* (London, Frank Cass), vol. 25, no. 2 (1989).
Karamchandani, K.P.: *Extension components of social forestry in Gujarat* (Vadodara, India: Gujarat State Forest Department, 1982).
Karshenas, M.: *Environment, employment and sustainable development*, mimeographed WEP working paper WEP 2-22/WP. 237 (Geneva: ILO, 1992).
Kate, A.T.; Joost, D.: *Environment and economic growth in Mexico*, paper presented at the WIDER Conference on Medium-Term Development Strategy, Helsinki, 15–17 April 1994.
Kates, R.W.; Johnson, D.L.; Haring, K.J.: 'Population, society and desertification', in United Nations Conference on Desertification: *Desertification: Its causes and consequences* (Oxford: Pergamon Press, 1977).
Khan, M.M.: 'Labour absorption and unemployment in rural Bangladesh', in *Bangladesh Development Studies* (Dhaka, Bangladesh Institute of Development Studies), vol. 13, no. 3–4 (1985).
Kolstad, C.D.; Braden, J.B.: 'Environmental demand theory', in J.B. Braden and C.D. Kolstad (eds.): *Measuring the demand for environmental quality* (Amsterdam: North Holland, 1991).
Konczacki, Z.A.: *The economics of pastoralism: A case study of sub-Saharan Africa* (London: Frank Cass, 1978).
Krutilla, J.V.: 'Conservation reconsidered', in *American Economic Review* (Nashville, Tennessee, American Economic Association), vol. LVII, no. 4 (1967).
Krutilla, J.V.; Fisher, A.C.: *The economics of natural environments*, 2nd edn. (Washington, DC: Resources for the Future, 1985).
Kulkarni, K.R.; Pandey, R.K.: 'Annual legumes for food and as green manure in a rice-based cropping system', in International Rice Research Institute: *Green manure in rice farming* (Los Banos, Philippines, 1988).
Kuznets, S.: *Economic growth and structural change* (New York: Norton, 1965).
Kuznets, S.: *Modern economic growth* (New Haven, Connecticut: Yale University Press, 1966).
Lau, Buong Tieng: 'The effects of shifting cultivation on sustainable yield management for Sarawak's national forests', in *Malaysian Forester* (Serdang, Malaysia, Universiti Pertanian Malaysia), vol. 42, no. 4 (1979).
Leach, M.; Mearns, R.: *Poverty and environment in developing countries: An overview study* (Brighton, United Kingdom: University of Sussex, Institute of Development Studies, 1992).
Leibenstein, H.: *Economic backwardness and economic growth* (New York: John Wiley, 1957).
Lipton, M.: 'The theory of the optimizing peasant', in *Journal of Development Studies* (London, Frank Cass), vol. 4, no. 3 (1968).

Lipton, M.: *Land assets and rural poverty*, World Bank Staff Working Paper No. 744 (Washington, DC: World Bank, 1985).

Lipton, M.: 'Limits of price policy for agriculture: Which way for the World Bank?', in *Development Policy Review* (London, Sage Publications), vol. 5, no. 2 (1987).

Lipton, M.; Longhurst, R.: *New seeds and poor people* (London: Unwin Hyman, 1989).

Livingstone, I.: 'Supply responses of peasant producers: The effect of own-account consumption on the supply of marketed output', in I. Livingstone (ed.): *Development economics and policy: Readings* (London: George Allen & Unwin, 1981).

López, R.: 'Environmental degradation and economic openness in LDCs: The poverty linkage', in *American Journal of Agricultural Economics* (Ames, Iowa, American Agricultural Economics Association), vol. 74, no. 5 (1992).

Lutz, E.; Daly, H.: *Incentives, regulations and sustainable land use in Costa Rica*, World Bank Environment Department Working Paper No. 34 (Washington, DC: World Bank, 1990).

Lutz, E.; Munasinghe, M.: *Environmental-economic analysis of projects and policies for sustainable development*, World Bank Environment Department Working Paper No. 42 (Washington, DC: World Bank, 1993).

Lutz, E.; Young, M.: *Agricultural policies in industrial countries and their environmental impacts: Applicability to and comparison with developing nations*, World Bank Environment Department Working Paper No. 25 (Washington, DC: World Bank, 1990).

Mabbutt, J.A.: 'A new global assessment of the status and trends of desertification', in *Environmental Conservation* (Lausanne, Switzerland, Elsevier Sequoia), vol. 11, no. 2 (1984), pp. 100–113

McCamley, F.; Kliebenstein, J.B.: 'Describing and identifying the complete set of target MOTAD solutions', in *American Journal of Agricultural Economics* (Ames, Iowa, American Agricultural Economics Society), vol. 69, no. 3 (1987).

Maclean, A.H.: *Moisture characteristics of some Zambian soils*, Soil Science Report No. 5 (Chilanga, Zambia: Mt Makulu Research Station, 1969).

McMillan, C.: *Mathematical programming* (New York: John Wiley, 1975).

Magrath, W.B.; Doolette, J.B.: 'Strategic issues in watershed development', in *Asean Economic Bulletin* (Singapore, Institute of Southeast Asian Studies), vol. 7, no. 2 (1990).

Mahar, D.J.: *Government policies and deforestation in Brazil's Amazonia Region* (Washington, DC: World Bank, 1989).

Majocchi, A.: *Energy/CO_2 tax and employment: Summary and conclusions*, paper presented at the Workshop on the Employment Potential of Sustainable Development Policies organized by the European Foundation for the Improvement of Living and Working Conditions (Dublin, 20–21 April 1994).

Malaysia, Department of Statistics: *Yearbook of Statistics 1989* (Kuala Lumpur, 1990).

Malaysia, Ministry of Finance: *Economic Report 1990–1991* (Kuala Lumpur, 1990).

Malaysia, Ministry of Primary Industries: *Profile of Primary Commodities 1989* (Kuala Lumpur, 1990).

Malaysia, Ministry of Primary Industries: *Forestry in Malaysia* (Kuala Lumpur, n.d.).
Malaysia, Prime Minister: *Fifth Five-Year Plan (1986-1990)* (Kuala Lumpur, 1986).
Maleka, P.T.: *Systems modelling approach to drought management strategies in the Lake Kariba District of Zambia*, unpublished PhD thesis (Wollongong, New South Wales, Australia: University of Wollongong, 1990).
Mäler, K.G.: *National accounts and environmental resources*, Beijer Reprint Series No. 4 (Stockholm: Beijer International Institute of Ecological Economics, 1990).
Malman, S.: *Socioeconomic foundations for and implications of national environmental policy in Chile*, mimeographed WEP working paper WEP 2-22/WP. 244 (Geneva: ILO, 1994).
Markandya, A.: *Criteria, instruments and tools for sustainable agricultural development*, paper prepared for the FAO/Netherlands Conference on Agriculture and the Environment, s'Hertogenbosch (Netherlands), 15-19 April 1991 (Rome: FAO, 1990).
Markandya, A.: *Technology, environment and employment: A survey*, mimeographed WEP workingpaper WEP 2-22/WP. 216 (Geneva: ILO, 1991).
Markandya, A.; Pearce, D.W.: 'Natural environments and the social rate of discount', in *Project Appraisal* (Guildford, United Kingdom, Beech Tree Publishing), vol. 3, no. 1 (1988).
Markandya, A.; Perrings, C.: *Accounting for ecologically sustainable development* (Rome: FAO, 1991).
Markandya, A.; Richardson, J.: *The debt crises, structural adjustment and the environment*, report prepared for the World Wide Fund for Nature (WWF) (Gland, Switzerland, 1990).
Mathur, H.S.; Yadav, S.D.: 'Communal diseases in a tribal eco-system', in R.M. Lodha (ed.): *Environmental ruin: The crisis of survival* (New Delhi: Indus Publishing Company, 1993).
Mattos, M.M.; Uhl, C.: 'Economic and ecological perspectives on ranching in the Eastern Amazon', in *World Development* (Oxford, Pergamon Press), vol. 22, no. 2 (1994).
Mehta, B.C.; Joshi, P.: 'Impact of IRDP on income of tribal beneficiaries: A case study on tribal sub-plan area in Rajasthan', in *Journal of Rural Development* (Hyderabad), vol. 2, no. 12 (1993).
Mellor, J.W.: *The new economies of growth: A strategy for India and the developing world* (Ithaca, New York: Cornell University Press, 1976).
Migot-Adholla, S.; Hazell, P.; Blarel, B.; Place, F.: 'Indigenous land rights in sub-Saharan Africa: A constraint on productivity?', in *World Bank Economic Review* (Washington, DC, World Bank), vol. 5, no. 1 (1991).
Monimart, M.: *Femmes du Sahel: La desertification au quotidien* (Paris: Karthala; Paris: OCDE/Club du Sahel, 1989).
Mosley, P.; Smith, L.: 'Structural adjustment and agricultural performance in sub-Saharan Africa 1980-1987', in *Journal of International Development* (Manchester, United Kingdom, Institute for Development Policy and Management), vol. 1, no. 3 (1989).
Moya, F.P.; Pingali, P.L.: *Can we close the yield gaps between the 'best' and 'ordinary' farmers in Luzon?*, Social Science Division Paper No. 89-03 (Los Banos, Philippines: International Rice Research Institute, 1989).

Munasinghe, M.: *Environmental economics and sustainable development*, World Bank Environment Department Working Paper No. 3 (Washington, DC: World Bank, 1993).
Nash, M.: 'The social context of economic choice in a small society', in G. Dalton (ed.): *Tribal and peasant economies* (Austin, Texas: University of Texas Press, 1967).
New Straits Times: 'Malacca plans tree-planting in barren areas', 7 February 1991; 'Heavy vehicles seized from illegal loggers', 26 February 1991; 'KL may ban Sabah timber exports due to excessive logging', 27 February 1991.
Nor, S.M.: 'Forestry in Malaysia', in *Journal of Forestry* (Bethesda, Maryland, Society of American Foresters), March 1983.
Norgaard, R.B.: *Sustainability as intergenerational equity: The challenge to economic thought and practice*, Asia Regional Series, Internal Discussion Paper IDP 97 (Washington, DC: World Bank, 1991).
Norman, D.W.: 'Economic rationality of traditional Hausa dryland farmers in the north of Nigeria', in R.D. Stevens (ed.): *Tradition and dynamics in small-farm agriculture* (Ames, Iowa: Iowa State University Press, 1977).
Nugent, S.: *Agrarian structure and environmental destruction in Brazilian Amazonia* (Geneva: ILO, 1994; draft).
Oniang'o, R.; Mukudi, E.: *Environment, technology and poverty linkages in Kenya* (Geneva: ILO, 1993; draft).
Oram, P.A.: 'Moving toward sustainability: Building the agroecological framework', in *Environment* (Washington, DC, Heldref Publications), vol. 30, no. 9 (1988).
Organization for Economic Cooperation and Development (OECD): *Economic instruments for environmental protection* (Paris, 1989).
Organization for Economic Cooperation and Development: *Technology and environment: Trade issues in the transfer of clean technologies* (Paris, 1992).
Organization for Economic Cooperation and Development: *The State of the Environment* (Paris), various issues.
Otsuka, K.; Cordover, V.G.; David, C.C.: 'Modern rice technology and regional wage differentials in the Philippines', in *Agricultural Economics* (Amsterdam, Elsevier Science Publishers, 1990).
Pachauri, R.K.: 'Energy efficiency and conservation in India', in *Industry and Environment* (Paris, United Nations Environment Programme, Industry and Environment Office), vol. 13, no. 2 (1990).
Pamecha, R.: *Elite in a tribal society* (Jaipur: Printwell Publishers, 1985).
Panayotou, T.: *Financing mechanisms for Agenda 21*, paper presented at the UNDP Asia and Pacific Resident Representatives Meeting (New Delhi, 14–15 November 1992).
Panayotou, T.; Sussangkarn, C.: 'Structural development and the environment: The case of Thailand', in D. Reed (ed.): *Structural adjustment and the environment* (Boulder, Colorado: Westview Press, 1992).
Parikh K.; Rabbas, F.: *Food for all in a sustainable world: The IIASA food and agricultural programme* (Luxembourg: IIASA, 1981).
Parry, M.: *Climate change and world agriculture* (London: Earthscan, 1990).
Pearce, D.W.: 'The economics of natural resource degradation in developing countries', in R.K. Turner (ed.): *Sustainable environmental management: Principles and practice* (London: Bellhaven Press, 1988).

Pearce, D.W.; Mäler, K.G.: 'Environmental economics and the developing world', in *Ambio* (Stockholm, Royal Swedish Academy of Science), vol. 20, no. 2 (1991).

Pearce, D.W.; Turner, R.K.: *The economics of natural resources and the environment* (Hemel Hempstead, United Kingdom: Harvester-Wheatsheaf, 1990).

Pearce, D.W.; Warford, J.J.: *World without end* (New York: Oxford University Press for the World Bank, 1993).

Pearce, D.W.; Barbier, E.B.; Markandya, A.: *Environmental economics and decision-making in sub-Saharan Africa*, London Environmental Economics Centre Paper 88-01 (London: International Institute for Economic Development, 1988a).

Pearce, D.W.; Barbier, E.B.: Markandya, A.: *Sustainable development and cost-benefit analysis*, paper presented at the Canadian Environmental Assessment Workshop in Integrating Economic and Environmental Assessment (Vancouver, Canada: Canadian Environmental Assessment Research Council, 1988b).

Pearce, D.W.; Barbier, E.B.; Markandya, A.: *Sustainable development: Economics and environment in the Third World* (Aldershot, United Kingdom, Edward Elgar; Brookfield, Vermont, Gower Publishing, 1990).

Pereira, A.F.: *Technology policy for environmental sustainability and for employment and income generation: Conceptual and methodological issues*, mimeographed WEP working paper WEP 2-22/WP. 215 (Geneva: ILO, 1991).

Perrings, C.: 'The natural economy revisited', in *Economic Development and Cultural Change* (Chicago, University of Chicago Press), vol. 33, no. 4 (1985).

Perrings, C.: 'An optimal path to extinction? Poverty and resource degradation in the open agrarian economy', in *Journal of Development Economics* (Amsterdam, Elsevier Science Publishers), vol. 30, no. 1 (1989a).

Perrings, C.: 'Debt and resource degradation in low income countries: The adjustment problem and the perverse effects of poverty in sub-Saharan Africa', in H. Singer and S. Sharma (eds.): *Economic development and world debt* (London: Macmillan, 1989b).

Perrings, C.: *Incentives for the ecologically sustainable use of human and natural resources in the drylands of sub-Saharan Africa: A review*, mimeographed WEP working paper WEP 2-22/WP. 219 (Geneva: ILO, 1991).

Perrings, C.: 'Sustainable livelihoods and environmentally sound technology', in *International Labour Review* (Geneva, ILO), vol. 133, no. 3 (1994).

Perrings, C.; Pearce, D.W.: 'Threshold effects and incentives for the conservation of biodiversity', in *Environmental and Resource Economics* (Dordrecht, Netherlands: Kluwer), vol. 4 (1994), pp. 13–28.

Perrings, C.; Folke, C.; Mäler, K.G.: 'The ecology and economics of biodiversity loss: The research agenda', in *Ambio* (Stockholm, Royal Swedish Academy of Science), vol. 30, no. 3 (1992).

Perrings, C.; Pearce, D.W.; Opschoor, J.B.; Arntzen, J.W.; Gilbert, A.J.: *Economics for sustainable development: Botswana – A case study* (Gland, Switzerland: International Union for Conservation of Nature and Natural Resources (IUCN), 1989).

Pezzey, J.: *Economic analysis of sustainable growth and sustainable development*, World Bank Environment Department Working Paper No. 15 (Washington, DC: World Bank, 1989).

Pezzey, J.: *Sustainable development concepts: An economic analysis*, World Bank Environment Department Working Paper No. 2 (Washington, DC: World Bank, 1992).

Philippines, Bureau of Agricultural Statistics: *Rice Statistics Handbook 1980–89* (Manila, 1990).

Pingali, P.L.; Bigot, Y.; Binswanger, H.B.: *Agricultural mechanization and the evolution of farming systems in sub-Saharan Africa* (Baltimore, Maryland: Johns Hopkins University Press for the World Bank, 1987).

Pingali, P.L.; Moya, P.F.; Velasco, L.E.: *The post Green Revolution bias in Asian rice production*, IRRI Social Division Paper No. 90-01 (Los Banos, Philippines: International Rice Research Institute, 1990).

Price, E.C.; Barker, R.Y.: 'The time distribution of crop labor in rice-based cropping patterns', in *Philippine Economic Journal* (Quezon City, Philippines, Philippine Economic Society), vol. 18, nos. 1 and 2 (1978).

Pryor, F.L.; Maurer, S.B.: 'On induced economic change in precapitalist societies', in *Journal of Development Economics* (Amsterdam, Elsevier Science Publishers), vol. 10 (1982), pp. 325–353.

Rahman, A.: 'Land concentration and dispossession in two villages of Bangladesh', in *Bangladesh Development Studies* (Dhaka, Bangladesh Institute of Development Studies), vol. 10, no. 2 (1982).

Ranadhawa, N.S.: *Green Revolution* (New Delhi: Allied Publishers, 1974).

Rao, J.M.: 'Agricultural supply response: A survey', in *Agricultural Economics* (Amsterdam, Elsevier Science Publishers), vol. 3, no. 1 (1989).

Rashid, Haroun Er: *Geography of Bangladesh*, 2nd revised edn (Dhaka: University Press, 1991).

Redclift, M.: 'Sustainable development and popular participation: A framework for analysis', in D. Ghai and J.M. Vivian (eds.): *Grassroots environmental action: People's participation in sustainable development* (London: Routledge, 1992).

Reed, D. (ed.): *Structural adjustment and the environment* (London: Earthscan, 1992).

Renner, M.: *Jobs in a sustainable economy*, Worldwatch Paper No. 104 (Washington, DC: Worldwatch Institute, 1991).

Repetto, R.: *World enough and time* (New Haven, Connecticut: Yale University Press, 1986).

Repetto, R.: 'Economic incentives for sustainable production', in G. Schramme and J.J. Warford (eds.): *Environmental management and economic development* (Baltimore, Maryland: Johns Hopkins University Press for the World Bank, 1989).

Repetto, R.; Holmes, T.: 'The role of population in resource depletion in developing countries', in *Population and Development Review* (New York, Population Council), vol. 9, no. 4 (1983).

Rosegrant, M.W.; Pingali, P.L.: *Sustaining rice production growth in Asia: A policy perspective*, IRRI Social Division Paper No. 91-01 (Los Banos, Philippines: International Rice Research Institute, 1991).

Rosegrant, M.W.; Roumasset, J.A.: 'Economic feasibility of green manure in a rice-based cropping system', in International Rice Research Institute: *Green manure in rice farming* (Los Banos, Philippines, 1988).

Rosegrant, M.W.; Roumasset, J.A.; Balisacan, A.M.: 'Biological technology and agricultural policy: An assessment in Philippine rice production', in *American*

Journal of Agricultural Economics (Ames, Iowa, American Agricultural Economics Association), vol. 67, no. 4 (1985).

Rosser, B.: *From catastrophe to chaos: A general theory of economic discontinuities* (Dordrecht, Netherlands: Kluwer Academic Publishers, 1991).

Ruddle, K.; Manshard, W.: *Renewable natural resources and the environment: Pressing problems in the developing world* (Dublin: Tycooly International for the United Nations University, 1981).

Rukuni, M.; Eicher, C.K. (eds.): *Food security for southern Africa* (Harare: University of Zimbabwe, Department of Agricultural Economics, 1987).

Ruttan, V.W.: 'The Green Revolution: Seven generalisations', in *International Development Review*, vol. 19, no. 4 (1977).

Sahlins, M.: *Stone Age economics* (London: Tavistock Press, 1974).

Schepers, J.S.; Hay, D.R.: 'Impacts of chemigation on groundwater contamination', in F.M. D'Itri and L.G. Wolfson (eds.): *Rural groundwater contamination* (Chelsea, Michigan: Lewis Publishers, 1988).

Schmidheiny, S.: *Changing course: A global business perspective on development and the environment* (Cambridge, Massachusetts: Massachusetts Institute of Technology Press, 1992).

Schultz, T.W.: *Transforming traditional agriculture* (New Haven, Connecticut: Yale University Press, 1964).

Scudder, T.: *The ecology of the Gwembe Tonga* (Manchester, United Kingdom: Manchester University Press, 1962).

Sen, A.K.: *Poverty and famines: An essay on entitlement and deprivation* (Oxford: Clarendon Press, 1981).

Serageldin, I.: 'Making development sustainable', in *Finance and Development* (Washington, DC, International Monetary Fund and the World Bank), vol. 30, no. 4 (1993).

Shafik, N.; Bandyopadhyay, S.: *Economic growth and environmental quality: Time series and cross-country evidence*, background paper for World Bank: World Development Report 1992 (Washington, DC, 1992).

Shankar, U.: 'Digging their own graves', in *Down to Earth* (New Delhi), 31 August 1993.

Sharma, T.C.; Nyumbu, I.L.: 'Some hydrology characteristics of the Upper Zambezi Basin', in W.L. Handlos and G.W. Howard (eds.): *Development prospects for the Zambezi Valley in Zambia* (Lusaka: University of Zambia, Kafue Basin Research Committee, 1985).

Singh, S.; Srivastava, K.: *Forests: People's access and rights*, report of the Third Training Programme with NGOs (Jaipur: Institute of Development Studies, 1993).

Smiddle, K.W.: 'The influence of the changing pattern of agriculture on fertiliser use', in *Proceedings of the Fertiliser Society*, No. 120 (1972).

Snijders, T.A.B.: 'Interstation correlation and non-stationarity of Burkino Faso rainfall', in *Journal of Climate and Applied Meteorology* (Boston, Massachusetts, American Meteorological Society), vol. 25, no. 4 (1986).

Solow, R.M.: 'Intergenerational equity and exhaustible resources', in *Review of Economic Studies*, (Oxford, Basil Blackwell), symposium reprint, pp. 29–46 (1974).

Solow, R.M.: 'On the intertemporal allocation of natural resources', in *Scandinavian Journal of Economics* (Oxford, Basil Blackwell), vol. 88, no. 1 (1986).

Speece, M.: 'Market performance of agricultural commodities in semi-arid South Kordofan, Sudan', in *Geoforum* (Oxford, Pergamon Press), vol. 20, no. 4 (1989).
Speece, M.; Wilkinson, M.J.: 'Environmental degradation and development of arid lands', in *Desertification Control Bulletin* (Nairobi, United Nations Environmental Programme), no. 7 (1982), pp. 2–9.
Sprenger, R.-U.: *Employment and environment: Facts and issues*, paper presented at the Workshop on the Employment Potential of Sustainable Development Policies organized by the European Foundation for the Improvement of Living and Working Conditions (Dublin, 20–21 April 1994).
Steele, J.H.: 'Marine functional diversity', in *BioScience* (Washington, DC, American Institute of Biological Sciences), vol. 41, no. 7 (1991).
Stern, D.I.; Common, M.S.; Barbier, E.B.: *Economic growth and environmental degradation: A critique of the environmental Kuznets curve*, Discussion Papers in Environmental Economics and Environmental Management No. 9409 (Heslington, United Kingdom, University of York), August. 1994.
Stewart, F.: *Technology and underdevelopment* (Boulder, Colorado: Westview Press, 1977; Oxford: Oxford University Press, 1978).
Stiles, D.: 'Arid land plants for economic development and desertification control', in *Desertification Control Bulletin* (Nairobi, United Nations Environmental Programme), no.17 (1988).
Tauer, I.W.: 'Target MOTAD', in *American Journal of Agricultural Economics* (Ames, Iowa, American Agricultural Economics Association), vol. 65, no. 3 (1983).
Tietenberg, T.: *Environmental and natural resource economics* (Glenview, Illinois: Scott Foresman, 1984).
Timberlake, L.: *Africa in crisis: The causes and cures of environmental bankrupcy* (London: Earthscan, 1988).
Tisdell, C.A.: 'Sustainable development: Differing perspectives of ecologists and economists, and relevance to LDCs', in *World Development* (Oxford, Pergamon Press), vol. 16, no. 3 (1988).
Tisdell, C.A.: *Natural resource, growth and development* (New York: Praeger, 1990).
Tisdell, C.A.: *Economics of environmental conservation* (Amsterdam: Elsevier Science Publishers, 1991).
Tisdell, C.A.: *Environmental economics: Policies for environmental management and sustainable development* (Aldershot, United Kingdom: Edward Elgar, 1993).
Tisdell, C.A.; Alauddin, M.: 'New crop varieties: Impact on diversification and stability of yields', in *Australian Economic Papers* (Adelaide, University of Adelaide), vol. 28, no. 52 (1989).
Tivy, J.: *Agricultural ecology* (Harlow, United Kingdom: Longman Scientific and Technical, 1990).
United Nations: *World Economic Survey 1986: Current trends and policies in the world economy* (New York, 1986).
United Nations Conference on Desertification: *Desertification: Its causes and consequences* (Oxford: Pergamon Press, 1977).
United Nations Conference on Environment and Development (UNCED): *Agenda 21: The United Nations Programme of Action from Rio* (New York: United Nations, 1993).

United Nations Development Programme (UNDP): *Human Development Report 1990* (Oxford: Oxford University Press, 1990).
United Nations, Economic Commission for Latin America and the Caribbean (ECLAC): *Sustainable development: Changing production patterns, social equity and the environment* (Santiago, Chile, 1991).
United Nations Environment Programme (UNEP): *Desertification control in Africa: Actions and directory of institutions* (Nairobi, 1985).
United Nations Industrial Development Organization (UNIDO): *Industrial Yearbook 1990*, Vol. 1, *General industrial statistics* (New York: United Nations, 1992a).
United Nations Industrial Development Organization: *Industry and development: Global Report, 1991/92* (Vienna, 1992b).
United Nations Industrial Development Organization: *Actes de la conférence sur un développement industriel écologiquement durable* (Vienna, 1992c).
United States Agency for International Development; Environmental Protection Agency (USAID/EPA): *Ranking environmental health risks in Bangkok, Thailand*, working paper prepared by Abt Associates and Sobotka & Co. (Washington, DC: USAID, Office of Housing and Urban Programmes, 1990).
United States, Congress, Office of Technology Assessment (OTA): *Industry, technology and the environment: Competitive challenges and business opportunities* (Washington, DC, 1993).
United States, House of Representatives: *The tropical timber industry in Sarawak, Malaysia*, staff mission report, 25 March – 2 April 1989.
van Ginneken, W.: *Trends in employment and labour incomes* (Geneva: ILO, 1988).
van Ginneken, W.; van der Hoeven, R.: 'Industrialisation, employment and earnings', in *International Labour Review* (Geneva, ILO), vol. 128, no. 5, September–October 1989.
von Amsberg, J.: *Project evaluation and the depletion of natural capital*, World Bank Environment Department Working Paper No. 56 (Washington, DC: World Bank, 1993).
Vyas, N.N.: *Bondage and exploitation in tribal India* (Jaipur and Delhi: Rawat Publications, 1980).
Vyas, N.N.: *Land and tribals* (Udaipur: MLV Tribal Research and Training Institute, 1983).
Vyas, V.S.: *Agrarian structure, environmental concerns and rural poverty*, VIth Elmhirst Memorial Lecture at the Twenty-First International Conference of Agricultural Economists (Tokyo, 1991).
Wade, R.: 'The management of common property resources: Finding a cooperative solution', in *World Bank Research Observer* (Washington, DC, World Bank), vol. 2, no. 2 (1987).
Wali, M.M.K.: *Tribal people in India*, A study undertaken for the International Labour Office (Geneva, ILO, December 1992).
Warford, J.: *Environment and development* (Washington, DC: World Bank/International Monetary Fund Development Committee, 1987).
Warford, J.: 'Environmental management and economic policy in developing countries', in G. Schramme and J.J. Warford (eds.): *Environmental management and economic development* (Baltimore, Maryland: Johns Hopkins University Press for the World Bank, 1989).

Watanabe, I.: 'Biological nitrogen fixation in sustainable rice farming', in S.K. Dutta and C. Sloger (eds.): *Biological nitrogen fixation associated with rice production* (New Delhi: Oxford and IBA Publishing Co., 1991).
Weinschenck, G.: 'The economic or the ecological way? Basic alternatives for the EC's agricultural policy', in *European Review of Agricultural Economics* (Amsterdam, Mouton de Gruyter), vol. 14 (1987), pp. 49–60.
Weisbrod, B.: 'Collective consumption services of individual consumption goods', in *Quarterly Journal of Economics* (Cambridge, Massachusetts, Harvard University, Department of Economics), vol. 78, no. 3 (1964).
World Bank: *Sudan forestry sector review* (Washington, DC, 1986).
World Bank: *World Bank experience with rural development, 1965–1986*, mimeographed (Washington, DC, 1987a).
World Bank: *World Development Report 1987* (Oxford: Oxford University Press for the World Bank, 1987b).
World Bank: *World Development Report 1988* (Oxford: Oxford University Press for the World Bank, 1988).
World Bank: *World Development Report 1990* (Oxford: Oxford University Press for the World Bank, 1990).
World Bank: *Development and the environment* (Oxford: Oxford University Press, 1992a).
World Bank: *World Development Report 1992* (Oxford: Oxford University Press for the World Bank, 1992b).
World Bank; USAID; The Land Tenure Center at the University of Wisconsin: *Proceedings of the Conference on Land Tenure Security and Tenure Reform in Africa, Nairobi, June 1990* (Washington, DC: World Bank, forthcoming).
World Commission on Environment and Development (WCED): *Our common future* (Oxford: Oxford University Press, 1987). (Also known as the Brundtland Report.)
World Resources Institute (WRI): *World Resources 1990–1991* (Oxford: Oxford University Press, 1990).
Wyatt-Smith, J.: *The agricultural system in the hills of Nepal: The ratio of agricultural to forest land and the problem of animal fodder*, Occasional Paper No. 12 (Kathmandu: Agricultural Projects Service Centre, 1982).
Yamey, B.S.: 'The study of peasant economic systems: Some concluding comments and questions', in R. Firth and B.S. Yamey (eds.): *Capital saving and credit in peasant societies* (London: George Allen & Unwin, 1964).
Younis, A.S.: *Soil conservation in developing countries* (Washington, DC: World Bank, 1987).
Yudelman, M.: *Africans on the land* (Cambridge, Massachusetts: Harvard University Press, 1964).
Zambia, National Commission for Development Planning: *Study on food strategy* (Lusaka, 1986).
Zambia, National Commission for Development Planning: *Economic Reports* (Lusaka, 1988).
Zimmerer, K.S.: 'Soil erosion and labour shortages in the Andes with special reference to Bolivia, 1953–91: Implications for "conservation-with-development"', in *World Development* (Oxford, Pergamon Press), vol. 21, no. 10 (1993).

Index

acid rain 15
Africa 8, 75, 79–80, 83, 85, 91, 113, 186, 359
see also specific countries
Africa, sub-Saharan
 agriculture 96, 97, 99, 101, 102–3, 111–30
 climate 96, 97, 98, 100–3, 113, 114, 122, 125
 consumption 95–6
 employment 125–7, 130
 environmental degradation 9, 77, 95–9, 103–6, 110–11, 114, 115, 122, 124–30
 government intervention 99
 population growth 97, 98–9, 103–6, 127
 poverty 9, 95–7, 99, 105–6, 122–5, 130
 property rights 106–11, 128–30
 risk management 99, 111–16, 130
 technology choice 99, 105–6, 111, 113–16, 129
agriculture
 and the environment 8, 13, 15, 20, 28, 32, 33, 34, 56, 70–87, 125, 354
 inputs 64, 69, 70, 81, 85, 86, 99, 112, 122, 307, 354, 361
 new technologies 65–6, 76–87, 92
 risk management in 111–16
 see also specific countries
agrochemicals 36, 70, 242, 254
aid 14, 34–6, 49, 87–8, 89, 90–1, 118, 200
air pollution 2, 14, 24–8, 72, 251, 302, 313, 338, 339, 340, 351, 355
 see also emissions
Amazon Forests 4, 20, 75, 80, 93, 257, 318, 319–20, 357, 361–2
Andes, the 75, 324
Argentina 6, 352, 355, 360

arid zones 3, 75–6, 77, 83
 see also sub-Saharan Africa
Asia 8, 72, 76, 78, 83, 113, 126, 195–6, 212, 214, 257, 320–1, 324, 360
 see also specific countries
Asian Developoment Bank (ADB) 200
Australia 88

Bangkok 13
Bangladesh
 agriculture 52, 83, 215, 224, 226–55, 358, 361
 employment 222, 227, 229, 233–4, 236, 238–40, 243–6, 248–51, 252–5
 environmental problems 221–4, 240–6, 251–5, 320
 forests 224–5, 246–53, 254
 Green Revolution 222, 223, 224, 226, 238, 240, 243, 244, 248–51, 253, 357
 health 224, 244, 245, 252
 HYVs 222, 223, 226, 227, 229–35, 238, 242, 248, 249–50, 253–4
 income 222, 226, 238–40, 244–5, 248–51, 252–4
 irrigation 222, 223, 224, 225–6, 238, 251
 population growth 222, 240, 242, 244–5, 251, 252, 253
 poverty 238, 244–6
 rice growing 225, 226, 227, 229–30, 232, 238, 240, 243, 245, 246, 248, 251, 320
 survey areas 224–7, 246–7
 technological change 221–2, 224, 226, 238, 240–5, 248–51, 253–5
Belgium 346, 353
Bolivia 324, 325
Botswana 90, 95, 109–10, 123

Index

Brazil 19, 80, 318, 349
'Brown' issues 2–3, 5, 7
Brundtland Report (WCED, 1987) 1, 38, 39, 40
Burkina Faso 75, 88, 131n, 135
Burundi 96

Canada 349
capital stocks 1, 5, 15, 17, 24, 38–40, 63
cement 15, 18, 349
Central African Republic 96, 97
Chad 91, 96, 101, 131n
chemical industry 5, 6, 15, 18, 342, 343, 344, 346, 350
children 43, 49, 63, 105, 226, 245, 253, 255
Chile
  agriculture 337–40, 353
  employment 335, 339–41, 352–4, 361
  environmental problems 335, 337–41
  fishing 335, 337–40
  forestry 335, 337–40
  growth rates 335
  mining 337–40, 354
China 19, 25, 91, 213, 214, 349
cities, pollution in 13, 15, 19, 24–8, 350, 351–2, 354
climate 45, 96, 97, 113, 137, 139–41, 161–2, 224, 299
  change 98, 100–3, 114, 122, 125, 133, 157, 161, 251, 257, 284
coal 25, 50–1, 343
cocoa 82, 132n, 263
coffee 82, 83, 85, 132n, 326
common property 9, 57, 106, 108, 109–10, 120–2, 128, 132n, 296, 307, 308, 312, 357, 362
construction 34, 337, 338, 339, 340, 341
consumption 4, 18, 23–5, 38–9, 42, 43, 44, 50, 55, 63, 113–14, 124, 166, 167, 184, 305
Convention on International Trade of Endangered Species (CITES) 273
Costa Rica 4, 85, 362n

cotton 82, 132n, 137, 138, 140–1, 143
crops 15, 45, 52, 70, 77–8
  environmental effects of new technology on 81–4
  see also specific crops

debt 4, 34, 134, 200, 211, 217, 284, 287
deforestation 3, 14, 15, 18, 75–6, 81, 88, 91–2, 318, 320
  and level of development 19–23, 28, 29
  see also specific countries
Denmark 346
desertification 73, 82, 83, 90, 95, 97, 115, 133, 159–61, 171, 183, 186–7, 189, 191
deserts 19, 131n
drought 101, 103, 113, 122, 132n, 133, 135, 138, 140, 141, 146, 160–1, 212, 222, 292, 328, 330

ecological systems 43, 44–7, 51–3, 55–6, 57, 60, 61, 63, 97, 143–4, 338
dryland 98, 101–3, 105–6, 117, 122, 131–2n
economic development
  environmental degradation and 13–19, 29–32, 317–18
  aid and 34–6
  deforestation 19–23, 28, 29
  employment and 32–5
  environmental pollution 23–9
  technology and 33–5
education 64, 236, 294–5, 298, 326, 331
Egypt 74
emissions
  industrial 15–19, 24, 25–9, 31, 33, 46, 50–2, 58, 61, 65, 100, 339, 349, 355
  lead 29–30, 350, 354, 355
  methane gas 302
employment
  and sustainable livelihood 322–4, 334

Index

and the environment 4–9, 14, 15, 32–5, 70–87, 92–3, 125–7, 130, 334–41
  in the environmental protection industry 3, 341–2, 346, 350–4, 355, 360
  technology and 69–71, 76–80, 84–7, 92–3
  see also specific countries
endangered species 4, 29, 273
energy 6, 7, 16, 23–4, 25, 31, 32, 81, 240, 284, 305, 361
environmental degradation 6, 9, 40–1, 54, 70, 284, 359
  and economic development 13–19, 29–32, 317–18
    aid and 34–6
    deforestation 19–23, 28, 29
    employment and 32–5
    environmental pollution 23–9
    technology and 33–5
  employment and 32–5, 70–7, 92–3
  government response to 87–92, 354–6
  technology and 33–5, 72–87, 92–3
  see also specific countries
Environmental Kuznets Curve 5, 8, 13–36, 317–18, 358–9
environmental protection 4, 5, 17–19, 31–2, 53, 58, 69–70, 86, 93, 346
  employment in 3, 341–2, 346, 350–4, 355, 360–1
environmental resources
  common property and 120–2
  demand for 42, 51, 52–3, 55–6
  depletion of 51–2, 60
  private costs of 42–3, 56
  use of 48, 55, 59–60, 63–5
  valuation of 40–2, 44–5, 55–6
  see also natural resources
environmental soundness 37, 44–8, 61–3
environmental sustainability
  and socioeconomic sustainability 324–34, 359
environmentally sound technology (EST) 37, 44, 47, 50–3, 359
  and Agenda 21 50–3, 57, 61, 63–4
  and the decision process 58–62, 64

operationalization of 57–8, 65–6
equity 2, 3
  intergenerational 38–9, 42–3, 54, 55
Ethiopia 22, 23, 73, 75, 76, 96
European Union 7, 9, 125
  employment in environmental protection 341–2, 346, 351–2, 353, 355, 360–1

famine 43, 63, 72, 96, 104, 105, 122, 133
FAO 76, 161, 261
fertilizers 46, 52, 72, 74, 77, 78, 81, 82, 164, 166, 180, 181–2, 197, 201, 202, 204–9, 238, 251, 254, 283, 294, 306, 324
  alternatives to chemical fertilizers 74, 211–17
  and environmental problems 70, 129, 209–11, 218–20, 222–4, 242–3, 320, 354–5
fish 14, 33, 58, 65, 240–1, 242, 245, 251, 254, 320, 335, 337–40
food security 9, 78, 127, 135, 142, 147, 160, 326, 330, 334–5, 359
forests 3, 4, 14, 15, 33, 83–4, 329, 330, 354, 359
  see also deforestation; reforestation; specific countries
France 346, 353

GATT 56
Germany 16, 341, 346, 349, 353
Ghana 131n
Global Environment Facility (GEF) 35, 36
Greece 26, 346
'green' issues 2–3, 5, 7, 18
Green Revolution 72, 77, 79, 99, 113, 114, 222
  see also specific countries
greenhouse effect 35, 100, 102–3, 161, 257, 299, 302
Guatemala 73
Guinea 112, 131n

health 18, 29–30, 209, 224, 244, 245, 252, 260, 262, 291, 294–5, 301, 302–3, 326

386 Index

Hong Kong 18–19
Horn of Africa 122, 126
Human Development Index 96, 131n

illiteracy and literacy 32, 131n, 227, 228, 236, 294–5
ILO 87–8, 89
IMF 118
incentive structures 4, 6–7, 9, 85, 159, 354
  economic 127–31
  financial incentives in Senegal 171–83, 192
  price incentives 116–22
incomes 9, 69, 70
  distribution 40, 49–50, 71, 122–3, 128–9
  inequality 13–14, 20
  see also specific countries
India
  agriculture 78, 79, 83, 89–90, 204, 214, 293–4, 296–306, 312–13
  deforestation 291–2, 295, 297, 299, 303, 308
  emissions 26, 27, 302
  employment 294, 295, 301–2, 313
  energy conservation 348
  environmental degradation 222, 289–90, 292, 295, 296–8, 299–300, 302–4, 311–14
  forests 91, 291–4, 296–7, 298, 301, 303, 304, 305–6, 308–12, 357–8, 362
  health 291, 294–5, 301, 302–3
  income 296, 297, 298, 302, 308, 313
  irrigation 294
  migration 297, 301–2, 357, 362
  natural resources 25, 290–2, 298–300, 303, 304–12, 314
  pollution control 349
  population growth 290, 292–3, 297, 300–2, 306, 307
  poverty 289–90, 292, 296–8, 300–2, 304, 314
  soil 73, 291–2, 299, 302, 306, 313
  survey area 289–95
  technology 290, 299–300, 302–4, 313–14

indigenous populations 4, 7, 9, 49, 80, 93, 258, 261, 262, 271, 272, 273, 280, 283, 284, 285, 286, 287, 318–20, 357, 358, 359, 361–2
Indonesia 19, 20, 35, 79, 83, 85, 266, 285, 321
industry
  emissions 15–19, 24, 25–9, 31, 33, 46, 50–2, 58, 61, 65, 100, 339, 349, 355
  polluting industries 2, 5, 13, 15–19, 25, 32, 81, 339, 342–6, 349–50, 355–6, 360–1
  pollution control 4, 7, 342–50, 359, 360
infrastructure 4, 7, 69, 81, 85, 120, 243
inputs 6, 16, 54, 81, 85, 86, 307, 354, 361
  prices 6, 16, 54, 69, 99, 122
integrated pest management (IPM) 74
International Rice Research Institute (IRRI) 77, 79, 195, 200–1, 203, 209, 210, 213, 216, 218
International Tropical Timber Organization (ITTO) 271, 274, 277
investment 43, 45, 49, 50, 63, 64, 65, 111, 122, 305, 306, 346
Ireland 346
irrigation 30, 35, 64, 72, 74, 77, 78, 83, 88, 89–90, 91, 354
  see also specific countries
Italy 26, 346, 353

Japan 13, 16, 18, 79, 200, 223, 254, 279, 349

Kenya 4, 83, 89, 90, 95, 156, 362n
  environmental and socioeconomic sustainability 324–34, 356
Korea, Republic of 19, 91, 279
Kuznets Curve, Environmental 5, 8, 13–36, 317–18, 358–9

labour 4, 32–3, 69, 71, 77, 78, 79, 84, 86, 126–7, 324, 325
  see also employment

land 7, 9, 32, 33, 77, 86, 107–11, 162–4, 171, 170–83, 188–92, 358, 359
see also common property
Latin America 4–5, 8, 72, 78–80, 83, 126, 324, 359, 360
see also specific countries
lead emissions 29–30, 350, 354, 355
lead-free petrol 19, 30
Lesotho 90, 95
livestock 80, 81–4, 85, 89, 90, 91, 97, 109–10, 113, 115, 123, 160, 161, 242, 243, 251, 253, 297, 299–300, 319–20, 357, 359
logging 4, 19, 75, 83, 223, 252, 253, 257, 258, 260, 261, 267, 268, 269–70, 271, 273, 274–7, 280, 282, 285, 287, 319–20, 336, 359
Lomé Convention 56
Luxembourg 346

Malawi 96
Malaysia
  agriculture 77, 263–4, 272, 274, 281, 282–3, 285, 287, 318–19, 357, 361
  forests
    deforestation 257, 258, 260, 270, 273, 274, 282–4, 287
    employment 4, 5, 257, 259–62, 265–7, 268, 270, 272, 274–7, 278, 280, 281, 283, 285, 334–5, 336, 359
    environmental concerns 4, 5, 19, 257–62, 267–8, 271, 273, 277–8, 285, 335
    forest resources 263–5, 336
    income 259, 262, 271, 335, 359
    indigenous population 258, 261, 262, 271, 272, 273, 280, 283, 284, 285, 286, 287, 318–19, 357, 361
    land use 263–5, 267, 284–7
    management of 259–62, 267–9, 271, 274, 280–2, 285
    sustainability 273–7, 278–81, 283, 284–7
    technology 258–62, 272, 275

  timber 257, 258–62, 265–7, 268, 271, 272–5, 277–81, 283, 284, 285, 286–7
  use of 258–60, 272
Mali 90, 131n
Mauritania 88, 96, 131n
Mexico 13, 19, 132n, 324, 350, 351, 354, 355, 360
Mexico City 13, 19, 350, 354
migration 70, 127, 132n, 147, 171, 173, 175, 179, 181, 183–6, 191–2, 204, 218, 255, 297, 301–2, 320–2, 324, 357, 359–60, 362
Mozambique 131n
multinational enterprises 136

natural capital 39–40, 55
natural disasters 30, 200, 222, 338
natural gas 25, 254, 298
natural resources 1–3, 4, 9, 14–19, 35, 36, 72, 84, 86, 93
  depletion 15, 18, 19–23, 28, 29, 34, 85, 240, 252
  overuse 9, 106–11, 354
  property rights over 6, 7, 31, 32, 34, 36, 53, 56–7, 63–4, 98–9, 105, 106–11, 128–30, 189–90, 280–1, 304, 306–11, 318, 356–7, 358–9, 362
  see also environmental resources
Nepal 73, 75–6, 222
Netherlands 346
New York City 351–2
newly industrializing economy (NIE) 18–19, 266
  see also specific countries
Niger 82, 90, 96, 101, 131n
Nigeria 85
noise pollution 2
North, the 5, 42, 72, 79, 342, 360

organic farming 74, 89
overgrazing 74–5, 80, 88, 90, 97, 160, 161, 171–2, 292
ozone layer 35

Pakistan 74, 79, 89
paper industry 5, 15, 83, 282, 342, 343, 344, 346, 350

Peru 324
pesticides 46, 52, 70, 74, 81, 82, 129, 164, 166, 175, 180, 181–2, 197, 202, 209, 223, 238, 242, 243, 354
petroleum refining 5, 342, 343, 344, 346
Philippines 216, 217–18, 220, 285, 321, 322–4, 354, 357, 360, 361
  employment 202–4
  environmental problems 209–11, 217–20, 322–4, 354–5, 360, 361
  fertilizers 197, 201, 202, 204–11, 218–20, 324, 354–5
    alternatives to chemical fertilizers 211–17
  forests 20, 83, 266, 279, 321
  Green Revolution 195–220
  irrigation 195, 197–201, 204, 205, 206, 208, 209, 211, 213, 214, 215, 220
  migration 204, 321–2, 359–60
  new technology 77, 202–3, 212
  pesticides 197, 202, 209
  population growth 195, 197, 210–11
  poverty 211, 218, 321–2, 360
  rice production 195–220
  soil conservation 321, 322–4
pollution
  and level of development 23–8, 34
  control 3, 4, 5, 7, 16, 23–4, 25, 31, 33, 341–52, 355, 359, 360, 362n
  industrial 2, 5, 13, 15–19, 25, 32, 81, 339, 342–6, 349–50, 355–6, 360–1
  see also air pollution; water pollution
population growth 52, 55, 69, 70, 72, 78, 82
  see also specific countries
Portugal 26, 346
poverty 3, 9, 17, 34, 69, 72, 80, 85, 354
  alleviation 2, 38, 40, 48–9, 71
  and the private valuation of environmental resources 40–2, 56
  see also specific countries

power generation 2, 34, 50–1, 135–6
prices 6, 24, 29, 33, 46–7, 54, 55, 56, 57, 61
  agricultural 56, 116–22
  input 6, 16, 54, 69, 99, 112, 122
  privatization 6, 111, 189–90, 281, 354, 358, 361
  productivity 18, 43, 46–7, 74, 80, 84
property rights
  and natural resources 6, 7, 31, 32, 34, 36, 53, 56–7, 63–4, 98–9, 105, 128–30, 189–90, 280–1, 304, 306–11, 318, 358–9, 362
  overutilization of 106–11, 356–7
  protection 17, 24, 33, 36, 125
pulp industry 5, 15, 83, 282, 319, 342, 343, 346, 349, 350

R&D 69, 85, 93, 304
radioactive contamination 29, 73, 338
reforestation 19, 20, 88, 171, 186–9
refugees 43, 49, 63, 96
Rio Earth Summit 1992
  Agenda 21 1, 35, 37, 38, 41, 42–3, 44, 47
  and environmentally sound technology (EST) 50–3, 57, 61, 63–4
  and sustainable livelihoods (SL) 48–50, 63–4
rural development projects 87–8
rural populations 2, 3, 15, 20, 49, 79, 95, 96, 159
Russia 79
Rwanda 88, 96

Sahel, the
  agriculture 82, 90, 112, 115, 122
  climatic change 100–1, 125, 161
  desertification 95, 131, 159–61, 183, 186–7, 189, 191
  irrigation 83
  population growth 105, 126, 160, 175, 178, 193
  poverty 96
  see also Africa, sub-Saharan; specific countries
sanitation 123, 328, 330, 332

semi-arid zones 3, 4, 77, 95, 97
  see also Africa, sub-Saharan;
    specific countries
Senegal
  agriculture 75, 90, 112, 161–71,
    191
  climate 101, 161–2
  consumption 96, 131n, 166, 167,
    184
  data collection 164–70
  farming and land tenure 162–4,
    188–92
  financial incentives to reduce peanut
    production 171–83, 191–2
  land conservation 170–83
  migration 171, 173, 175, 179, 181,
    183–6, 191–2
  reforestation 186–9
  regional development 186
  Social Accounting Matrix (SAM)
    159, 165, 168, 170–1, 171–2,
    184–6, 192–3
  soil 162, 175, 177–8, 180–3, 191,
    192, 193n
Seoul 13
service industries 15, 24, 28, 32–3,
  126, 349
Sierra Leone 131n
Singapore 19, 26
socioeconomic sustainability
  and environmental sustainability
    324–34, 359
soil
  conservation 88–9, 177–8, 191,
    193n, 283, 302, 313, 321,
    322–5, 329–31, 354
  degradation 14, 52, 117, 125,
    159–61, 175, 178, 180–3, 192,
    201, 210–11, 217, 291, 299,
    306, 338
  erosion 3, 15, 18, 30, 70, 73, 74,
    76, 85, 89, 104, 246, 251, 252,
    254, 257, 261, 264, 358
  fertility 240–1, 242–3, 245–6, 251,
    254, 264, 291
  fertilizer effect on 209, 210–11,
    223
  in India 73, 291–2, 299, 302, 306,
    313
  in Senegal 162, 175, 177–8,
    180–3, 191, 192, 193n
  in Zambia 136, 140, 157
Somalia 89, 96, 131n
South, the 5, 79, 342, 360
South Africa 107, 126
Spain 27, 346, 353
Sri Lanka 79, 90
steel 5, 6, 15, 18, 349
Sudan
  agriculture 75, 76, 82, 131n, 135
  climate 101
  desertification 82, 95, 159–61
  population growth 105
  poverty 96, 123
sustainable development 1–3, 5, 8
  and access to technology 57–8
  and environmental safety and
    soundness 44–8, 62–3
  concepts of 37–43, 87, 221
  failures in 6–7, 42–3
sustainable livelihoods (SL) 37, 38,
  47, 221, 354, 359–60
  and Agenda 21 48–50, 63–4
  and the decision process 58–62, 64
  destruction of 318–34
  destruction versus creation of 3–4
  employment intensity and 322–4,
    334
  'green' versus 'brown' issues in
    2–3
  innovative approaches for 7–8
  institutional dimensions of 356–8
  operationalization of 54–7, 65–6
Switzerland 22, 23, 26, 27

Taiwan 19, 79, 279
Tanzania 79, 88, 90, 96, 97, 107,
  131n
technology
  and employment 69–71, 76–80,
    84–7, 92–3
  and environmental degradation
    13–14, 15–16, 18, 24, 25, 33–5,
    51–2, 72–87, 92–3
  choice of 69–71, 99, 105–6, 111,
    113–16, 129
  environmentally sound (EST) 37,
    44, 47, 50–3, 350, 359

technology – *continued*
  and Agenda 21 50–3, 57, 61, 63–4
  and the decision process 58–62, 64
  operationalization of 57–8, 65–6
  government policies and 81, 85–92
  transfer 47, 50, 53, 57, 65–6, 71, 76–87
  see also specific countries
textiles 5, 34, 342, 343, 344, 349
Thailand 13, 19, 20, 25, 26, 34, 35, 279
timber 28, 83, 97, 104, 117, 242, 257, 258–62, 265–7, 268, 271, 272–5, 277–81, 283, 284, 285, 286–7, 297, 304, 305, 306, 308, 319, 328

Uganda 96 97, 131n
unemployment 71–2, 126, 135, 146–7, 210–11, 218, 301, 313, 320–1, 353
United Kingdom 346, 349, 353
United Nations Conference on Desertification (1977) 95, 97
United States 4, 9, 18, 20, 79, 88, 125, 200, 284, 349
  pollution control 341, 346–8, 350–2, 355, 360, 362n
  pollution in cities 13, 352
  water pollution 209–10
urban areas 2, 15, 32, 33, 49, 70, 79, 96, 222, 255
  pollution 4, 15, 19, 24–8
  see also cities

Venezuela 22
Vietnam 213

waste disposal 2, 4, 13, 16, 17, 18, 22, 29, 31, 45, 50, 121, 329, 333, 350–2, 355, 359, 360
water 6, 16, 33, 35, 45, 46, 64, 81, 86, 89, 90, 222, 240–1, 246, 251, 271, 299, 342

in dryland environments 114–16, 123, 124, 128
pollution 2, 209–10, 217, 219–20, 222, 224, 244, 257, 261–2, 302, 305, 313, 338, 354, 355
  table 157, 161, 292
  see also irrigation
watersheds 30, 72, 258–9, 303, 304
welfare 42, 54, 77, 103, 159, 209, 221, 305, 308
women 49, 56, 77, 88, 123, 179, 226, 245, 253, 255, 294–5, 302, 326, 333–4
World Bank 54, 88, 90, 118, 165, 200, 304
World Trade Organization (WTO) 56

Zaire 96, 97, 112
Zambia
  agriculture 131n, 133–8, 140–4, 154, 155, 156, 157, 334
  analytical models for 138–57
  climate 137, 139–41
  employment 134–5, 137, 141, 142, 144–51, 152, 155, 156–7, 334, 359
  environment issues 133–8, 140–2, 143–58, 334, 359
  food 133–5, 137, 141, 142, 146–51, 152, 155, 156, 334, 359
  income 134–5, 137, 140–1, 142–3, 144–6, 147–51, 154–6, 158, 334
  irrigation 133–6, 137–8, 139–48, 151, 153–4, 156
  Kariba Dam 135–6
  Lake Kariba 133, 136, 147, 153
  pollution control 349
  soil 136, 140, 157
  tourism 152–3, 154, 155, 156, 334
  Zambesi River Authority 136
Zimbabwe 135–6, 156, 358, 361